PROGRESS IN ORGANIC AND PHYSICAL CHEMISTRY

Structures and Mechanisms

PROGRESS IN ORGANIC AND PHYSICAL CHEMISTRY

Structures and Mechanisms

Edited by

Gennady E. Zaikov, DSc, Alexander N. Goloshchapov, PhD, and Anton V. Lobanov, PhD

Apple Academic Press

TORONTO NEW JERSEY

Apple Academic Press Inc.
3333 Mistwell Crescent
Oakville, ON L6L 0A2
Canada

Apple Academic Press Inc.
9 Spinnaker Way, Waretown, NJ 08758
USA

Exclusive worldwide distribution by CRC Press, a Taylor & Francis Group

International Standard Book Number: 978-1-926895-40-6 (Hardback)

Library of Congress Control Number: 2012951938

Library and Archives Canada Cataloguing in Publication

Progress in organic and physical chemistry: structures and mechanisms/edited by Gennady E. Zaikov, Alexander N. Goloshchapov, and Anton V. Lobanov.

Includes bibliographical references and index.
ISBN 978-1-926895-40-6
1. Chemistry, Physical and theoretical. 2. Chemistry, Organic. I. Zaikov, G. E. (Gennadiˇi Efremovich), 1935- II. Goloshchapov, Alexander N III. Lobanov, Anton V

QD453.3.P76 2013 541 C2012-906396-7

Apple Academic Press also publishes its books in a variety of electronic formats. Some content that appears in print may not be available in electronic format. For information about Apple Academic Press products, visit our website at **www.appleacademicpress.com**

About the Editors

Gennady E. Zaikov, DSc

Gennady E. Zaikov, DSc, is Head of the Polymer Division at the N. M. Emanuel Institute of Biochemical Physics, Russian Academy of Sciences, Moscow, Russia, and a professor at Moscow State Academy of Fine Chemical Technology, Russia, as well as a professor at Kazan National Research Technological University, Kazan, Russia. He is also a prolific author, researcher, and lecturer. He has received several awards for his work, including the the Russian Federation Scholarship for Outstanding Scientists. He has been a member of many professional organizations and on the editorial boards of many international science journals.

Alexander N. Goloshchapov, PhD

Alexander N. Goloshchapov, PhD, is the Deputy of the N. M. Emanuel Institute of Biochemical Physics and Director of the Russian Academy of Sciences, Moscow. He is Head of the laboratory and a specialist in the field of chemical kinetics, biochemical kinetics, chemical physics, and biochemical physics. He has published four books and about 100 original papers in the field of biochemistry and biology (cancer, gerontology, biochemical reactions in the living body).

Anton V. Lobanov, PhD

Anton V. Lobanov, PhD, graduated from Lomonosov Moscow State Academy of Fine Chemical Technology (MIFCT) with honors in 2001, and he received his PhD degree in chemistry in 2005. He is Deputy to the Head of the Laboratory of the Semenov Institute of Chemical Physics of the Russian Academy of Sciences (ICP RAS), Deputy to the Head of Department of MIFCT, and a member of the Scientific Council of Scientific Education Centre of Lomonosov Moscow State University. He is member of the Mendeleev Russian Chemical Society and the Russian Photobiology Society, as well as on the organizing committees of several international conferences. He is the author of the lecture courses "Photochemistry of Bioactive Compounds" and "Supramolecular Photochemistry". He is a reviewer for several professional journals. He is the author of more than 40 articles, eight book chapters, two patents, two patent requests, and 90 conference reports. He was a recipient of a scholarship from the President of Russia (2000), the Prize of "MAIK-Nauka/Interperiodika" in chemistry (2004), a grant from the government of Moscow (2004), and several prestigious scholarships. His scientific interests are focused on photochemistry and coordination chemistry of metal complexes, mechanistic studies of supramolecular assemblies of metal complexes, and the development of new methods in diagnosis and pharmacology.

About the Editors

Contents

List of Contributors

M. A. Akhmanova
Departement of Biophysics, Faculty of Physics, Lomonosov Moscow State University, Moscow, Russia.
E-mail: maria.akhmanova@gmail.com

O. M. Alekseeva
Institute of Biochemical Physics N. M. Emanuel Academy of Sciences, Russia, 119334, Moscow, st. Kosygin d.4. (495) 939-74-09, olgavek@yandex.ru

J. Aneli
1R.Dvali Institute of Machine Mechanis
JimAneli@yahoo.com

Esen Arkış
Institute of Technology
Gulbahce Campus Urla
Izmir, Turkey 35430
† devrimbalkose@iyte.edu.tr

M. I. Artsis
General Institute of Chemical Physics, Russian Academy of Sciences, Kosygin str. 4, 119991 Moscow, Russia; ioran@chph.ras.ru

Devrim Balköse
Institute of Technology
Gulbahce Campus Urla
Izmir,Turkey 35430
† devrimbalkose@iyte.edu.tr

P. Yu. Barzilovich
Institute of Problem of Chemical Physics, Russian Academy of Sciences,
1 Semenov pr., 142432 Chernogolovka, Moscow region, Russia

S. A. Bekusarova
Doctor of Biological Science, professor
Gorsky State Agrarian University, Kirov str., 37, Vladikavkaz, Republic of North Ossetia Alania, Russia, ggau@globalalania.ru

E. B. Blokhina
N. M. Emanuel Institute of Biochemical Physics, Russian Academy of Sciences, Moscow, Russia
E-mail: groshotan@gmail.com

N. A. Bome
Doctor of Agricultural Science, professor, head of the Cathedra of botany and plant biotechnology
Tyumen State University, Semakova str., 1, 625003, Tyumen, Russia, +7(3452) 46-40-61, 46-81-69. president@utmn.r

A. P. Bonartsev
A. N. Bach's Institute of Biochemistry, Russian Academy of Sciences, Leninskii prosp. 33, 119071 Moscow, Russia bonar@inbi.ras.ru bonar@inbi.ras.ru
Faculty of Biology, Moscow State University, Leninskie gory 1-12, 119992 Moscow, Russia; ant_bonar@mail.ru

G. A. Bonartseva
A. N. Bach's Institute of Biochemistry, Russian Academy of Sciences, Leninskii prosp. 33, 119071 Moscow, Russia bonar@inbi.ras.ru bonar@inbi.ras.ru

E. B. Burlakova
N. M. Emanuel Institute of Biochemical Physics, Russian Academy of Sciences, Moscow, Russia
E-mail: groshotan@gmail.com

Hugh D. Burrows
Department of Chemistry, University of Coimbra, 3004 - 535 Coimbra, Portugal
rpereira@qui.uc.pt, avalente@ci.uc.pt, burrows@ci.uc.pt, burrows@ci.uc.pt, vlobo@ci.uc.pt

S. P. Domogatsky
Institute of Experimental Cardiology, Russian Cardiology Research Center MHSD RF, Moscow, Russia

M. M. Feldshtein
A.V. Topchiev Institute of Petrochemical processing Russian Academy of Sciences
117912 Moscow, Lenin avenue, 29

G. V. Fetisov
Moscow M.V. Lomonosov State University, Chemistry Department,
Leninskie Gory, 1, str. 3

N. Yu. Gerasimov
N. M. Emanuel Institute of Biochemical Physics, Russian Academy of Sciences, Moscow, Russia
E-mail: groshotan@gmail.com

M. A. Goldschtrakh
Moscow M.V. Lomonosov State Academy of Fine Chemical Technology, Moscow, Vernadskogo prosp. 86.
E-mail: aolkhov72@yandex.ru

A. N. Goloshchapov
N. M. Emanuel Institute of Biochemical Physics, Russian Academy of Sciences, Moscow, Russia
E-mail: groshotan@gmail.com

A. K. Haghi
University of Guilan, Rasht, Iran
Haghi@Guilan.ac.ir

A. A. Ischenko
Moscow M.V. Lomonosov State Academy of Fine Chemical Technology, Moscow, Vernadskogo prosp. 86.
E-mail: aolkhov72@yandex.ru

A. L. Iordanskii
General Institute of Chemical Physics, Russian Academy of Sciences, Kosygin str. 4, 119991 Moscow, Russia; ioran@chph.ras.ru
N.N. Semenov Institute of Chemical Physics Russian Academy of Sciences 119991 Moscow, street Kosygina, 4

I. A. Kish
Moscow State University of Food Production
33 Talalihin st. Moscow Russia 109316
kaf.vms@rambler.ru

E. Klodzinska
Institute for Engineering of Polymer Materials and Dyes, 55 M. Sklodowskiej-Curie str., 87-100 Torun, Poland,
E-mail: S.Kubica@impib.pl

G. G. Komissarov
Semenov Institute of Chemical Physics, Russian Academy of Sciences,
4 Kosygin st., 119991 Moscow, Russia

G. A. Korablev
Izhevsk State Agricultural Academy
Basic Research and Educational Center of Chemical Physics and Mesoscopy, URC, UrD, RAS
Russia, Izhevsk, 426000, e-mail: korablev@mail.ru,biakaa@mail.ru

N. N. Kononov
Institute of General Physics by the name of A.M. Prokhorov, RAS.
Moscow, Vavilova ul. 38

A. A. Krutikova
Moscow M.V. Lomonosov State Academy of Fine Chemical Technology, Moscow, Vernadskogo prosp. 86.
E-mail: aolkhov72@yandex.ru

G. V. Luschenko
PhD student
North Caucasus Research Institute of mountain and foothill agriculture, Williams's str, 1, 391502, Mikhailovskoye, Republic of North Ossetia Alania, Russia, t/f +7(8672) 73-03-40

D. J. Liaw
National Taiwan University of Science and Technology, 43, Sec.4, Keeling RD., Taipei, 106

A. V. Lobanov
Semenov Institute of Chemical Physics, Russian Academy of Sciences,
4 Kosygin st., 119991 Moscow, Russia

Victor M. M. Lobo
Department of Chemistry, University of Coimbra, 3004 - 535 Coimbra, Portugal
rpereira@qui.uc.pt, avalente@ci.uc.pt, burrows@ci.uc.pt, burrows@ci.uc.pt, vlobo@ci.uc.pt
Departamento de Química, Universidade de Coimbra, 3049 Coimbra, Portugal
anacfrib@ci.uc.pt

E. Markarashvili
I. Javakhishvili Tbilisi State University Faculty of Exact and Natural Sciences, Department of Macromolecular Chemistry. I. Chavchavadze Ave., 3, Tbilisi 0128, Republic of Georgia
OmariMui@yahoo.com

O. Mukbaniani
I. Javakhishvili Tbilisi State University Faculty of Exact and Natural Sciences, Department of Macromolecular Chemistry. I. Chavchavadze Ave., 3, Tbilisi 0128, Republic of Georgia
OmariMui@yahoo.com

Joaquim J. S. Natividade
Departamento de Química, Universidade de Coimbra,
3049 Coimbra, Portugal
anacfrib@ci.uc.pt

N. I. Neshev
Institute of Problems of Chemical Physics, Russian Academy of Sciences.
1 Academician Semenov avenue, 142432, Moscow region, Chernogolovka, Russia.
E-mail: Cyber-shop@mail.ru

O. V. Nevrova
Semenov Institute of Chemical Physics, Russian Academy of Sciences,
4 Kosygin st., 119991 Moscow, Russia

A. A. Olkhov
Moscow M.V. Lomonosov State Academy of Fine Chemical Technology, Moscow, Vernadskogo prosp. 86.
E-mail: aolkhov72@yandex.ru

Rui F. P. Pereira
Department of Chemistry, University of Coimbra, 3004 - 535 Coimbra, Portugal
rpereira@qui.uc.pt, avalente@ci.uc.pt, burrows@ci.uc.pt, burrows@ci.uc.pt, vlobo@ci.uc.pt

D. A. Pomogova
Moscow State University of Food Production
33 Talalihin st. Moscow Russia 109316
kaf.vms@rambler.ru

B. L. Psikha
Institute of Problems of Chemical Physics, Russian Academy of Sciences.
1 Academician Semenov avenue, 142432, Moscow region, Chernogolovka, Russia.
E-mail: Cyber-shop@mail.ru

Ana C. F. Ribeiro
Departamento de Química, Universidade de Coimbra,
3049 Coimbra, Portugal
anacfrib@ci.uc.pt

T. N. Rudneva
Institute of Problems of Chemical Physics, Russian Academy of Sciences.
1 Academician Semenov avenue, 142432, Moscow region, Chernogolovka, Russia.
E-mail: Cyber-shop@mail.ru

N. A. Roubtsova
Semenov Institute of Chemical Physics, Russian Academy of Sciences,
4 Kosygin st., 119991 Moscow, Russia

N. A. Sanina
Institute of Problems of Chemical Physics, Russian Academy of Sciences.
1 Academician Semenov avenue, 142432, Moscow region, Chernogolovka, Russia.
E-mail: Cyber-shop@mail.ru

D. A. Sogrina
Moscow State University of Food Production
33 Talalihin st. Moscow Russia 109316
kaf.vms@rambler.ru

E. M. Sokolova
Institute of Problems of Chemical Physics, Russian Academy of Sciences.
1 Academician Semenov avenue, 142432, Moscow region, Chernogolovka, Russia.
E-mail: Cyber-shop@mail.ru

P. A. Storozhenko
GNC FGUP GNIIKHTEOS, Moscow, Entuziastov, 38

F. T. Tzomatova
Ph.D. student
North Caucasus Research Institute of mountain and foothill agriculture, Williams's str, 1, 391502, Mikhailovskoye, Republic of North Ossetia Alania, Russia, t/f +7(8672) 73-03-40

Artur J. M. Valente
Department of Chemistry, University of Coimbra, 3004 - 535 Coimbra, Portugal
rpereira@qui.uc.pt, avalente@ci.uc.pt, burrows@ci.uc.pt, burrows@ci.uc.pt, vlobo@ci.uc.pt
Departamento de Química, Universidade de Coimbra, 3049 Coimbra, Portugal
anacfrib@ci.uc.pt

L. I. Weisfeld
Senior researcher
N.M. Emanuel Institute of Biochemical Physics of Russian Academy of Sciences, Kosygina str., 4, 119334, Moscow, Russia, chembio@sky.chph.ras.ru;

Larissa I. Weisfeld
N.M. Emanuel Institute of Biochemical Physics RAS, Moscow, Russia, e-mail liv11@yandex.ru

G. E. Zaikov
N.M. Emanuel Institute of Biochemical Physics, RAS, Russia, Moscow 119991, 4 Kosygina St., E-mail: chembio@sky.chph.ras.ru
Institute of Biochemical Physics after N.M. Emanuel, Kosygina 4, Moscow, 119991, Russia, E-mail: chembio@sky.chph.ras.ru

Russian Academy of Sciences, Moscow, Russia
GEZaikov@Yahoo.com

N. Emanuel Institute of Biochemical Physics, Kosigin 4, Moscow , Russia
Chembio@sky.chph.ras.ru

General Institute of Chemical Physics, Russian Academy of Sciences, Kosygin str. 4, 119991 Moscow, Russia; ioran@chph.ras.ru

N. M. Emanuel Institute of Biochemical physics Russian Academy of Sciences
119991 Moscow, street Kosygina, 4

List of Abbreviations

ATP	Adenosine triphosphate
AFM	Atomic force microscopy
BTz	Benzthiazol tetranotrosyl iron complex
BSA	Bovine serum albumin
BE	Bulbectomized
CMC	Critical micella concentration
Cys	Cysteinamine-contained tetranitrosyl iron complex
DNA	Deoxyribonucleic acid
DSK	Differential scanning microcalorymetry
DMPC	Dimyristoilphosphatidylcholine
EAC	Ehrlich ascetic carcinoma
ESCA	Electron spectroscopy of chemical analysis
EN	Electronegativity
eV	Electron-Volts
ER	Endoplasmic reticulum
IGF	Like growth factors
LPO	Lipid peroxide oxidation
LPO	Lipids peroxidation
MDA	Malondialdehyde
MR	Methyl red dye
Mim	Methylimidazol tetranitrosyl iron complex
MC	Moisture content
nc-Si	Nanocrystalline silicon
NADP	Nicotinamidadenindinucleotid phosphate
NO	Nitric oxide
OD	Optical density
ODEs	Ordinary differential equations
PABA	Para-Aminobenzoic acid
PHB	Poly((R)-3-hydroxybutyrate)
LDPE	Polyethylene of low density
PET	Polyethylene terephthalate
PLA	Polylactic acid
PP	Polypropylene
PDSC	Pressurized differential scanning calorimetry
Pym	Pyrimidin-contained tetranitrosyl iron complex
REV	Representative element volume
RNA	Ribonucleic acid
ShO	Sham operated
SNIC	Sulfur nitrosyl iron complexes
TNIC	Tetranitrosyl iron complex
TPP	Tetraphenylporphyrin

THP	Theophylline
TC	Thermomechanical curves
TEM	Transmission electron microscopy
TN	Turnover numbers

Preface

This volume includes information about very important research in the field of organic and physical chemistry: physical principles of the conductivity of electrical conducting polymer compounds, dependence of constants of electromagnetic interactions upon electron spatial-energy characteristic, effect of chitosan molecular weight on rheological behavior of chitosan-modified nanoclay at highly hydrated state bio-structural energy criteria of functional states in normal and pathological conditions, potentiometric study on the interactions between divalent cations and sodium carboxylates in aqueous solution, diffusion of electrolytes and non-electrolytes in aqueous solutions, structural characteristics changes in erythrocyte membranes of mice bearing Alzheimer's-like disease caused by the olfactory bulbectomy, structural state of erythrocyte membranes from humans with Alzheimer's disease, the rheology of particulate-filled polymer nanocomposites, gradiently oriented state of polymers: formation and investigation, organosilicon polymers with photo switchable fragments in the side chain and complexes of iron and cobalt porphyrins for controlled radical polymerization of methyl methacrylate and styrene, and more.

This volume will be very useful for students, engineers, teachers, and researchers who are working in the field of organic and physical chemistry.

The editors and contributors will be happy to receive comments from the readers that we can take in account into our future research.

— Gennady E. Zaikov

1 Dependence of Constants of Electromagnetic Interactions Upon Electron Spatial Energy Characteristic

G. A. Korablev and G. E. Zaikov

CONTENTS

1.1 INTRODUCTION

By the analogy with basic equation of electron spectroscopy of chemical analysis (ESCA) method, the simple dependence of magnetic and electric constants upon electron spatial energy parameter is obtained with semi-empirical method. The relative calculation error is under 0.1%.

1.2 INITIAL CRITERIA

The comparison of multiple regularities of physical and chemical processes allows to assume that in such and similar cases the principle of adding reciprocals of volume energies or kinetic parameters of interacting structures are realized.

Thus, the lagrangian equation for relative motion of the system with two interacting material points with masses m_1 and m_2 in coordinate x with reduced mass m_r, can be written down as follows:

$$\frac{1}{\Delta U} \approx \frac{1}{\Delta U_1} + \frac{1}{\Delta U_2}$$

where ΔU_1 and ΔU_2 = potential energies of material points, ΔU = resultant potential energy of the system.

The electron with mass m, moving near the proton with mass M, is equivalent to the particle with mass $m_r = \frac{mM}{m+M}$.

Therefore, assuming that the energy of atom valence orbitals (responsible for interatomic interactions) can be calculated by following the principle of adding the reciprocals of some initial energy components. The introduction of P-parameter as an averaged energy characteristic of valence orbitals is assigned [1] based on the following equations:

$$\frac{1}{q^2/r_i} + \frac{1}{W_i n_i} = \frac{1}{P_E} \text{ or } \frac{1}{P_0} = \frac{1}{q^2} + \frac{1}{(Wrn)_i} ; P_E = P_0/r \qquad (1), (2), (3)$$

where W_i = orbital energy of electrons [2], r_i = orbital radius of i–orbital [3], $q = Z^*/n^*$, n_i = number of electrons of the given orbital, Z^* and n^* = nucleus effective charge and effective main quantum number [4, 5], r = bond magnitude.

In Equations (1, 2) the parameters q^2 and Wr can be considered as initial (primary) values of P_0-parameter that are tabulated constants for the electrons of the given atom orbital. For its dimensionality it can be written down as follows:

$$[P_0] = \left[q^2\right] = [E]\cdot[r] = [h]\cdot[\upsilon] = \frac{kg\cdot m^3}{s^2} = Jm$$

where [E], [h], and [υ] = dimensionality of energy, plank constant, and velocity.

At the same time, for like-charged systems (for example—orbitals in the given atom) and homogeneous systems, the principle of algebraic addition of such parameters are preserved:

$$\sum P_E = \sum (P_0/r_i), \sum P_E = \frac{\sum P_0}{r} \qquad (4), (4a)$$

Applying the Equation (4, 4a) to hydrogen atom for initial values of P-parameters we can obtain the following:

$$K\left(\frac{e}{n}\right)_1^2 = K\left(\frac{e}{n}\right)_2^2 + mc^2\lambda \qquad (5)$$

where: e = elementary charge, n_1 and n_2 = main quantum numbers, m = electron mass, c = electromagnetic wave velocity, λ = wave length, K = constant.

Using the known correlations $\nu = C/\lambda$ and $\lambda = h/mc$ (where h = Plank constant, ν = wave frequency) from the Equation (5), we can obtain the equation of spectral regularities in hydrogen atom, in which $2\pi^2 e^2/hc = K$.

Taking into account the main quantum characteristics of atom sublevels, and based on the Equation (3) we have the following:

$$\Delta P_{\text{Å}} \approx \frac{\Delta P_0}{\Delta x} \text{ or } \partial P_{\text{Å}} = \frac{\partial P_0}{\partial x},$$

where the value ΔP_0 is equal to the difference between P_0-parameter of *i*-orbital and P_{cn}-parameter of counting (parameter of the main state at the given set of quantum numbers).

According to the rule of adding P-parameters of like-charged or homogeneous systems for two orbitals in the given atom with different quantum characteristics and in accordance with energy conservation law we have:

$$\Delta P''_{Э} - \Delta P'_{Э} = P_{Э,\lambda}$$

где $P_{E,\lambda}$ = spatial energy parameter of quantum transition.

Taking the interaction $\Delta\lambda = \Delta x$ as a magnitude we have:

$$\frac{\Delta P''_0}{\Delta\lambda} - \frac{\Delta P'_0}{\Delta\lambda} = \frac{P_0}{\Delta\lambda} \text{ or } \frac{\Delta P'_0}{\Delta\lambda} - \frac{\Delta P''_0}{\Delta\lambda} = -\frac{P_0}{\Delta\lambda}$$

Let us divide again by term $\Delta\lambda$: $\dfrac{\left(\dfrac{\Delta P'_0}{\Delta\lambda} - \dfrac{\Delta P''_0}{\Delta\lambda}\right)}{\Delta\lambda} = -\dfrac{P_0}{\Delta\lambda^2}$

where $\dfrac{\left(\dfrac{\Delta P'_0}{\Delta\lambda} - \dfrac{\Delta P''_0}{\Delta\lambda}\right)}{\Delta\lambda} \approx \dfrac{d^2 P_0}{\Delta\lambda^2}$, that is: $\dfrac{d^2 P_0}{d\lambda^2} + \dfrac{P_0}{\Delta\lambda^2} \approx 0.$

Taking into account only those interactions when $2\pi\Delta x = \Delta\lambda$ (closed oscillator), we get the equation similar to Schroedinger Equation for stationary state in coordinate x:

$$\frac{d^2 P_0}{dx^2} + 4\pi^2 \frac{P_0}{\Delta\lambda^2} \approx 0$$

Since $\Delta\lambda = \dfrac{h}{mv}$, then:

$$\frac{d^2 P_0}{dx^2} + 4\pi^2 \frac{P_0}{h^2} m^2 v^2 \approx 0 \text{ or } \frac{d^2 P_0}{dx^2} + \frac{8\pi^2 m}{h^2} P_0 E_k = 0$$

where $E_k = \dfrac{mv^2}{2}$ = electron kinetic energy.

Besides we obtained the correlation [1] between the values of P_E-parameter and energy of valence electrons in atom statistic model, which allows to assume that P-parameter is the direct characteristic of electron charge density. Therefore, the exchange spatial energy interactions based on the equalization of electron densities apparently have broad manifestation in physical-chemical processes [6-9].

1.3 ANALOG COMPARISONS

The value of the relative difference of P-parameters of atoms components was used as a complex characteristic of structural interactions:

$$\alpha = \frac{P_1 - P_2}{(P_1 + P_2)/2} \text{ or } \Delta P = \frac{\alpha}{2}\sum P \tag{6}$$

The equalization of electron densities of interacting components brings their unbalanced system into the equalized state which is similar to diffusion processes of transfer. The general equation of mass transfer (m) in x direction (diffusion) is as follows:

$$m = -D\frac{\Delta\rho}{\Delta x}St$$

where D = diffusion coefficient, S = transfer square, t = transfer time, and $\frac{\Delta\rho}{\Delta x}$ = density gradient. Or:

$$m\Delta x = -D\Delta\rho St.$$

As applicable to elementary particles, we can consider that $\Delta\rho$ = value characterizing the energy mass density change, where $mc^2 = E$ (MeV).

As $Er = P_e$ = electron parameter, then there must be a functional bond between P_e-parameter and constants of electromagnetic interactions.

The electron classical radius is:

$$r = \frac{e^2}{mc^2},\ P_e \sim e^2$$

The value of initial P_e-parameter for the electron:

$$P_e = 0.5110 \cdot 2.8179 = 1.43995 MeVfm.$$

In experimental and theoretical investigations in ESCA the following equation is used as the basic one [10]:

$$\Delta W_i = e\sum\frac{1-K}{d_i}q_i \tag{7}$$

where ΔW_i = "shift" in the bond energy (change in the bond energy), d_i = inter nuclear distances, K = constant characterizing the element (approximately equal to the electrostatic interaction between the skeleton and valence electrons in a free atom), q_i = effective transferred charge. As applicable to the binary bond, we have:

$$\Delta W_i d_i = \Delta P = (1-K)q^2 \tag{8}$$

Equation (8) is similar to Equation (6), but the calculations in it are made for initial (primary) P-parameters q^2 and Wd.

Considering the electric charge transfer as the electric current generating the magnetic field under vacuum with magnetic constant μ_0, we can assume by analogy with Equation (8), that it is more an elementary processes:

$$(1-\kappa)q^2 \sim (k\mu_0)^2$$

then $P_e \sim (k\mu_0)^2$,
where k = proportionality coefficient.

The calculations revealed that in these cases the simple correlation for the magnetic constant is fulfilled:

$$(2\pi\mu_0)^2 = 3P_e c^2 \rightarrow 4\pi^2\mu_0^2 = 3P_e c^2 \rightarrow k\mu_0 = P_e^{1/2} c \tag{9}$$

where $k = \frac{2\pi}{\sqrt{3}}$, number 3 is apparently determined by the number of electrons interacting with three quarks or three protons. Transfer coefficient from MeVfm into Jm: $10^6 \cdot 1.602 \cdot 10^{-19} \cdot 10^{-15} = 1.602 \cdot 10^{-28}$ Jm.

Then: $P_e = 1.43995 \cdot 1.602 \cdot 10^{-28} = 2.3068 \cdot 10^{-28}$ Jm.

Calculation of μ_0 by Equation (9) gives $\mu_0 = 1.2554 \cdot 10^{-6} \frac{Gn}{m}$. The relative error in comparison with the actual value of μ_0 is below 0.1%. Value k^2 gets the dimensionality Jm^3/Gn^2, where the numerator characterizes the volume values of P-parameter.

As $P_e \sim e^2$, then the dimensionality $[P_e^{1/2}c]$ gives the physical sense of magnetic constant as the current element for the elementary charge: [Am].

The known correlation between three fundamental constants of electromagnetic interactions can be represented as follows:

$$\mu_0 \tilde{n} = 1/(\varepsilon_0 c)$$

where ε_0 = electric constant.

Using this dependence together with Equation (9), and after the relevant calculations we can obtain the equation P-parameter bond with constants of electromagnetic interactions:

$$k\mu_0 c = \frac{k}{\varepsilon_0 c} = P_e^{1/2} c^2 \approx 1366(Am)\frac{m}{c} \qquad (10)$$

1.4 CONCLUSION

The simple dependence of electromagnetic constants upon initial electron spatial energy characteristics is obtained with semi-empirical method.

KEYWORDS

- **Electromagnetic constants**
- **Electromagnetic interactions**
- **Electron spectroscopy of chemical analysis method**
- **Spatial energy parameter**
- **Semi-empirical method**

REFERENCES

1. Korablev, G. A. *Spatial Energy Principles of Complex Structures Formation*. Brill Academic Publishers and VSP, Netherlands, Leiden, p. 426 (Monograph) (2005).
2. Fischer, C. F. Average-energy of Configuration Hartree-Fock Results for the Atoms Helium to Radon. *Atomic Data*, **4**, 301 (1972).
3. Waber, J. T. and Cromer, D. T. Orbital Radii of Atoms and Ions. *J. Chem. Phys.*, **42**(12), 4116 (1965).
4. Clementi, E. and Raimondi, D. L. Atomic screening constants from S. C. F. functions 1. *J. Chem. Phys.*, **38**(11), 2686 (1963).
5. Clementi, E. and Raimondi, D. L. Atomic screening constants from S. C. F. functions. II. *J. Chem. Phys.*, **47**(4), 1300–1307 (1967).
6. Korablev, G. A. and Zaikov, G. E. Energy of chemical bond and spatial energy principles of hybridization of atom orbitals. *J. of Applied Polymer Science*, USA **101**(3), 2101 (2006).
7. Korablev, G. A. and Zaikov, G. E. Spatial energy interactions of free radicals. *Successes in gerontology*, **21**(4), 535 (2008).
8. Korablev, G. A. and Zaikov, G. E. Spatial energy parameter as a materialized analog of wave function. *Progress on Chemistry and Biochemistry* Nova Science Publishers, Inc. New York, pp. 355–376 (2009).
9. Korablev, G. A. and Zaikov, G. E. Chemical bond energy and spatial energy principles of atom orbital hybridization. *Chemical physics. RAS. M.*, **25**(7), 24 (2006).
10. Zigban, K., Norling, K., Valman, A., et al. *Electron spectroscopy*, (M. Mir), p. 493 (1971).

2 Bio-structural Energy Criteria of Functional States in Normal and Pathological Conditions

G. A. Korablev and G. E. Zaikov

CONTENTS

2.1 INTORDUCTION

With the help of spatial energy notions it is demonstrated that molecular electronegativity (EN) and energy characteristics of functional states of biosystems are defined basically by P-parameter values of atom first valence electron. The principles of stationary biosystems formation are similar to the conditions of wave processes in the phase.

2.2 SPATIAL ENERGY P-PARAMETER

The analysis of kinetics of various physical and chemical processes demonstrates that in many cases the reverse values of velocities, kinetic or energy characteristics of corresponding interactions are added.

A few examples: ambipolar diffusion, summary rate of topochemical reaction, and change in the light velocity when moving from vacuum to a given medium, effective penetration of biomembranes.

In particular, such assumption is confirmed by the probability formula of electron transport (W_∞) due to the overlapping of wave functions 1 and 2 (in stationary state) at electron conformation interactions:

$$W\infty = \frac{1}{2}\frac{W_1 W_2}{W_1 + W_2} \tag{1}$$

Equation (1) is used [1] when assessing the characteristics of diffusion processes accompanied with electron nonradiating transport in proteins.

The modified Lagrangian equation is also illustrative. For the relative movement of isolated system of two interacting material points with the masses m_1 and m_2 in coordinate x it looks as follows:

$$m_{eq}X'' = -\frac{\partial U}{\partial x} \qquad \frac{1}{m_{eq}} = \frac{1}{m_1} + \frac{1}{m_2}$$

where U = mutual potential energy of points and m_{eq} = equivalent mass. Here $x'' = a$ (characteristic of system acceleration). For interaction elementary areas Δx can be taken as follows:

$$\frac{\partial U}{\partial x} \approx \frac{\Delta U}{\Delta x} \quad \text{that is} \quad m_{\text{гд}} a\Delta x = -\Delta U$$

Then: $$\frac{1}{1/(a\Delta x)\left(1/m_1 + 1/m_2\right)} \approx -\Delta U \qquad \frac{1}{1/(m_1 a\Delta x) + 1/(m_2 a\Delta x)} \approx -\Delta U$$

Or: $$\frac{1}{\Delta U} \approx \frac{1}{\Delta U_1} + \frac{1}{\Delta U_2} \tag{2}$$

where ΔU_1 and ΔU_2 = potential energies of material points on the interaction elementary area and ΔU = resultant (mutual) potential energy of these interactions.

"Electron with mass m moving near the proton with mass M is equivalent to the particle with mass $m_{eq} = \frac{mM}{m+M}$" [2].

In this system the energy characteristics of subsystems are: electron orbital energy (W_i) and effective nucleus energy that takes screening effects into account и (by Clementi).

Therefore, assuming that the resultant interaction energy of the system orbital nucleus (responsible for interatomic interactions) can be calculated by the following principle of adding the reverse values of some initial energy components, we substantiate the introduction of P-parameter [3] as averaged energy characteristics of valence orbitals according to the following equations:

$$\frac{1}{q^2/r_i} + \frac{1}{W_i n_i} = \frac{1}{P_E} \tag{3}$$

$$P_E = \frac{P_0}{r_0} \tag{4}$$

$$\frac{1}{P_0} = \frac{1}{q^2} + \frac{1}{(Wrn)_i} \tag{5}$$

$$q = \frac{Z^*}{n^*} \tag{6}$$

where W_i = electron bond energy [4], or (E_i) = atom ionization energy [5], r_i = orbital radius of *i*–orbital [6], n_i = number of electrons of the given orbital, Z^* and n^* = nucleus effective charge and effective main quantum number [7, 8].

P_0 is called the spatial energy parameter, and P_E = the effective P-parameter. The effective P_E-parameter has a physical sense of some averaged energy of valence electrons in the atom and is measured in energy units, for example electron volts (eV).

The values of P-parameters are calculated based on the Equations (3) and (6), some results are given in Table 1. At the same time, the values of parameters during the orbital hybridization are used (marked with index "H") [9]. The calculations are carried out taking into account the possibility of fold bond formation (single, double, and triple). For carbon atom the covalent radius of triple bond (0.60 Å) nearly equals its orbital radius of 2P-orbital (0.59 Å), therefore, the possibility of carbon compound formation becomes more effective.

And now we will briefly discuss the reliability of such approach. According to the calculations [3] the values of P_E-parameters numerically equal (in the limits of 2%) the total energy of valence electrons (*U*) by the atom statistic model. Using a well-known correlation between the electron density (β) and interatomic potential by atom statistic model, we can obtain the direct dependence of P_E-parameter upon the electron density at the distance r_i from the nucleus:

$$\beta_i^{2/3} = \frac{AP_0}{r_i} = AP_E$$

where A = constant.

The rationality of this equation can be confirmed when calculating the electron density using Clementi's wave functions [10] and comparing it with the value of electron density calculated *via* the values of P_E-parameter.

The modules of maximum values of radial part of Ψ function are correlated with the values of P_0-parameter and the linear dependence between these values is established.

Using some properties of wave function as applied to P-parameter, we obtain the wave equation of P-parameter that is formally analogous with Ψ function equation.

Based on the calculations and comparisons, two principles of adding spatial energy criteria depending on wave properties of P-parameter and system character of interactions and particle charges are substantiated:

(1) Interaction of oppositely charged (heterogeneous) systems containing I, II, and III, atom grades is satisfactorily described by the principle of adding their reverse energy values based on Equations (5) and (8), thus corresponding to the minimum of weakening the oscillations in antiphase.
(2) When similarly charged (homogeneous) subsystems are interacted, the principle of algebraic addition of their P-parameters is followed based on the equations:

$$\sum_{i=1}^{m} P_0 = P_0^{'} + P_0^{''} + \ldots + P_0^{m} \tag{7}$$

$$\sum P_E = \frac{\sum P_0}{R} \tag{8}$$

where R = dimensional characteristic of an atom (or chemical bond) thus corresponding to the maximum of enhancing the oscillations in phase.

2.3 P-PARAMETER AS AN OBJECTIVE ELECTRONEGATIVITY ASSESSMENT

The notion of EN was introduced by Poling in 1932s, as a quantitative characteristic of atoms' ability to attract electrons in a molecule. Currently there are a lot of methods to calculate the EN (thermal chemical, geometrical, spectroscopic, etc) producing comparable results. The notion of EN is widely applied in chemical and crystal chemical investigations.

Taking into account Sandersen's idea that EN changes symbately with atom electron density and following the physical sense of P-parameter as a direct characteristic of atom electron density at the distance r_i from the nucleus, we assume that EN for the lowest stable oxidation degree of the element equals the averaged energy of one valence electron:

$$X = \frac{Po}{3R} \tag{9}$$

or for the first valence electron:

$$X = \frac{Po}{3R} \tag{9a}$$

where $\sum P_0$ = total of P_0-parameters for n-valence electrons, R = atom radius (depending on the bond type—metal, crystal or covalent), value of P_0-parameter is calculated *via* the electron bond energy [4]. Digit 3 in the denominator (9, 9a) demonstrates that probable interatomic interaction is considered only along the bond line that is in one of the three spatial directions. The calculation of molecular EN for all elements by these equations is given in Table 2 there is no table for metal one.

For some elements (characterized by the availability of both metal and covalent bonds) the EN is calculated in two variants—with the use of the values of atomic and covalent radii.

The deviations in calculations of EN values by Batsanov [11, 12] and Allred-Rokhov in most cases do not exceed 2–5% from those generally accepted.

Thus, simple correlations (9 and 9a) quite satisfactorily assess the EN value in the limits of its values based on reference data.

The advantage of this approach—greater possibilities of P-parameter to determine the EN of groups and compounds, as P-parameter quite simply (based on the initial rules) can be calculated both for simple and complex compounds.

At the same time, the individual features of structures can be considered, and consequently, important physical-chemical properties of these compounds can be not only characterized but also predicted (isomorphism, mutual solubility, eutectic temperature, etc). For instance, with the help of the notion of P-parameter we can evaluate the EN of not only metal but also crystal structures.

From the data obtained we can conclude that molecular EN for the majority of elements numerically equals the P_E-parameter of the first valence electron divided by three.

It should be noted that there is a significant difference between the notions of EN and P_E-parameter: The EN—stable characteristic of an atom (or radical), and the value of P_E-parameter depends not only on quantum number of valence orbital, but also on the bond length and bond type. Thus, P_E-parameter is an objective and most differentiated energy characteristic of atomic structure.

2.4 SPATIAL ENERGY CRITERIA OF FUNCTIONAL STATES OF BIOSYSTEMS

With the help of P-parameter methodology the spatial energy conditions of isomorphic replacement are found based on experimental data of about a thousand different systems [3]:

TABLE 1 P-parameters of atoms calculated *via* the electron bond energy.

Atom	Valence electrons	W (eV)	r_i (Å)	q^2_0 (eVÅ)	P_0 (eVÅ)	R (Å)	P_0/R (eV)
	$1S^1$	13.595	0.5292	14.394	4.7969	0.5292	9.0644
H						0.375	12.792
						0.28	17.132

TABLE 1 *(Continued)*

Atom	Valence electrons	W (eV)	r_i (Å)	q^2_0 (eVÅ)	P_0 (eVÅ)	R (Å)	P_0/R (eV)
	$2P^1$	11.792	0.596	35.395	5.8680	0.77	7.6208
						0.67	8.7582
						0.60	9.780
	$2P^2$	11.792	0.596	35.395	10.061	0.77	13.066
						0.67	15.016
						0.60	16.769
C	$2P^3_r$				13.213	0.77	17.160
	$2S^1$	19.201	0.620	37.240	9.0209	0.77	11.715
	$2S^2$				14.524	0.77	18.862
	$2S^1+2P^3_r$				22.234	0.77	28.875
	$2S^1+2P^1_r$				13.425	0.77	17.435
	$2S^2+2P^2$				24.585	0.77	31.929
					24.585	0.67	36.694
						0.60	40.975
	$2P^1$	15.445	0.4875	52.912	6.5916	0.70	9.4166
						0.55	11.985
	$2P^2$				11.723	0.70	16.747
						0.63	18.608
N	$2P^3$				15.830	0.70	22.614
						0.55	28.782
	$2S^2$	25.724	0.521	53.283	17.833	0.70	25.476
	$2S^2+2P^3$				33.663	0.70	48.090
	$2P^1$	17.195	0.4135	71.383	6.4663	0.66	9.7979
	$2P^1$					0.55	11.757
	$2P^2$	17.195	0.4135	71.383	11.858	0.66	17.967
						0.59	20.048
O	$2P^4$	17.195	0.4135	71.383	20.338	0.66	30.815
						0.59	34.471
	$2S^2$	33.859	0.450	72.620	21.466	0.66	32.524
	$2S^2+2P^4$				41.804	0.66	63.339
						0.59	70.854

(1) Complete (100%) isomorphic replacement and structural interaction take place at approximate equality of P-parameters of valence orbitals of exchangeable atoms: $P'_E \approx P''_E$

(2) P-parameter of the least value determines the orbital, which is mainly responsible for the isomorphism and structural interactions.

Modifying the rules of adding the reverse values of the energy magnitudes of subsystems as applicable to complex structures, we can obtain the equations to calculate a P_C-parameter of a complex structure [13]:

$$\frac{1}{P_c} = \left(\frac{1}{NP_E}\right)_1 + \left(\frac{1}{NP_E}\right)_2 + \ldots \tag{10}$$

where N_1 and N_2 = number of homogeneous atoms in subsystems.

The calculation results in this equation for some atoms and radicals of biosystems are given in Table 3.

When forming the solution and other structural interactions, the same electron density should be set in the contact spots of atom components. This process is accompanied with the redistribution of electron density between the valence areas of both particles and transition of a part of electrons from some external spheres into the neighboring ones.

Apparently, with the approximation of electron densities in free atom components, the transition processes between boundary atoms of particles are minimal, thus facilitating the formation of a new structure. So, the task of evaluating the degree of structural interactions mostly comes to the comparative assessment of electron density of valence electrons in free atoms (on averaged orbitals) participating in the process based on the following equation:

$$\alpha = \frac{P_E' - P_E''}{P_E' + P_E''} 200\% \tag{11}$$

The degree of structural interactions (ρ) in many (over a thousand) simple and complex systems is evaluated following this technique. The nomogram of ρ dependence on the coefficient of structural interaction (α) is created (Figure 1).

Isomorphism as a phenomenon is usually considered as applicable to crystal structures. But, obviously, similar processes can also flow between molecular compounds, where the bond energy can be evaluated *via* the relative difference of electron densities of valence orbitals of interacting atoms. Therefore, the molecular EN is quite easily calculated *via* the values of the corresponding P-parameters.

Since P-parameter possesses wave properties (similar to Ψ' function), mainly the regularities in the interference of the corresponding waves should be fulfilled.

The interference minimum, oscillation weakening (in anti-phase) takes place if the difference in wave move (Δ) equals the odd number of semi-waves:

$$\Delta = (2n+1)\frac{\lambda}{2} = \lambda(n + \frac{1}{2}), \quad \text{where } n = 0, 1, 2, 3, \tag{12}$$

As applicable to P-parameters, this rule means that the minimum of interactions take place if P-parameters of interacting structures are also "in anti-phase" either oppositely charged systems or heterogeneous atoms are interacting (e.g. during the formation of valence-active radicals CH, CH_2, CH_3, NO_2, etc).

In this case, P-parameters are summed up based on the principle of adding the reverse values of P-parameters—Equations (3–5).

The difference in wave move (Δ) for P-parameters can be evaluated *via* their relative value ($\gamma = \frac{P_2}{P_1}$) or *via* relative difference of P-parameters (coefficient α), which give an odd number at minimum of interactions:

$$\gamma = \frac{P_2}{P_1} = \left(n + \frac{1}{2}\right) = \frac{3}{2}; \frac{5}{2} \ldots \text{ At n = 0 (basic state) } \frac{P_2}{P_1} = \frac{1}{2} \qquad 12(a)$$

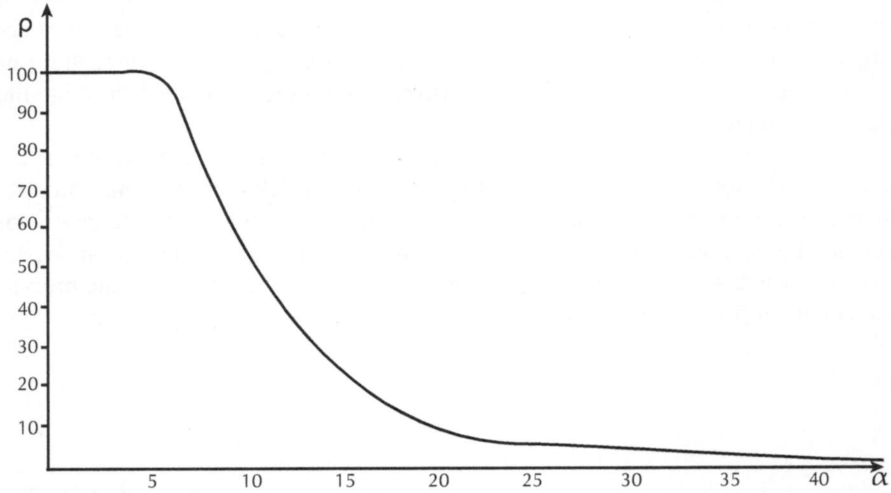

FIGURE 1 Dependence of the degree of structural interactions on coefficient.

Let us point out that for stationary levels of one-dimensional harmonic oscillator the energy of theses levels $\varepsilon = h\nu(n+\frac{1}{2})$, therefore, in quantum oscillator, as opposed to the conventional one, the least possible energy value does not equal zero.

In this model the interaction minimum does not provide the energy equaled to zero corresponding to the principle of adding the reverse values of P-parameters.

Interference maximum, oscillation enhancing (in phase) takes place if the difference in wave move equals an even number of semi-waves:

$$\Delta = 2n\frac{\lambda}{2} = \lambda n \quad \text{or } \Delta = \lambda(n+1) \qquad (13)$$

As applicable to P-parameters, the maximum enhance of interaction in the phase corresponds to the interactions of similarly charged systems or systems homogeneous by their properties and functions (e.g. between the fragments or blocks of complex organic structures, such as CH_2 and N-NO_2 in octogene).

Then:

$$\gamma = \frac{P_2}{P_1} = (n+1) \qquad 13(a)$$

In the same way, for "singular" systems (with similar values of functions) of two-dimensional harmonic oscillator the energy of stationary states is found as follows:

$$\varepsilon = h\nu(n+1)$$

By this model the maximum of interactions corresponds to the principle of algebraic addition of P-parameters Equations (7) and (8). When n = 0 (basic state), we have $P_2 = P_1$, or: the maximum of structure interaction occurs if their P-parameters are equal. This postulate and Equation (13(a)) are used as basic conditions for the formation of stable structures [13].

Hydrogen atom, element No 1 with orbital $1S^1$ defines the main energy criteria of structural interactions (their "ancestor").

Table 1 shows its three P_E-parameters corresponding to three different characteristics of the atom.

TABLE 2 Calculation of molecular EN.

Atom	Orbital	W (eV)	r_i (Å)	q^2 (eVÅ)	P_0 (eVÅ)	R_K (Å)	$P_0/3Rn$ (eV)	X (Batsanov)
1	2	3	4	5	6	7	8	9
Li	$2S^1$	5.3416	1.586	5.8902	3.475	1.33	0.87	0.98
Be	$2S^1$	8.4157	1.040	13.159	5.256	1.13 (M)	1.55	1.52
B	$2P^1$	8.3415	0.770	21.105	4.965	0.81	2.04	2.03
C	$2P^1$	11.792	0.596	35.395	5.868	0.77	2.54	2.56
N	$2P^1$	15.445	0.4875	52.912	6.5903	0.74	2.97	3.05
O	$2P^1$	17.195	0.4135	71.383	6.4660	0.66	3.27	3.42
F	$2S^22P^5$				50.809	0.64	3.78	3.88
Na	$3S^1$	4.9552	1.713	10.058	4.6034	1.54	1.00	0.98
Mg	$3S^1$	6.8859	1.279	17.501	5.8588	1.6 (M)	1.22	1.28
Al	$3P^1$	5.7130	1.312	26.443	5.8401	1.26	1.55	1.57

TABLE 2 *(Continued)*

Atom	Orbital	W (eV)	r_i (Å)	q^2 (eVÅ)	P_0 (eVÅ)	R_K (Å)	$P_0/3Rn$ (eV)	X (Batsanov)
1	2	3	4	5	6	7	8	9
Si	$3P^1$	8.0848	1.068	29.377	6.6732	1.17	1.90	1.89
P	$3P^1$	10.659	0.9175	38.199	7.7862	1.10	2.36	2.19
S	$3P^1$	11.901	0.808	48.108	8.432	1.04	2.57	2.56
Cl	$3P^1$	13.780	0.7235	59.844	8.546	1.00	2.85	2.89
K	$4S^1$	4.0130	2.162	10.993	4.8490	1.96	0.83	0.85
Ca	$4S^1$	5.3212	1.690	17.406	5.9290	1.74	1.14	1.08
Sc	$4S^1$	5.7174	1.570	19.311	6.1280	1.44	1.42	1.39
Ti	$4S^1$	6.0082	1.477	20.879	6.227	1.32	1.57	1.62
V	$4S^1$	6.2755	1.401	22.328	6.3077	1.22	1.72	(1.7)
						1.34	1.57	1.54
Cr	$4S^23d^1$				17.168	1.19	1.60	1.63
Mn	$4S^1$	6.4180	1.278	25.118	6.4180	1.18	1.81	1.73
Fe	$4S^1$	7.0256	1.227	26.572	6.5089	1.17	1.85	1.82
						1.26	1.72	1.74
Co	$4S^1$	7.2770	1.181	27.983	6.5749	1.16	1.89	1.88
Ni	$4S^1$	7.5176	1.139	29.348	6.6226	1.15	1.92	1.98
Cu	$4S^1$	7.7485	1.191	30.117	7.0964	1.31	1.81	(1.8)
	$4S^13d^1$				13.242	1.31	1.68	1.64
Zn	$4S^1$	7.9594	1.065	32.021	6.7026	1.31	1.71	1.72
Ga	$4P^1$	5.6736	1.254	34.833	5.9081	1.25	1.58	(1.7)
	$4S^24P^1$				20.760	1.25	1.82	1.87
Ge	$4P^1$	7.819	1.090	41.372	7.0669	1.22	1.93	2.08
	$4S^24P^2$				30.370	1.22	2.07	

TABLE 2 *(Continued)*

Atom	Orbital	W (eV)	r_i (Å)	q^2 (eVÅ)	P_0 (eVÅ)	R_K (Å)	$P_0/3Rn$ (eV)	X (Batsanov)
1	2	3	4	5	6	7	8	9
As	$4P^1$	10.054	0.9915	49.936	8.275	1.21	2.28	2.23
Se	$4P^1$	10.963	0.909	61.803	8.5811	1.17	2.44	2.48
Br	$4P^1$	12.438	0.8425	73.346	9.1690	1.11	2.75	2.78
Rb	$5S^1$	3.7511	2.287	14.309	5.3630	2.22	0.81	0.82
Sr	$5S^1$	4.8559	1.836	21.224	6.2790	2.00	1.05	1.01
Y	$5S^1$	6.3376	1.693	22.540	6.4505	1.69	1.27	1.26
Zr	$5S^1$	5.6414	1.593	23.926	6.5330	1.45	1.50	1.44
Nb	$5S^1$	5.8947	1.589	20.191	6.3984	1.34	1.52	1.56
Mo	$5S^1$	6.1140	1.520	21.472	6.4860	1.29	1.68	1.73
Tc	$5S^1$	6.2942	1.391	30.076	6.7810	1.27	1.78	1.76
Ru	$5S^1$	6.5294	1.410	24.217	6.6700	1.24	1.79	1.85
Ru (II)	$5S^14d^1$				15.670	1.24	2.11	(2.1)
Rh	$5S^1$	6.7240	1.364	33.643	7.2068	1.25	1.92	1.96
Rh ($5S^14d^8$)	$5S^1$	6.7240	1.364	25.388	6.7380	1.25	1.87	(1.82)
Pd	$5S^24d^2$				30.399	1.28	1.98	1.95

TABLE 2 *(Continued)*

Atom	Orbital	W (eV)	r_i (Å)	q^2 (eVÅ)	P_0 (eVÅ)	R_K (Å)	$P_0/3Rn$ (eV)	X (Batsanov)
1	2	3	4	5	6	7	8	9
Pd	$5S^1$	6.9026	1.325	35.377	7.2670	1.28	1.89	1.95; 1.85
Ag^* ($5S^2 4d^9$)	$5S^1$	7.0655	1.286	37.122	6.9898	1.25	1.86	1.90
Ag ($5S^1 4d^{10}$)	$5S^1$	7.0655	1.286	26.283	6.7520	1.34	1.68	(1.66)
Cd	$5S^1$	7.2070	1.184	38.649	6.9898	1.38	1.69	(1.68); 1.62
Jn	$5S^2 5P^1$				21.841	1.42	1.71	1.76; 1.68
Sn	$5P^1$	7.2124	1.240	47.714	7.5313	1.40	1.79	(1.80); 1.88

R_1 = 0.5292 Å—orbital radius – quantum-mechanical characteristic gives the initial main value of P_E-parameter equaled to 9.0644 eV;

R_2 = 0.375 Å—distance equaled to the half of the bond energy in H_2 molecule. But if hydrogen atom is bound with other atoms, its covalent radius is ≈ 0.28 Å. Let us explain the reason:

In accordance with Equation (13(a)) $P_2 = P_1$ (n+1), therefore $P_1 \approx 9.0644$ eV, $P_2 \approx 18.129$ eV.

These are the values of possible energy criteria of stable (stationary) structures. The dimensional characteristic 0.375 Å does not satisfy them; therefore, there is a transition into the covalence radius ≈ 0.28 Å, which provides the value of P-parameter approximately equaled to P_2.

From a big number of different combinations of interactions we can obtain series with approximately equal values of P-parameters of atoms (or radicals). Such series, by initial values of hydrogen atom, are given in Table 4 (at $\alpha < 7.5\%$).

First series for P_E = 9.0644 eV—the main, initial, where H, C, O, and N atoms have P_E-parameters only of the first electron and interactions proceed in the phase.

Second series for P'_E = 12.792 eV is the nonrational, pathological as it more corresponds to the interactions in antiphase: by Equation (12(a)) P''_E = 13.596 eV.

Coefficient α between the parameters P'_E and P''_E equals 6.1%, thus defining the possibility of forming "false" bio-structures containing the molecular hydrogen H_2. Coefficient α between series I and II is 34.1%, thus confirming the irrationality of series II.

Third series for P'_E = 17.132 eV—stationary as the interactions are in the phase: by Equation (13(a)) P''_E = 18.129 eV (α = 5.5%).

With specific local energy actions (electromagnetic fields, radiation, etc), the structural formation processes can grow along the pathological series II. Maybe this is the reason of oncological diseases? If this is so, the practical recommendations can be given. Some of them are simple and common but are now being substantiated. They propose to transform the molecular hydrogen H_2 into the atomic one. In former times the ashen alkaline water was used in Russian sauna that is OH^- hydroxyl groups. A so-called "live" alkaline water fraction successfully used for the treatment of a number of diseases has the same value. The water containing fluorine and iodine ions is similar.

During the transplantation and use of stem cells the condition of approximate equality of P-parameters of the corresponding structures should be observed (not by the series II).

From Table 4 it is seen that the majority of atoms and radicals, depending on the bond types and bond lengths, have P_E-parameters of different series. When introducing the stem cells, it is important for the molecular hydrogen not to be present in their structures. Otherwise atoms and radicals can transfer into the series II and disturb the vital functions of the main first system.

Tables 2–4 contain only those atoms and radicals that have the main value in the formation of the molecules of DNA, RNA, and nitrogenous bases of nucleic acids (Ц-Г, A-Г). For these pairs the calculations give α equaled to 0.3%. The average values of P-parameters of contacting hydrocarbon rings Ц-$Г_1$ and A-T_1 are also nearly the same. The calculations demonstrate that the systems with P-parameters approximately two times less than in system II are also stable, as in the pair of structural formations they produce the nominal value of the parameter close to the initial one (9.0644 eV).

TABLE 3 Structural P_C-parameters calculated *via* the bond energy of electrons.

Radicals, molecule fragments	P_i' (eV)	P_i'' (eV)	P_C (eV)	Orbitals
	9.7979	9.0644	4.7084	O ($2P^1$)
OH	30.815	17.132	11.011	O ($2P^4$)
	17.967	17.132	8.7710	O ($2P^2$)
H_2O	2·9.0644	17.967	9.0237	O ($2P^2$)
	2·17.132	17.967	11.786	O ($2P^2$)
	17.160	2·9.0644	8.8156	C ($2S^12P^3_г$)
CH_2	31.929	2·17.132	16.528	C ($2S^22P^2$)
	36.694	2·9.0644	12.134	C ($2S^12P^3_г$)
	31.929	3·17.132	19.694	C ($2S^22P^2$)
CH_3	15.016	3·9.0644	9.6740	C ($2P^2$)
	40.975	3·9.0644	16.345	C ($2S^22P^2$)

TABLE 3 *(Continued)*

Radicals, molecule fragments	P_i' (eV)	P_i'' (eV)	P_C (eV)	Orbitals
	36.694	17.132	11.679	C ($2S^22P^2$)
CH	31.929	12.792	9.1330	C ($2S^22P^2$)
	40.975	17.132	12.081	C ($2S^22P^2$)
	16.747	17.132	8.4687	N($2P^2$)
NH	19.538	17.132	9.1281	N($2P^2$)
	48.090	17.132	12.632	N($2S^22P^3$)
	19.538	2·9.0644	9.4036	N($2P^2$)
NH_2	16.747	2·17.132	12.631	N($2P^2$)
	28.782	2·17.132	18.450	N($2P^3$)
C_2H_5	2·31.929	5·17.132	36.585	C ($2S^22P^2$)
NO	19.538	17.967	9.3598	N($2P^2$)
	28.782	20.048	11.817	N($2P^3$)
CH_2	31.929	2·9.0644	11.563	C ($2S^22P^2$)
CH_3	16.769	3·17.132	12.640	C ($2P^2$)
CH_3	17.160	3·17.132	12.865	C ($2P^3_r$)
CO–OH	8.4405	8.7710	4.3013	C ($2P^2$)
CO	31.929	20.048	12.315	C ($2S^22P^2$)
C=O	15.016	20.048	8.4405	C ($2P^2$)
C=O	31.929	34.471	16.576	O ($2P^4$)
CO=O	36.694	34.471	17.775	O ($2P^4$)
C–CH_3	31.929	19.694	12.181	C ($2S^22P^2$)
C–CH_3	17.435	19.694	9.2479	–
C–NH_2	31.929	18.450	11.693	C ($2S^22P^2$)
C–NH_2	17.435	18.450	8.8844	–

TABLE 4 Bio-structural spatial energy parameters (eV).

Series number	H	C	N	O	CH	CO	NH	C–NH$_2$	C–CH$_3$	$\langle P_E \rangle$	α
I	9.0644 ($1S^1$)	8.7582 ($2P^1$) 9.780 ($2P^1$)	9.4166 ($2P^1$)	9.7979 ($2P^1$)	9.1330 ($2S^22P^2$–$1S^1$)	8.4405 ($2P^2$–$2P^2$)	8.4687 ($2P^2$–$1S^1$) 9.1281 ($2P^2$–$1S^1$)	8.8844 $2S^12P^1_r$– ($2P^3$–$1S^1$)	9.2479 $2S^12P^1_r$– ($2S^2 2P^2$–$1S^1$)	9.1018	0.34–7.54
II	12.792 ($1S^1$)	13.066 ($2P^2$) 11.715 ($1S^1$)	11.985 ($2P^1$)	11.757 ($2P^1$)	11.679 ($2S^22P^2$–$1S^1$) 12.081 ($2S^22P^2$–$1S^1$)	12.315 ($2S^22P^2$–$2P^2$)	12.632 ($2S^22P^3$–$1S^1$)	11.693 $2S^22P^2$–($2P^3$–$1S^1$)	12.181 $2S^22P^2$–($2S^22P^2$–$1S^1$)	12.173	0.07–7.08
III	17.132 ($1S^1$)	16.769 ($2P^2$) 17.435 ($2S^12P^1$)	16.747 ($2P^2$)	17.967 ($2P^2$)	C and H blocks	16.576 ($2S^22P^2$–$2P^4$)	N and H blocks	C and NH$_2$ blocks	C and NH$_2$ blocks	17.104	0.16–4.92

Note: The designations of interacting orbitals are given in brackets.

2.5 CONCLUSION

Based on spatial energy notions it is shown that:

(1) Molecular EN of the majority of elements numerically equals the P-parameter of the first valence electron (divided by three).

(2) P-parameters of the first valence electron of atoms define the energy characteristics of stationary states (in normal state) under the condition of the maximum of wave processes.

(3) Under the condition of the minimum of such interactions, the pathological (but not stationary) bio-structures containing the molecular hydrogen can be formed.

KEYWORDS

- **Biosystems**
- **Electronegativity**
- **Lagrangian equation**
- **Spatial energy parameter**
- **Stationary and pathological states**

REFERENCES

1. Rubin, A. B. *Biophysics.* Book 1. Theoretical Biophysics. Vysshaya shkola, Moscow p. 319 (1987).
2. Aring, G., Walter, D., and Kimbal, D. *Quantum Chemistry.* Inostrannaya Literatura, Moscow, p. 528 (1948).
3. Korablev, G. A. *Spatial energy Principles of Complex Structures Formation.* Brill Academic Publishers and VSP, Netherlands, p. 426 (Monograph) (2005).
4. Fischer, C. F. Average-energy of Configuration Hartree-Fock Results for the Atoms Helium to Radon. *Atomic Data*, **4**, 301–399 (1972).
5. Allen, K. U. *Astrophysical magnitudes.* M. Mir, p. 446 (1977).
6. Waber, J. T. and Cromer, D. T. Orbital Radii of Atoms and Ions. *J. Chem. Phys.*, **42**(12), 4116–4123 (1965).
7. Clementi, E. and Raimondi, D. L. Atomic Screening constants from S. C. F. Functions, 1. *J. Chem. Phys.*, **38**(11), 2686–2689 (1963).
8. Clementi, E. and Raimondi, D. L. Atomic Screening constants from S. C. F. Functions, 1. *J. Chem. Phys.*, **47**(14), 1300–1307 (1967).
9. Korablev, G. A. and Zaikov, G. E. Energy of chemical bond and spatial energy principles of hybridization of atom orbitals. *J. of Applied Polymer Science*, **101**(3), 2101–2107 (August 5, 2006)
10. Clementi, E. Tables of atomic functions. *J. B. M. S. Res. Develop/Suppl.*, **9**(2), 76 (1965).
11. Batsanov, S. S. and Zvyagina, R. A. *Overlap integrals and problem of effective charges.* Nauka, Novosibirsk, p. 386 (1966).
12. Batsanov, S. S. *Structural refractometry.* Vysshaya shkola, Moscow, p. 304 (1976).
13. Korablev, G. A. *Exchange spatial energy interactions.* Publishing house of Udmurt State University, p. 530 (Monograph) (2010).

3 Potentiometric Study on the Interactions between Divalent Cations and Sodium Carboxylates in Aqueous Solution

Rui F. P. Pereira, Artur J. M. Valente, Hugh D. Burrows, and Victor M. M. Lobo

CONTENTS

3.1 INTRODUCTION

The interaction of cations with anionic surfactants in aqueous solutions is of both considerable theoretical [1-5] and practical [6, 7], importance, and can have dramatic effects on the mixed solution phase behavior. The factors responsible for the formation of aggregates by amphiphilic molecules in aqueous solutions are well established [6-9], and there is a vast literature of experimental data which supports theoretical predictions.

The charge of these cations suggests that they are likely to bind anions strongly. In addition, aqueous solutions, metal ions are susceptible to hydrolysis [10], which can lead to particularly rich phase behavior. Various methods have been used to study the interaction of cations with anionic surfactants, including surface tension [11-13], electrical conductivity [14-16], potentiometry [17, 18], nuclear magnetic resonance [19,

20], and electron spin resonance spectroscopy [21], self-diffusion measurements [22], ultrafiltration [23], ellipsometry [13], and time-resolved fluorescence quenching [24].

Although these systems have been widely studied and a number of reviews have been reported [25-29], there is not much information about the formation and structure of the resulting aggregates, which complicates the development of models of such systems in solution.

The long chain carboxylates of divalent metal ions (metal soaps) are an important group of compounds, which find applications as emulsifiers, paint driers, grease thickeners, dispersant agents, and so on [30-33]. They are also used in solvent extraction procedures [34], and may find interesting material applications in metal organic mesogen systems [35, 36]. They are in general insoluble in water, but dissolve in a variety of organic solvents.

Of particular importance is the interaction of divalent metal ions, such as calcium(II) and lead(II) with carboxylates. Lead and calcium are ions of high interest due to their industrial, environmental, and biological importance, and the carboxylates of these ions play an important role on, for example, lipid organization in human hair [37] or in aged traditional oil paint [38]. In addition, the two ions have contrasting complexation behavior, with the chemically "harder" [39] calcium(II) forming a more ionic bond with the carboxylate group.

In this chapter, the effect of divalent cations, Pb^{2+} and Ca^{2+}, and carboxylate anions with different alkyl chain lengths (octanoate, decanoate, and dodecanoate) on solution properties, at 25°C, is studied in water using conductimetric and potentiometric techniques. The effect of metal ion hydration on the interaction process is also analyzed. With the present study, we feel that we can contribute to a deeper understanding of the solution structure of these salts which will be valuable in the development of new potential applications for these systems.

The effect of the divalent cations Pb^{2+} and Ca^{2+} on the properties of carboxylates with different alkyl chain lengths (octanoate, decanoate, and dodecanoate) in aqueous solution at 25°C, is evaluated using conductimetric and potentiometric data. Electrical conductance data show strong interaction for concentration ratio (carboxylate divalent ion) up to around two, suggesting an interaction stoichiometry of 2:1. With lead(II), the concentration of free metal ion in solution has also been determined by using a selective electrode, and stability constants are proposed. In the presence of lead ions, the length of the surfactant alkyl chain has a direct influence in the hydrolysis process, followed by pH measurements, and this effect becomes more pronounced upon increasing the number of methylene groups of the carboxylate.

3.2 EXPERIMENTAL

Lead (98%) and calcium nitrate tetrahydrates were purchased from Panreac Quimica SA and Riedel-de Haën, respectively. Sodium octanoate (99%), sodium decanoate (98%), and sodium dodecanoate (99–100%) were purchased from Sigma. These reagents were used as received, and all solutions were prepared using Millipore-Q water. No control was made on the pH, which was the natural value for each solution.

Electrical conductance measurements were carried out with a Wayne-Kerr model 4265 automatic LCR meter at 1 kHz, through the recording of solution electrical resistances, measured by a conductivity cell with a constant of 0.1178 cm^{-1}, uncertainty 0.02% [40]. Cell constant was determined from electrical resistance measurements with KCl (reagent grade, recrystallized, and dried) using the procedure and data of Barthel et al. [41]. Measurements were taken at 25.00°C (±0.02°C) in a Thermo Scientific Phoenix II B5 thermostat bath. Solutions were always prepared immediately before the experiments. In a typical experiment, 20 ml of metal divalent solution 1 mM were placed in the conductivity cell then aliquots of the surfactant solution were added at 4 min intervals by a Gilson Pipetman micropipette. The specific conductance of the solution was measured after each addition and corresponds to the average of three ionic conductances (uncertainty less than 0.2%), determined using homemade software. The specific electrical conductance of the solutions, κ, is calculated from the experimental specific conductance, κ_{exp}, and corrected for the specific conductance of water, κ_0:$\kappa = \kappa_{exp} - \kappa_0$.

Potentiometric measurements were carried out with a pH Radiometer PHM 240. For Pb(II), the metal ion concentration present in solution was measured using a lead selective electrode (WTW, Pb 500) and an Ag/AgCl reference electrode (Ingold). The pH measurements were carried out with a pH conjugated electrode (Ingold U457-K7), the pH was measured on fresh solutions, and the electrode was calibrated immediately before each experimental set of solutions using IUPAC-recommended pH 4 and 7 buffers. In a typical experiment, using a Gilson Pipetman micropipette, aliquots of the surfactant solution were added to 20 ml of metallic divalent solution. In both cases, the electrode potential was recorded after signal stabilization, and all measurements were carried out at 25.00°C (±0.02°C).

3.3 DISCUSSION AND RESULTS

Figure 1 shows the effect of adding sodium carboxylates ($C_nH_{2n+1}COO^-$, which for simplicity we will designate as C_nCOO^-) on the specific conductance of aqueous solutions of lead and calcium nitrate at surfactant concentrations below their critical micelle concentration (cmcs: C_7COO^- 0.34 M; C_9COO^- 94 mM, $C_{11}COO^-$ 24 mM [42]). With the aqueous Pb^{2+} solution, addition of sodium carboxylate leads to a decrease in the specific conductance until around 2 mM carboxylate. Since the specific conductance gives a measure of the ion concentration in solution, this observation indicates a strong interaction between the ions of the two salts, with the consequent charge neutralization [16]. We can also conclude that the change in the k = f(c) behavior occurs for a molar ratio $[C_nCOO^-]/[Pb^{2+}]$ of 2, indicating an interaction stoichiometry is 2:1 (C_nCOO^-:Pb^{2+}), consistent with the charge neutralization. We can also see that in the case of Pb^{2+}, the effect of alkyl chain length is not significant for the solutions behavior [32]. For molar ratios above 2, the specific conductance increases with increasing concentration of carboxylate. This behavior can readily be understood since after all the Pb^{2+} has been "consumed" by the carboxylate, the excess of surfactant produces an increase in the solution conductivity, similar to what occurs in the absence of the divalent ions.

However, the carboxylate behavior, in presence of Ca^{2+} is significantly different from that observed for Pb^{2+}. With calcium(II), it is possible to distinguish three different zones of k as a function of the surfactant concentration, depending on the alkyl chain length.

The addition of sodium octanoate leads to an increasing electrical conductivity of the resulting solution, which indicates the absence of carboxylate calcium(II) interactions, or, if they exist they are so weak and cannot be detected by electrical conductivity. For sodium decanoate, the electrical conductivity behavior is very similar to what occurs in solutions for example, sodium dodecyl sulfate and trivalent ions [43]. That is, the addition of decanoate to a solution containing Ca^{2+} leads to an initial increase of solution electrical conductivity because there is no interaction between species in this zone, however, for a molar ratio $[C_9COO^-]/[Ca^{2+}]$ around 1, the specific conductance tends to stabilize, which can be interpreted either as a critical aggregation concentration or ion pair formation, after this, the Ca^{2+} species interact with the decanoate, until the stoichiometric a molar ratio of 2.0, the specific conductance increases again.

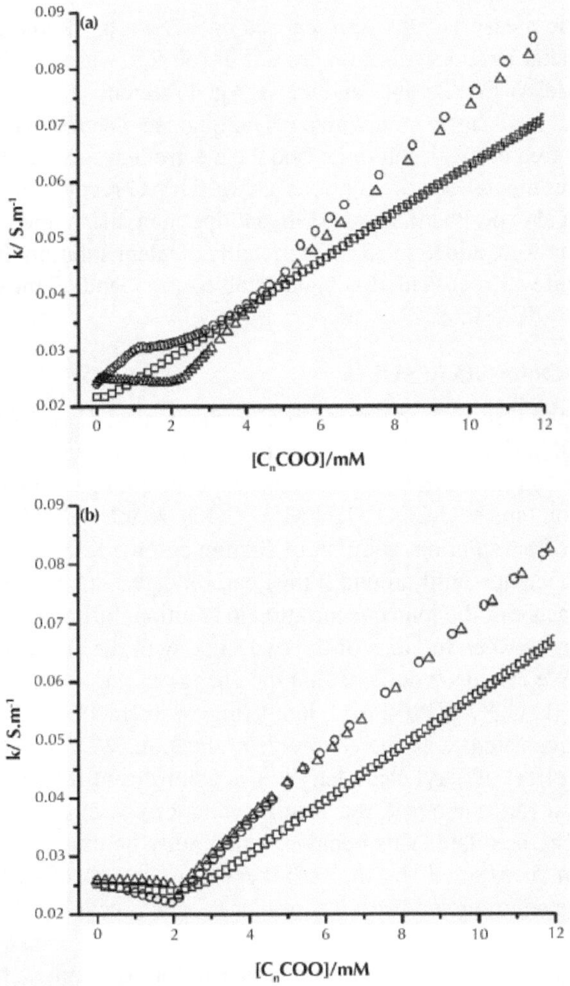

FIGURE 1 Effect of Ca^{2+} 1 mM (a) and Pb^{2+} 1 mM and (b), in the specific conductivity of sodium octanoate (□, n = 7), sodium decanoate (○, n = 9), and sodium dodecanoate (Δ, n = 11) at 25°C.

However, in the case of sodium dodecanoate, interaction with Ca^{2+} occurs from the first addition of carboxylate to the divalent ion solution. From these results we can conclude that, in the case of Ca^{2+}, the interaction with carboxylates is not only affected by ionic interactions, but also by the interactions involving the hydrophobic chain of the amphiphilic ion. The differences in behavior can be justified by the higher charge density of the hydrated ions of Pb^{2+} when compared with Ca^{2+}, because their radii of hydration are 4.01 Å and 4.12 Å [44], respectively. In addition, crystallographic evidence has been presented that the lead(II) carboxylate bond has appreciable covalent character [45] here, as would be expected from it being a softer acid than Ca(II). Another important point that cannot be neglected in the analysis of the experimental results presented in Figure 1 is related with the possible influence of hydrolysis processes in the interaction between the carboxylate and divalent metal ions [10].

Figure 2 shows the variation of pH in a titration of a 1 mM solution of divalent ion with sodium carboxylates having different alkyl chain lengths. It is interesting to note that in the case of solutions containing Ca^{2+}, there is a pH increase with the addition of octanoate and decanoate, following a behavior similar to what occurs in the absence of divalent ions (Figure 3), and that the hydrolysis of alkanoates increases with increasing the alkyl chain length. This behavior, which is consistent with that observed by analysis of electrical conductivity measurements, suggests that in this concentration range Ca^{2+}/carboxylate interactions do not exist or are very weak.

The addition of $C_{11}COO^-$ to solutions of Ca^{2+} or Pb^{2+}, initially leads to a pH decrease, which can be explained by the breakup of dimers/oligomers of carboxylates following the interaction with divalent metals, with the consequent release of hydronium ions [46]. As was observed in the conductivity measurements, the presence of Pb^{2+} leads to more significant changes in the macroscopic physical-chemical properties of those mixed solutions.

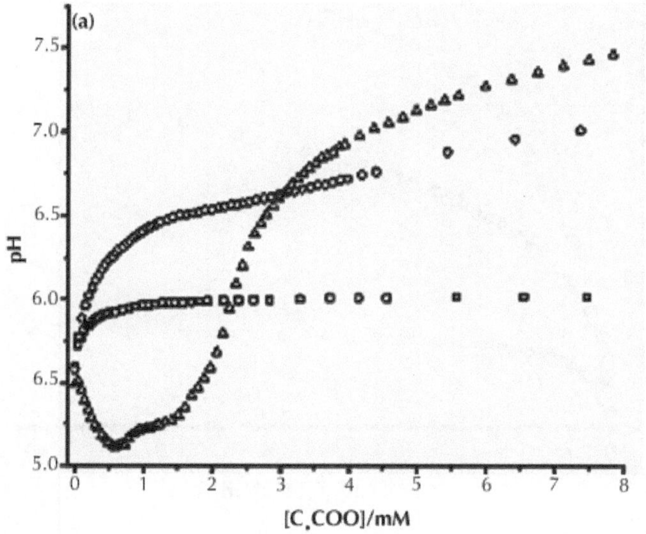

FIGURE 2 *(Continued)*

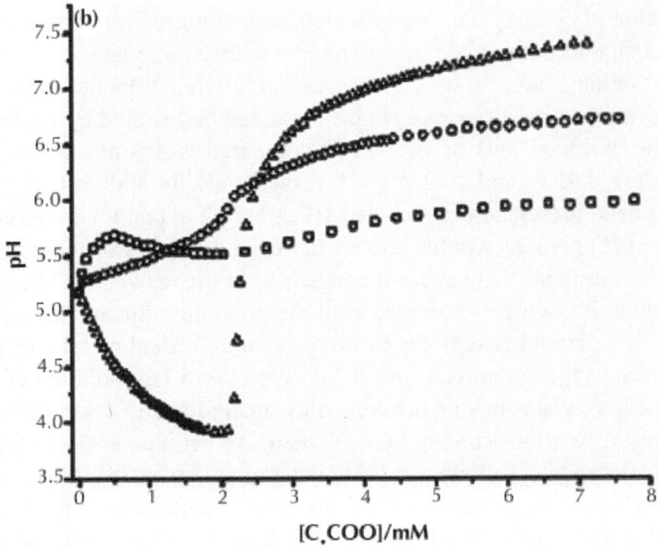

FIGURE 2 Effect of the addition of sodium alkyl carboxylates on the pH of 1 mM $Ca(NO_3)_2$ (a) and $Pb(NO_3)_2$ (b) solutions, at 25°C, (□) sodium octanoate, (○) sodium decanoate, and (Δ) sodium dodecanoate.

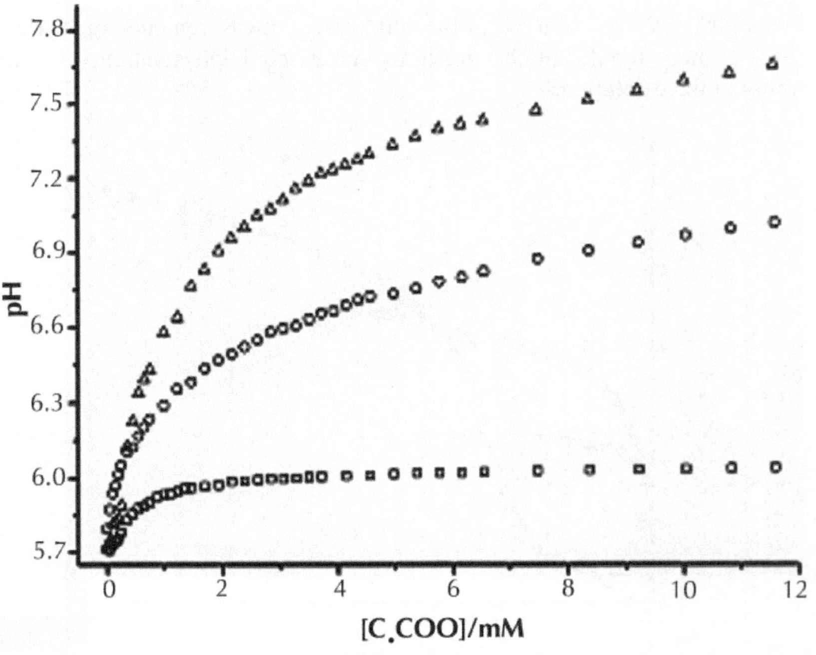

FIGURE 3 Dependence of pH on the concentration of sodium octanoate (□), sodium decanoate (○), and sodium dodecanoate (Δ) aqueous solution at 25°C.

From analysis of Figure 2(b), we can also conclude that the addition of C_7COO^- and C_9COO^- to Pb^{2+} solutions, leads to an increase in pH values but with different trends of those obtained in aqueous solutions in the absence of divalent ion.

From the analysis of Figures 1, 2, and 4 it is still possible to obtain the interaction stoichiometry between the divalent ions and different alkanoates. The calculated values, by using different techniques, are summarized in Table 1. It appears that the interaction of $C_nCOO^-:Pb^{2+}$ and $C_{11}COO^-:Ca^{2+}$ occurs with a stoichiometry of 2:1, which is consistent with charge neutralization [45-47].

As observed, the interaction between Pb^{2+} and the alkanoates studied is independent of the size of the alkyl chain, and as such it is relevant to quantify the association constant. For this purpose, potentiometric measurements were carried out to quantify the concentration of free lead in solutions containing alkanoates (Figure 4). From analysis of Figure 4, it appears that the concentration of free Pb^{2+} in solution decreases with increasing concentration of alkanoates, reaching a minimum at a concentration of alkanoates approximately of 2 mM, for an initial 1 mM Pb^{2+} solutions, which is again consistent with the values of stoichiometry association calculated by electrical conductivity.

TABLE 1 Interaction stoichiometry of divalent ions (Ca^{2+} and Pb^{2+}) and sodium carboxylates obtained by different techniques.

	$[C_nCOO^-]/[M^{2+}]$		
	Conductivity	**pH measurements**	**Selective electrode**
Ca^{2+}/C_7COO^-	Not detected	Not detected	-
Ca^{2+}/C_9COO^-	2.6 ± 0.2	–	-
$Ca^{2+}/C_{11}COO^-$	2.15 ± 0.02 2.26 ± 0.01[d]	2.35 ± 0.01[a] 2.304 ± 0.009[c]	-
Pb^{2+}/C_7COO^-	2.49 ± 0.03 2.36 ± 0.06[d]	1.85 ± 0.01[b]	2.35 ± 0.05
Pb^{2+}/C_9COO^-	1.88 ± 0.04 1.917 ±0.008[d]	2.08 ± 0.05[a] 2.15 ± 0.01[c]	1.89 ± 0.15
$Pb^{2+}/C_{11}COO^-$	2.18 ± 0.06 2.08 ± 0.01[c]	2.38 ± 0.03[a] 2.307 ± 0.009[c]	2.08 ± 0.05

[a] Boltzmann fit; [b] Gauss fit; [c]First derivative; [d] Second derivative.

It is worth noticing that the interaction between Pb^{2+} and alkanoates (Figure 2(b)) is quite different from that observed with, for example, sodium dodecyl sulfate and trivalent ions [48]. In that case, it is observed that under surfactant excess conditions, the precipitate (aggregate formed by dodecyl sulfate and the metal ion) redissolves, due to preferential hydrophobic interactions between the alkyl chains of the surfactant, destabilizing the aggregates of dodecyl sulfate/metal ion. The non-occurrence of redissolution of lead alkanoates, under conditions of excess alkanoate is in agreement with the lead(II) carboxylate bond having some covalency, and may indicate the existence of weak interactions between the alkyl chains of the carboxylates.

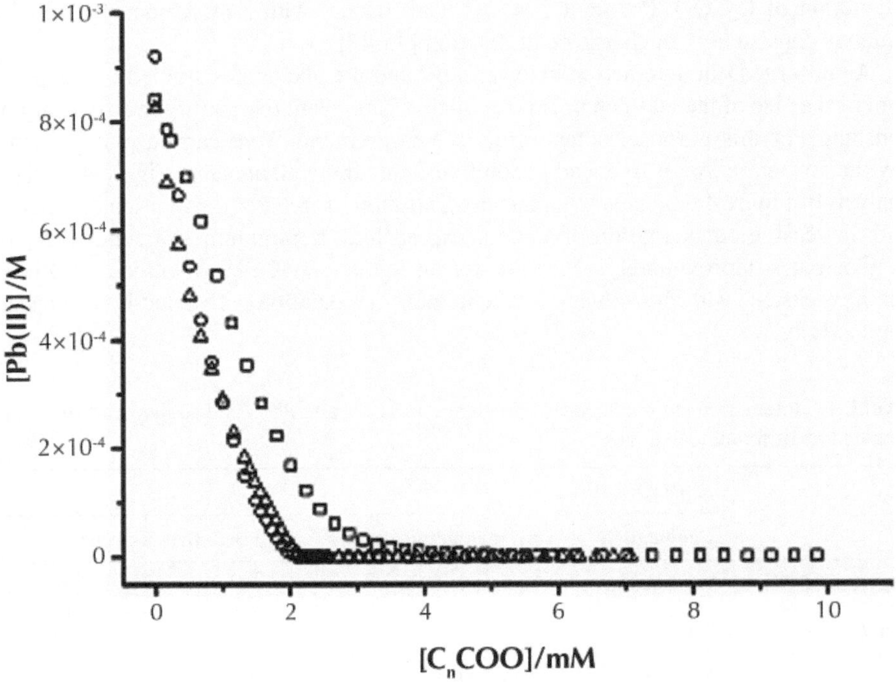

FIGURE 4 Effect of sodium octanoate (□), sodium decanoate (○), and sodium dodecanoate (Δ) in the concentration of free Pb^{2+} (in equilibrium) at 25°C.

Assuming that the reaction between Pb^{2+} and the carboxylates (C_nCOO^-, with n = 7, 9, 11) can be described by Equation (1), the association constant, K_a, can be defined by Equation (2):

$$Pb^{2+} + 2\,C_nCOO^- \rightleftharpoons Pb(C_nCOO)_2 \tag{1}$$

$$K_a = [Pb(C_nCOO)_2]_{eq}/([C_nCOO^-]_{eq}^{\,2}\,[Pb^{2+}]_{eq}) \tag{2}$$

Considering the mass balance equations,

$$[Pb(C_nCOO)_2]_{eq} = [Pb^{2+}]_i - [Pb^{2+}]_{eq} \tag{3}$$

$$[C_nCOO^-]_{eq} = [C_nCOO^-]_i - 2[Pb^{2+}]_{eq} \quad (4)$$

where subscripts *eq* and *i* indicate the concentrations of species in equilibrium and initially, respectively; from (3) and (4) and (2) we have:

$$K_a = \frac{[Pb^{2+}]_i - [Pb^{2+}]_{eq}}{[Pb^{2+}]_{eq}([C_nCOO^-]_i^2 + 4[Pb^{2+}]_{eq}^2 - 4[Pb^{2+}]_{eq}[C_nCOO^-]_i)} \quad (5)$$

Using the experimental values of $[Pb^{2+}]_{eq}$, shown in Figure 4, and the initial concentration of carboxylate, it is possible to calculate the association constants (K_a). The K_a values obtained were analyzed using the Grubbs method resulting the values 1.8 (±0.5) × 10^6 M^{-2}, 5.2 (±0.6) × 10^6 M^{-2} and 3.8 (±0.7) × 10^6 M^{-2} for the formation of lead octanoate, decanoate, and dodecanoate, respectively.

3.4 CONCLUSION

The interactions between the divalent cations Ca^{2+} and Pb^{2+}, and sodium carboxylates with different alkyl chain lengths were studied in water using conductimetric and potentiometric techniques. In the case of Ca^{2+}, the interaction with carboxylates is not only affected by ionic interactions, but also by the interactions involving the hydrophobic chain of the amphiphilic ion. The interaction between Ca^{2+} and sodium octanoate does not exist or is very weak. The Pb^{2+} carboxylate interactions are not significantly affected by the surfactant chain length, in agreement with the idea of some specific interactions in the complexation due to the bonds having substantial covalent character. However, the presence of Pb^{2+} markedly affects the physico-chemical properties of the solutions when compared with Ca^{2+}. The interactions C_nCOO^- Pb^{2+} and $C_{11}COO^-Ca^{2+}$ occur with a stoichiometry of 2:1, which is consistent with charge neutralization. Considering this interaction stoichiometry, association constant values are proposed for the lead carboxylate system.

KEYWORDS

- **Calcium(II)**
- **Electrical conductivity**
- **Lead(II)**
- **Potentiometry**
- **Sodium carboxylates**

ACKNOWLEDGMENT

Rui F. P. Pereira thanks FCT for a PhD. grant (SFRH/BD/38696/2007).

REFERENCES

1. Stigter, D. *J. Phys. Chem.*, **79**, 1008–1014 (1975).
2. Stigter, D., *J. Phys. Chem.*, **79**, 1015–1022 (1975).
3. Gunnarsson, G., Jonsson, B., and Wennerstrom, H. *J. Phys. Chem.*, **84**, 3114–3121 (1980).

4. Evans, D. F., Mitchell, D. J., and Ninham, B. W. *J. Phys. Chem.*, **88**, 6344–6348 (1984).
5. Evans, D. F. and Ninham, B. W. *J. Phys. Chem.*, **90**, 226–234 (1986).
6. Holmberg, K., Jönsson, B., Kronberg, B., and Lindman, B. *Surfactants and Polymers in Aqueous Solution* 2nd ed., Wiley Chichester, UK (2003).
7. Bunton, C. A., Nome, F., Quina, F. H., and Romsted, L. S. *Accounts Chem. Res.*, **24**, 357–364 (1991).
8. Israelachvili, J. N., Mitchell, D. J., and Ninham, B. W. *J. Chem. Soc., Faraday Trans. II*, **72**, 1525–1568 (1976).
9. Israelachvili, J. N. *Intermolecular and Surface Forces*. Academic Press, London (1985).
10. Baes, C. F. and Mesmer, R. E. *The Hydrolysis of Cations*. Wiley, New York (1976).
11. Miyamoto, S. *Bull. Chem. Soc. Jpn.*, **33**, 375–379 (1960).
12. Hato, M. and Shinoda, K. *J. Phys. Chem.*, **77**, 378–381 (1973).
13. Teppner, R., Haage, K., Wantke, D., and Motschmann, H. *J. Phys. Chem. B*, **104**, 11489–11496 (2000).
14. Bunton, C. A., Ohmenzetter, K., and Sepulveda, L. *J. Phys. Chem.*, **81**, 2000–2004 (1977).
15. Pereira, R. F. P., Valente, A. J. M., and Burrows, H. D. *J. Mol. Liq.*, **176**, 109–114 (2010).
16. Valente, A. J. M., Burrows, H. D., Pereira, R. F., Ribeiro, A. C. F., Pereira, J., and Lobo, V. M. M. *Langmuir*, **22**, 5625–5629 (2006).
17. Cutler, S. G., Meares, P., and Hall, D. G. *J. Chem. Soc., Faraday Trans. I*, **74**, 1758–1767 (1978).
18. Rathman, J. F. and Scamehorn, J. F. *J. Phys. Chem.*, **88**, 5807–5816 (1984).
19. Robb, I. D. *J. Colloid Interface Sci.* **37**, 521–527 (1971).
20. Rosenholm, J. B. and Lindman, B. *J. Colloid Interface Sci.*, **57**, 362–378 (1976).
21. Stilbs, P. and Lindman, B. *J. Colloid Interface Sci.*, **46**, 177–179 (1974).
22. Clifford, J. and Pethica, B. A. *Trans. Faraday Soc.*, **60**, 216–224 (1964).
23. Hafiane, A., Issid, I., and Lemordant, D. *J. Colloid Interface Sci.*, **142**, 167–178 (1991).
24. Okano, L. T., Alonso, E. O., Waissbluth, O. L., and Quina, F. H. *Photochem. Photobiol.*, **63**, 746–749 (1996).
25. Pilpel, N. *Chem. Rev.*, **63**, 221–234 (1963).
26. Winsor, P. A. *Chem. Rev.*, **68**, 1–40 (1968).
27. Burrows, H. D. In *The Structure Dynamics and Equilibrium Properties of Colloidal Systems*. D. M. Bloor and E. Wyn-Jones (Eds.). Kluwer Dordrecht p. 415 (1990).
28. Lyklema, J. *Adv. Colloid Interface Sci.*, **2**, 65 (1968).
29. Kertes, A. S. and Gutman, H. *Surface and Colloid Science*. Vol **8**, Plenum, New York (1975).
30. Buono, F. J. and Feldman, M. L. In *Kirk–Othmer Encyclopedia of Chemical Technology*. Vol. 8, Wiley, New York, p. 34 (1979).
31. Parfitt, G. D. and Peacock, J. *Surface and Colloid Science*. Vol. 10, Plenum, New York, p. 163 (1978).
32. Akanni, M. S., Okoh, E. K., Burrows, H. D., and Ellis, H. A. *Thermochim. Acta,* **208**, 1–41 (1992).
33. Bossert, R. G. *J. Chem. Educ.*, **27**, 10–14 (1950).
34. Yamada, H. and Tanaka, M. *Adv. Inorg. Chem.*, **29**, 143–168 (1985).
35. Binnemans, K. and Gorller-Walrand, C. *Chem. Rev.*, **102**, 2303–2345 (2002).
36. Donnio, B. *Curr. Opin. Colloid Interface Sci.*, **7**, 371–394 (2002).
37. Bertrand, L., Doucet, J., Simionovici, A., Tsoucaris, G., and Walter, P. *Biochim. Biophys. Acta-Gen. Subj.*, **1620**, 218–224 (2003).
38. Plater, M. J., De Silva, B., Gelbrich, T., Hursthouse, M. B., Higgitt, C. L., and Saunders, D. R. *Polyhedron,* **22**, 3171–3179 (2003).
39. Pearson, R. G. *Inorg. Chem.*, **27**, 734–740 (1988).
40. Ribeiro, A. C. F., Valente, A. J. M., Lobo, V. M. M., Azevedo, E. F. G., Amado, A. M., da Costa, A. M. A., Ramos, M. L., and Burrows, H. D. *J. Mol. Struct.*, **703**, 93–101 (2004).
41. Barthel, J., Feuerlein, F., Neueder, R., and Wachter, R. *J. Solut. Chem.*, **9**, 209–219 (1980).
42. Mukerjee, P. and Mysels, K. J. *Critical Micelle Concentrations of Aqueous Surfactatns Systems.* NBS Wasington DC, p. 222 (1971).

43. Pereira, R. F. P., Valente, A. J. M., Burrows, H. D., Ramos, M. L., Ribeiro, A. C. F. and Lobo, V. M. M. *Acta Chim. Slov.*, **56**, 45–52 (2009).
44. Nightingale, E. R. *J. Phys. Chem.*, **63**, 1381–1387 (1959).
45. Lacouture, F., Francois, M., Didierjean, C., Rivera, J. P., Rocca, E., and Steinmetz, J. *Acta Crystallogr. Sect. C-Cryst. Struct. Commun.* **57**, 530–531 (2001).
46. Kralchevsky, P. A., Danov, K. D., Pishmanova, C. I., Kralchevska, S. D., Christov, N. C., Ananthapadmanabhan, K. P., and Lips, A. *Langmuir,* **23**, 3538–3553 (2007).
47. Ellis, H. A., White, N. A., Hassan, I., and Ahmad, R. *J. Mol. Struct.*, **642**, 71–76 (2002).
48. Pereira, R. F. P., Tapia, M. J., Valente, A. J. M., Evans, R. C., Burrows, H. D., and Carvalho, R. A. *J. Colloid Interface Sci.* **354**, 670–676 (2011).

4 Diffusion of Electrolytes and Non-electrolytes in Aqueous Solutions: A Useful Strategy for Structural Interpretation of Chemical Systems

Ana C. F. Ribeiro, Joaquim J. S. Natividade, Artur J. M. Valente, and Victor M. M. Lobo

CONTENTS

4.1 INTRODUCTION

In the last years, the diffusion Coimbra group, headed by Prof. Lobo, has been particularly dedicated to the study of mutual diffusion behavior of binary, ternary, and quaternary solutions [1-19], involving electrolytes and non-electrolytes, helping to go deeply in the understanding of their structure, aiming at practical applications in fields as diverse as corrosion studies occurring in biological systems or therapeutic uses. In fact, the scarcity of diffusion coefficients data in the scientific literature, due to the difficulty of their accurate experimental measurement and impracticability of their determination by theoretical procedures, allied to their industrial and research need, well justify our efforts in accurate measurements of such transport property.

This transport property has been measured in different conditions (several electrolytes, concentrations, temperatures, and techniques used), having in mind a contribution to a better understanding of the structure of those solutions, behavior of electrolytes or non-electrolytes in solution and, last but not the least, supplying the scientific and technological communities with data on this important parameter in solution transport processes. Whereas an open ended capillary cell developed by Lobo has been used to obtain mutual diffusion coefficients of a wide variety of electrolytes [1, 2], the Taylor technique has been used mainly for ternary and quaternary systems with non-electrolytes (e.g., [11-19]). From comparison between our experimental results and those obtained from different models, for example Nernst, Nernst-Hartley, Stokes, Onsager, Fuoss, and Pikal theoretical equations, and from our semiempirical equations, and Gordon's and Agar's as well, it has been possible to obtain some structural information, such as diffusion coefficient at infinitesimal concentration, ion association, complex formation, hydrolysis, hydration, or estimations of the mean distance of closest approach involving ions as diffusing entities.

4.1.1 Concepts of Diffusion in Solutions

Many techniques are used to study diffusion in aqueous solutions. It is very common to find misunderstandings concerning the meaning of a parameter, frequently just denoted by D and merely called diffusion coefficient, in the scientific literature, communications, meetings, or simple discussions among researchers. In fact, it is necessary to distinguish self-diffusion (interdiffusion, tracer diffusion, single ion diffusion, ionic diffusion) and mutual diffusion (interdiffusion, concentration diffusion, salt diffusion) [20, 21]. The NMR and capillary tube, the most popular methods, can only be used to measure interdiffusion coefficients [20, 21]. In our case, the mutual diffusion is analyzed [20, 21].

Mutual diffusion coefficient, D, in a binary system, may be defined in terms of the concentration gradient by a phenomenological relationship, known as Fick's first law.

$$J = -D\frac{\partial c}{\partial x} \tag{1}$$

where J represents the flow of matter across a suitable chosen reference plane per area unit and per time unit, in a one-dimensional system, and c is the concentration of solute in moles per volume unit at the point considered: (1) may be used to measure D. The diffusion coefficient may also be measured considering Fick's second law.

$$\frac{\partial c}{\partial t} = \frac{\partial}{\partial x}\left(D\frac{\partial c}{\partial x}\right) \tag{2}$$

In general, the available methods are assembled into two groups: steady and unsteady-state methods, according to Equations (1) and (2). In most of the processes, diffusion is a three-dimensional (3D) phenomenon. However, many of the experimental methods used to analyze diffusion restrict it to a one-dimensional process, making it much easier to study its mathematical treatments in one dimension (which then may be generalized to a 3D space).

The resolution of Equation (2) for a unidimensional process is much easier if we consider *D* as a constant. This approximation is applicable only when there are small differences of concentration, which is the case of our open-ended conductimetric technique, and of the Taylor technique [20, 21]. In these conditions, it is legitimate to consider that our measurements of differential diffusion coefficients obtained by the techniques are parameters with a well defined thermodynamic meaning [20, 21].

In our research group, we also have measured mutual diffusion for multicomponent systems, that is, for ternary and, more recently, quaternary systems.

Diffusion in a ternary solution is described by the diffusion equations (Equations 3 and 4).

$$-(J_1) = (D_{11})_V \frac{\partial c_1}{\partial x} + (D_{12})_V \frac{\partial c_2}{\partial x} \tag{3}$$

$$-(J_2) = (D_{21})_V \frac{\partial c_1}{\partial x} + (D_{22})_V \frac{\partial c_2}{\partial x} \tag{4}$$

where J_1, J_2, $\frac{\partial c_1}{\partial x}$ and $\frac{\partial c_2}{\partial x}$ are the molar fluxes and the gradients in the concentrations of solute 1 and 2, respectively. The index V represents the volume fixed frame of the reference used in these measurements. Main diffusion coefficients give the flux of each solute produced by its own concentration gradient. Cross diffusion coefficients D_{12} and D_{24} give the coupled flux of each solute driven by the concentration gradient in the other solute. A positive D_{ik} cross coefficient ($i \neq k$) indicates co-current coupled transport of solute i from regions of higher to lower concentrations of solute k. However, a negative D_{ik} coefficient indicates counter-current coupled transport of solute i from regions of lower to higher concentration of solute k.

Recently, diffusion in a quaternary solution has been described by the diffusion equations[1] (Equations 5-7) [19].

---

[1]An aqueous quaternary system, which for brevity we will indicate with ijk, not indicating the solvent 0, has three corresponding aqueous ternary systems (ij, ik, and jk), and three corresponding aqueous binary systems (i, j, and k). The main term quaternary diffusion coefficients can then be compared with two ternary values, ${}^{ij}D_{ii}$ and ${}^{ik}D_{ii}$, and with one binary value; similarly for the other two main terms ${}^{ijk}D_{jj}$ and ${}^{ijk}D_{kk}$. The quaternary cross diffusion coefficient ${}^{ijk}D_{ij}$ can be compared only with one ternary diffusion coefficient ${}^{ij}D_{ij}$; this is also true for all the other cross terms. The comparison between the diffusion coefficients of system ijk with those of the systems ij, ik, and jk, permits to obtain information on the effect of adding each solute to the other two. The comparison between the diffusion coefficients of the quaternary system with those of the systems ijk, and with those of the systems i, j, and k, permits to obtain information on the effect of adding each couple of solutes to the other one.

$$-(J_1) = {}^{123}(D_{11})_v \frac{\partial c_1}{\partial x} + {}^{123}(D_{12})_v \frac{\partial c_2}{\partial x} + {}^{123}(D_{13})_v \frac{\partial c_3}{\partial x} \tag{5}$$

$$-(J_2) = {}^{123}(D_{21})_v \frac{\partial c_1}{\partial x} + {}^{123}(D_{22})_v \frac{\partial c_2}{\partial x} + {}^{123}(D_{23})_v \frac{\partial c_3}{\partial x} \tag{6}$$

$$-(J_3) = {}^{123}(D_{31})_v \frac{\partial c_1}{\partial x} + {}^{123}(D_{32})_v \frac{\partial c_2}{\partial x} + {}^{123}(D_{33})_v \frac{\partial c_3}{\partial x} \tag{7}$$

where the main diffusion coefficients ${}^{123}D_{11}$, ${}^{123}D_{22}$ and ${}^{123}D_{33}$, give the flux of each solute produced by its own concentration gradient. Cross diffusion coefficients, ${}^{123}D_{12}$, ${}^{123}D_{13}$, ${}^{123}D_{21}$, ${}^{123}D_{23}$, ${}^{123}D_{31}$, and ${}^{123}D_{32}$ give the coupled flux of each solute driven by a concentration gradient in the other solute.

4.2 EXPERIMENTAL TECHNIQUES: CONDUCTIMETRIC AND TAYLOR DISPERSION TECHNIQUES

Experimental methods that can be employed to determine mutual diffusion coefficients [20, 21]: Diaphragm cell (inaccuracy 0.5-1%), conductimetric (inaccuracy 0.2%), Gouy and Rayleigh interferometry (inaccuracy <0. 1%), and Taylor Dispersion (inaccuracy 1-2%). While the first and second method consume days in experimental time, the last one imply just hours. The conductimetric technique follows the diffusion process by measuring the ratio of electrical resistances of the electrolyte solution in two vertically opposed capillaries as time proceeds. Despite this method has given us reasonably precise and accurate results, it is limited to studies of mutual diffusion in electrolyte solutions, and like in diaphragm cell experiments, the run times are inconveniently long (~days). The Gouy method also has high precision, but when applied to microemulsions they are prone to gravitational instabilities and convections. Thus, the Taylor dispersion has become increasingly popular for measuring diffusion in solutions, because of its experimental short time and its major application to the different systems (electrolytes or non electrolytes). In addition, with this method it is possible to measure multicomponent mutual diffusion coefficients.

Mutual differential diffusion coefficients of binary (e.g. [6-8]) and pseudo binary systems (such as, e.g., cobalt chloride in aqueous solutions of sucrose [9]), have been measured using a conductimetric cell and an automatic apparatus to follow diffusion. This cell uses an open ended capillary method and a conductimetric technique is used to follow the diffusion process by measuring the resistance of the solution inside the capillaries, at recorded times. Figure 1 shows a schematic representation of the open-ended capillary cell.

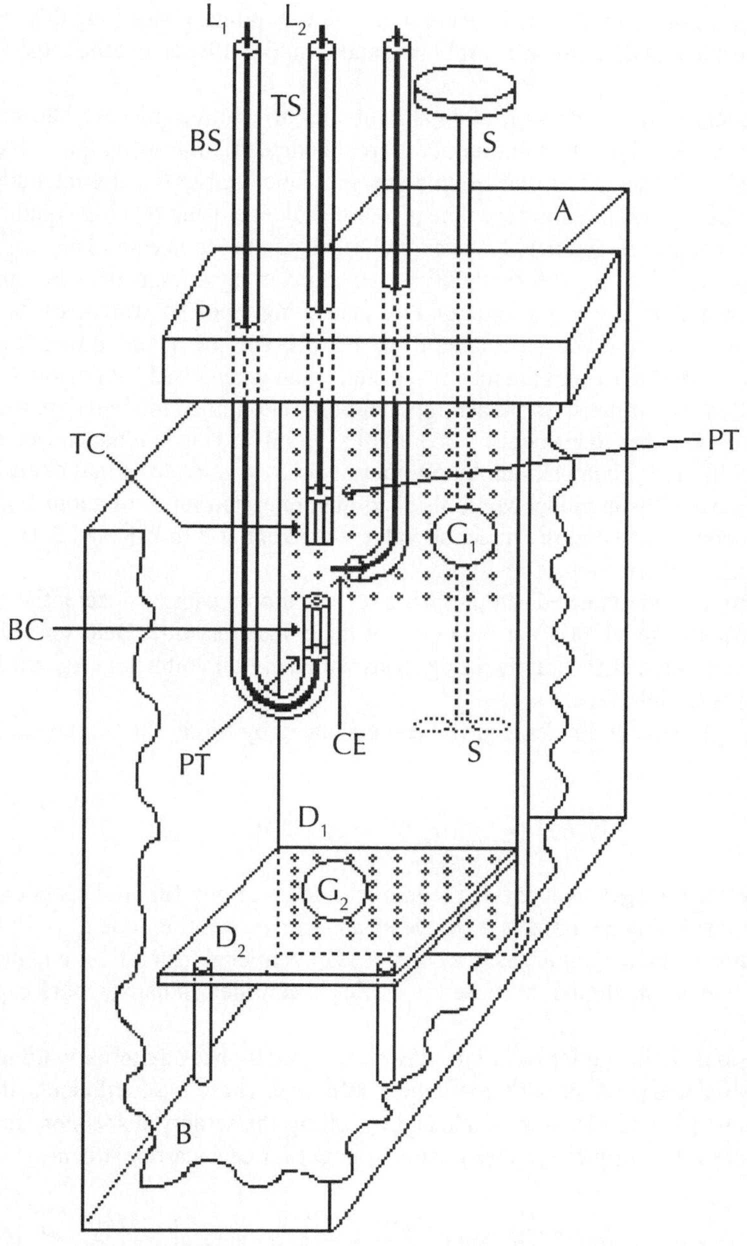

FIGURE 1 The TS and BS: support capillaries, TC and BC: top and bottom diffusion capillaries, CE: central electrode, PT: platinum electrodes, D1 and D2: perspex sheets, S: glass stirrer, P: perspex block, G1 and G2: perforations in perspex sheets, A, B: sections of the tank, and L1 and L2: small diameter coaxial leads [1].

The theory of the Taylor dispersion technique is well described in the literature [11-20], and so the authors only indicate some relevant points concerning this method on the experimental determination of binary and ternary diffusion coefficients (Figure 2).

It is based on the dispersion of small amounts of solution injected into laminar carrier streams of solvent or solution of different composition, flowing through a long capillary tube. The length of the Teflon dispersion tube used in the present study was measured directly by stretching the tube in a large hall and using two high quality theodolytes and appropriate mirrors to accurately focus on the tube ends. This technique gave a tube length of 3.2799 (± 0.0001) $\times 10^4$ mm, in agreement with less precise control measurements using a good quality measuring tape. The radius of the tube, 0.5570 (± 0.0003) mm, was calculated from the tube volume obtained by accurately weighing (resolution 0.1 mg) the tube when empty and when filled with distilled water of known density. At the start of each run, a 6-port Teflon injection valve (Rheodyne, model 5020) was used to introduce 0.063 ml of solution into the laminar carrier stream of slightly different composition. A flow rate of 0.17 ml min^{-1} was maintained by a metering pump (Gilson model Minipuls 3) to give retention times of about 1.1×10^4 s. The dispersion tube and the injection valve were kept at 298.15K and 310.15K (± 0.01K) in an air thermostat.

Dispersion of the injected samples has been monitored using a differential refractometer (Waters model 2410) at the outlet of the dispersion tube. Detector voltages, $V(t)$, were measured at accurately 5 s intervals with a digital voltmeter (Agilent 34401 A) with an IEEE interface.

Binary diffusion coefficients have been evaluated by fitting the dispersion Equation (8)

$$V(t) = V_0 + V_1 t + V_{max} (t_R/t)^{1/2} \exp[-12D(t-t_R)^2/r^2 t] \tag{8}$$

to the detector voltages, where r is the internal radius of our Teflon dispersion tube. The additional fitting parameters were the mean sample retention time t_R, peak height V_{max}, baseline voltage V_0, and baseline slope V_1. Gravitational instabilities and unwanted convection are negligible because the carrier is confined to narrow bore capillary tubing.

Extensions of the Taylor technique have been used to measure ternary mutual diffusion coefficients (D_{ik}) for multicomponent solutions. These D_{ik} coefficients, defined by Equations (3) and (4), were evaluated by fitting the ternary dispersion equation (Equation 9) to two or more replicate pairs of peaks for each carrier stream.

$$V(t) = V_0 + V_1 t + V_{max}(t_R/t)^{1/2}\left[W_1 \exp\left(-\frac{12D_1(t-t_R)^2}{r^2 t}\right) + (1-W_1)\exp\left(-\frac{12D_2(t-t_R)^2}{r^2 t}\right)\right] \tag{9}$$

Two pairs of refractive index profiles, D_1 and D_2, are the eigenvalues of the matrix of the ternary D_{ik} coefficients. In these experiments, small volumes of ΔV of solution, of composition $\bar{c}_1 + \overline{\Delta c_1}$ and $\bar{c}_2 + \overline{\Delta c_2}$ are injected into carrier solutions of composition $\bar{c}_1$ and $\bar{c}_2$, at time t = 0.

Extensions of the Taylor technique have been used to measure quaternary mutual diffusion coefficients $^{ijk}(D_{ij})$ for multicomponent solutions. These $^{ijk}(D_{ij})$ coefficients, defined by Equations (5-7), were evaluated by fitting the quaternary dispersion equation (Equation 10),

$$V(t) = V_0 + V_1 t + K \sum_{i=1}^{3} R_i \left[c_i(t) - \bar{c}_i \right] \tag{10}$$

where $K = dV/dn$ is the sensitivity of the detector, being n the refractive index, $R_i = dn/d\bar{c}_i$ measures the change in the detected property per unit change in the concentration of solute, and $c_i(t) - \bar{c}_i$ represents the dispersion solute average concentrations given by:

$$c_i(t) = \bar{c}_i + \frac{2\Delta v}{\pi r^3 u}\left(\frac{3}{\pi t}\right)^{1/2} \sum_{k=1}^{3}\sum_{p=1}^{3} A_{ik} B_{kp} \Delta C_p D_k^{1/2} \exp\left[-12 D_k (t - t_R)^2 / r^2 t\right] \tag{11}$$

D_k are the eigenvalues of the matrix D of quaternary diffusion coefficients. The columns of matrix A are independent eigenvectors of D, and B is the inverse of A.

The terms $V_0 + V_1 t$ are often included in practice to allow for small drifts in the detector signal. In these experiments, small volumes of ΔV of solution of composition $\bar{c}_1 + \overline{\Delta c}_1$, $\bar{c}_2 + \overline{\Delta c}_2$, and $\bar{c}_3 + \overline{\Delta c}_3$ are injected into carrier solutions of composition, $\bar{c}_1$, $\bar{c}_2$, and $\bar{c}_3$ at time $t = 0$.

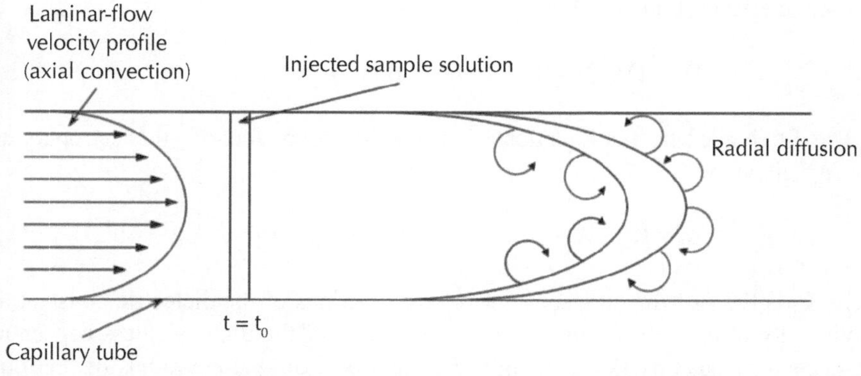

FIGURE 2 Schematic representation of the Taylor dispersion technique [22].

4.3 SOME EXPERIMENTAL DISCUSSION AND RESULTS

Mutual differential diffusion coefficients of several electrolytes 1:1, 2:2, and 2:1, in different media (considering these systems as binary or pseudo binary systems, depending on the circumstances) have been measured using a conductimetric cell [1]. The already published mutual differential diffusion coefficients data are average results of at least three independent measurements. The imprecision of such average results is, with few exceptions, lower than 1%.

The calculation of diffusion coefficients from equations based on some models describing the movement of matter in electrolyte solutions is, in the end, a process contributing to the knowledge of their structure, provided we have accurate experimental data to test these equations. Thus, to understand the behavior of transport process of these aqueous systems, experimental mutual diffusion coefficients have been compared with those estimated using several equations, resulting from different models.

Assuming that each ion of the diffusing electrolyte can be regarded as moving under the influence of two forces: (i) a gradient of the chemical potential for that ionic species and (ii) an electrical field produced by the motion of oppositely charged ions, we come up to the Nernst-Hartley equation [20, 21].

$$D = [(v_1+v_2)\,\lambda_1^\circ\,\lambda_2^\circ / (v_1\,|Z_1|\,(\lambda_1^\circ + \lambda_2^\circ))]\,(R\,T / F^2)\,[1 + (d\,\ln g_\pm / d\,\ln c)] \qquad (12)$$

where λ° are the limiting ionic conductivities of the ions (subscripts 1 and 2 for cation and anion, respectively), Z is the algebraic valence of the ion, v is the number of ions formed upon complete ionization of one solute "molecule", T is the absolute temperature, R and F are the gas and Faraday constants, respectively, and $g_\pm$ is the mean molar activity coefficient.

Equation (12) is often written as:

$$D = D^\circ\,[1 + (d\,\ln\gamma_\pm / d\,\ln c)] \qquad (13)$$

where D° is the Nernst limiting value of the diffusion coefficient.

Onsager and Fuoss [23] improved Equation (13) by taking into account the electrophoretic effects (Equation 14):

$$D = (D^\circ + \Sigma\Delta_n)[1 + (d\,\ln\gamma_\pm / d\,\ln c)] \qquad (14)$$

The difference between Equations (13) and (14) can be found in the lectrophoretic term, Δ_n, given by:

$$\Delta_n = K_B\,T\,A_n\,(Z_1^{\,n}\,t_2^\circ + Z_2^{\,n}\,t_1^\circ)^2 / (a^n\,|Z_1\,Z_2|) \qquad (15)$$

where K_B is the Boltzmann's constant, A_n are functions of the dielectric constant, of the viscosity of the solvent, of the temperature, and of the dimensionless concentration-dependent quantity (k *a*), being k the reciprocal of average radius of the ionic atmosphere; t_1° and t_2° are the limiting transport numbers of the cation and anion, respectively.

Since the expression for the electrophoretic effect has been derived on the basis of the expansion of the Boltzmann exponential function, because that function had been consistent with the Poisson equation, we only, in major cases, would have to take into account the electrophoretic term of the first order (n = 1). For symmetrical electrolytes we can consider the second term. Thus, the experimental data can be compared with the calculated *D* on the basis of Equations (16) and (17) for symmetrical and non-symmetrical electrolytes, respectively.

$$D = (D^{\circ} + \Delta_1 + \Delta_2)\,[1 + c\,(d\,\ln g_{\pm} / d\,c)] \qquad (16)$$

$$D = (D^{\circ} + \Delta_1)\,[1 + c\,(d\,\ln g_{\pm} / d\,c)] \qquad (17)$$

The theory of mutual diffusion in binary electrolytes, developed by Pikal [24], includes the Onsager-Fuoss equation, but has new terms resulting from the application of the Boltzmann exponential function for the study of diffusion. The eventual formation of ion pairs is taken into account in this model, not considered in the Onsager-Fuoss.

$$D = \frac{1}{\frac{1}{M^0}(1 - \frac{\Delta M}{M^0})}\,(10^3\,R\,T\,v)\,[1 + c\,(d\,\ln g_{\pm} / d\,c)] \qquad (18)$$

Data from these models for different types of electrolytes in dilute aqueous solutions have been presented in the literature [25, 26]. From those data we conclude that for symmetrical uni-univalent, both theories (Onsager and Pikal) give similar results, and they are consistent with experimental ones. In fact, if Pikal's theory is valid, ΔM^{OF} must be the major term, all other terms are much smaller and they partially cancel each other. Concerning symmetrical but polyvalent electrolytes [25, 26], we can well see that Pikal's theory is a better approximation than the Onsager-Fuoss'. The ion association, taken into account in this model [27] can justify this behavior.

In polyvalent nonsymmetrical electrolytes, agreement between experimental data and Pikal calculations is not so good, eventually because of the full use of Boltzmann's exponential in Pikal's development.

Although no theory on diffusion in electrolyte solutions is capable of giving generally reliable data on D, we suggest, for estimating purposes, when no experimental data are available, the calculations of D_{OF} and D_{Pikal} for hundreds of electrolytes already made by Lobo et al. [25, 26]. That is, for symmetrical uni-univalent electrolyte (1:1) we suggest the application of Onsager-Fuoss equation with any *a* (ion size) from the literature (e.g., Lobo's publication), because *a* parameter has little effect on final conclusions of D_{OF}; for symmetrical polyvalent (basically 2:2), we suggest the application of Pikal equation. In this case, because D_{Pikal} is strongly affected by the choice of *a,* we suggest calculation with two (or more) reasonable values of *a*, assuming that the actual value of *D* should lie between them, for nonsymmetrical polyvalent, we suggest both Onsager-Fuoss and Pikal theories, assuming the actual value of *D* should lie between them. Now, the choice of *a* is irrelevant, within reasonable limits.

Concerning more concentrated solutions, no definite conclusion is possible. In fact, the results predicted from these models differ from experimental observation (ca>4%). This is not surprising if we take into account the change with concentration of parameters such as viscosity, dielectric constant, hydration, and hydrolysis, which are not taken into account in these models [20-28].

For example, the experimental diffusion coefficient values of $CrCl_3$ in dilute solutions at 298.15K [6] are higher than the calculated ones (D_{OF}).This can be explained not only by the initial $CrCl_3$ gradient and the formation of complexes between chloride and chromium(III), but also by a further hydrogen ion flux, according to Equation. (19).

$$x\,Cr^{3+} + 2y\,H_2O \leftrightarrows Cr_x(OH)_y^{(3x-y)+} + y\,H_3O^+ \tag{19}$$

Similar situations have been found for the systems $BeSO_4$/water [3], $CoCl_2$ chloride/water [4], and $Pb(NO_3)_2$/water [5].

In fact, comparing the estimated diffusion coefficients of $Pb(NO_3)_2$, D_{OF}, with the related experimental values [5], an increase in the experimental D values is found in lead(II) nitrate concentrations below 0.025 M. This can be explained not only by the initial $Pb(NO_3)_2$ gradient, but also by a further H_3O^+ flux. Consequently, as H_3O^+ diffuses more rapidly than NO_3^- or Pb^{2+}, the lead(II) nitrate gradient generates "its own" HNO_3 flux. Thus, the $Pb(NO_3)_2$/water mixture should be considered a ternary system. However, in the present experimental conditions we may consider the system as pseudobinary, mainly for c≥0.01 M, and, consequently, take the measured parameter, D_{av}, as the main diffusion coefficient, D_{11}.

For c<0.01 M, we can estimate the concentration of H_3O^+ produced by hydrolysis of Pb(II) using Equations (20) and (21), assuming that: (a) the fluxes of the species, HNO_3 and $Pb(NO_3)_2$, are independent, (b) the values of the diffusion coefficients, D_{OF}, come from Equation (3). The percentages of H_3O^+ (or the amount of acid that would be necessary to add to one solution of $Pb(NO_3)_2$ in the absence of hydrolysis, resulting in this way a simulation of a more real system) are estimated from the following equations [5].

$$\alpha\, D_{of}(HNO_3) + \beta\, D_{of}(Pb(NO_3)_2) = D_{av} \tag{20}$$

$$\alpha + \beta = 1 \tag{21}$$

where $\alpha \times 100$ and $\beta \times 100$ are the percentages of nitric acid and lead nitrate, respectively. From those data, we can conclude that, for c≥0.01 M, α becomes very low, suggesting that either the hydrolysis effect or the contribution of $D_{of}(Pb(NO_3)_2)$ to the whole diffusion process, can be neglected. Tables (1-3) give the estimated percentage of hydrogen ions, α, resulting from the hydrolysis of Pb^{2+}, Be^{2+}, and Co^{2+} ions in aqueous solutions of lead(II) nitrate, beryllium sulfate, and cobalt chloride, respectively.

TABLE 1 Estimated percentage of hydrogen ions, α, resulting from the hydrolysis of Pb^{2+} in aqueous solutions of lead (II) nitrate at 298.15K, using Equations (20) and (21) [5].

$[Pb(NO_3)_2]/(mol\ dm^{-3})$	**α/%**
0.001	26.0
0.005	7.2
0.01	2.5
0.05	α)

(a) For this concentration we can consider α as nonrelevant.

It what concerns binary systems involving non-electrolytes, we have been measuring mutual diffusion coefficients of some cyclodextrins (α-CD, β-CD, HP-α-CD, and HP-β-CD) [12, 13] and some drugs (e.g., caffeine and isoniazid [14]) in aqueous solutions. Also, from comparison of these experimental diffusion coefficients with the related calculated values, it is possible to give some structural information, such as diffusion coefficients at infinitesimal concentration at different temperatures, estimation of activity coefficients by using equations of Hartley and Gordon, estimation of hydrodynamic radius and estimation of activation energies, *Ea*, of the diffusion process at several temperatures.

TABLE 2 Estimated percentage of hydrogen ions, α, resulting from the hydrolysis of Be^{2+} in aqueous solutions of beryllium sulfate at 298.15K, using similar equations to the Equations (20) and (21) [3].

$[BeSO_4]/(mol\ dm^{-3})$	$\alpha/\%$
0.003	69.0
0.005	48.0
0.008	26.0
0.010	24.0

TABLE 3 Estimated percentage of hydrogen ions, α, resulting from the hydrolysis of Co^{2+} in aqueous solutions of cobalt (II) chloride at 298.15K, using equations similar to Equations (20) and (21) [4].

$[CoCl_2]/(mol\ dm^{-3})$	$\alpha/\%$
0.001	6.0
0.003	6.0
0.005	6.0
0.008	8.0
0.010	5.0

Also, from the measurements of diffusion coefficients of the ternary systems already studied (e.g., *β*-cyclodextrin plus caffeine [15], 2-hydroxypropyl-β-cyclodextrin plus caffeine [16], $CuCl_2$ (1) plus caffeine [10], and KCl plus theophylline (THP) [18]), it is possible to give a contribution in understanding the structure of electrolyte solutions and their thermodynamic behavior. For example, by using Equations (22) and (23), and through the experimental tracer ternary diffusion coefficients of KCl dissolved in supporting THP solutions, D_{11}^{0} ($c_1/c_2 = 0$) and tracer ternary diffusion coefficients of THP dissolved in supporting KCl solutions, D_{22}^{0} ($c_2/c_1 = 0$) [18], it will be possible to estimate some parameters, such as the diffusion coefficient of the aggregate between KCl and THP [18] and the respective association constant.

$$D_{11}^{0}(c_1/c_2=0)=2D_{\text{Cl}^-}\frac{(X_1)\,D^0_{\text{complex}^+}\,(1-X_1)\;D^0_{\text{K}^+}}{D_{Cl^-}+(X_1)\,D^0_{\text{complex}^+}\,(1-X_1)\;D^0_{\text{K}^+}} \tag{22}$$

$$D_{11}^{0}(c_1/c_2=0)=2D_{\text{Cl}^-}\frac{(X_1)\,D^0_{\text{complex}^+}\,(1-X_1)\;D^0_{\text{K}^+}}{D_{Cl^-}+(X_1)\,D^0_{\text{complex}^+}\,(1-X_1)\;D^0_{\text{K}^+}} \tag{23}$$

4.4 CONCLUSION

The study of diffusion processes of electrolytes and non-electrolytes in aqueous solutions is important for fundamental reasons, helping to understand the nature of aqueous electrolyte structure, for practical applications in fields such as corrosion, and provide transport data necessary to model diffusion in pharmaceutical applications. Although no theory on diffusion in electrolyte or non-electrolyte solutions is capable of giving generally reliable data on *D*, there are, however, estimating purposes, whose data, when compared with the experimental values will allow us to take off conclusions on the nature of the system.

KEYWORDS

- **Aqueous solutions**
- **Diffusion coefficients**
- **Electrolytes**
- **Non-ionic solutes**
- **Transport properties**

REFERENCES

1. Agar, J. N. and Lobo, V. M. M. *J. Chem. Soc., Faraday Trans. I*, **71**, 1659–1666 (1975).
2. Lobo, V. M. M. *Handbook of Electrolyte Solutions*. Elsevier, Amsterdam (1990).
3. Lobo, V. M. M., Ribeiro, A. C. F., and Veríssimo, L. P. *J. Chem. Eng. Data*, **39**, 726–728 (1994).
4. Ribeiro, A. C. F., Lobo, V. M. M., and Natividade, J. J. S. *J. Chem. Eng. Data*, **47**, 539–541 (2002).
5. Valente, A. J. M., Ribeiro, A. C. F., Lobo, V. M. M., and Jiménez, A. *J. Mol. Liq.*, **111**, 33–38 (2004).
6. Ribeiro, A. C. F., Lobo, V. M. M., Oliveira, L. R. C., Burrows, H. D., Azevedo, E. F. G., Fangaia, S. I. G., Nicolau, P. M. G., and Guerra, F. A. D. R. A. *J. Chem. Eng. Data*, **50**, 1014–1017 (2005).
7. Ribeiro, A. C. F., Lobo, V. M. M., Valente, A. J. M., Simões, S. M. N., Sobral, A. J. F. N., Ramos, M. L., and Burrows, H. D. *Polyhedron*, **25**, 3581–3587 (2006).
8. Ribeiro, A. C. F., Esteso, M. A., Lobo, V. M. M., Valente, A. J. M., Sobral, A. J. F. N., and Burrows, H. D. *Electrochim. Acta*, **52**, 6450–6455 (2007).
9. Ribeiro, A. C. F., Costa, D. O., Simões, S. M. N., Pereira, R. F. P., Lobo, V. M. M., Valente, A. J. M., and Esteso, M. A. *Electrochim. Acta*, **55**, 4483–4487 (2010).
10. Ribeiro, A. C. F., Esteso, M. A., Lobo, V. M. M., Valente, A. J. M., Simões, S. M. N., Sobral, A. J. F. N., Ramos, L., Burrows, H. D., Amado, A. M., and Amorim da Costa, A. M. *J. Carbohydr. Chem.*, **25**, 173–185 (2006).

11. Ribeiro, A. C. F., Lobo, V. M. M., Leaist, D. G., Natividade, J. J. S., Veríssimo, L. P., Barros, M. C. F., and Cabral, A. M. T. D. P. V. *J. Sol. Chem.*, **34**, 1009–1016 (2005).
12. Ribeiro, A. C. F., Leaist, D. G., Esteso, M. A., Lobo, V. M. M., Valente, A. J. M., Santos, C. I. A. V., Cabral, A. M. T. D. P. V., and Veiga, F. J. B. *J. Chem. Eng. Data*, **51**, 1368–1371 (2006).
13. Ribeiro, A. C. F., Valente, A. J. M., Santos, C. I. A. V., Prazeres, P. M. R. A., Lobo, V. M. M., Burrows, H. D., Esteso, M. A., Cabral, A. M. T. D. P. V., and Veiga, F. J. B. *J. Chem. Eng. Data*, **52**, 586–590 (2007).
14. Ribeiro, A. C. F., Santos, A. C. G., Lobo, V. M. M., Veiga, F. J. B., Cabral, A. M. T. D. P. V., Esteso, M. A., and Ortona, O. *J. Chem. Eng Data*, **54**, 3235–3237 (2009).
15. Ribeiro, A. C. F., Santos, C. I. A. V., Lobo, V. M. M., Cabral, A. M. T. D. P. V., Veiga, F. J. B., and Esteso, M. A. *J. Chem. Eng. Data,* **54**, 115–117 (2009).
16. Ribeiro, A. C. F., Santos, C. I. A. V., Lobo, V. M. M., Cabral, A. M. T. D. P. V., Veiga, F. J. B., and Esteso, M. A. *J. Chem. Thermodynamics*, **41**, 1324–1328 **(2009).**
17. Ribeiro, A. C. F., Simões, S. M. N., Lobo, V. M. M., Valente, A. J. M., and Esteso, M. A. *Food Chem.*, **118**, 847–850 (2010).
18. Santos, C. I. A. V., Lobo, V. M. M., Esteso, M. A., and Ribeiro, A. C. F. *Effect of Potassium Chloride on Diffusion of Theophylline at T = 298.15K.* In press.
19. Ribeiro, A. C. F., Santos, C. I. A. V., Lobo, V. M. M., and Esteso, M. A. *J. Chem. Eng Data*, **55**, 2610–2612 (2010).
20. Robinson, R. A. and Stokes, R. H. *Electrolyte Solutions* 2nd Ed. Butterworths, London (1959).
21. Tyrrell, H. J. V. and Harris, K. R. *Diffusion in Liquids* 2nd Ed. Butterworths, London (1984).
22. Callendar, R. and Leaist, D. G. *J. Sol. Chem.*, **35**, 353–379 (2006).
23. Onsager, L. and Fuoss, R. M. *J. Phys. Chem.*, **36**, 2689–2778 (1932).
24. Pikal, M. J. *J. Phys. Chem.*, **75**, 663–680 (1971).
25. Lobo, V. M. M., Ribeiro, A. C. F., and Andrade, S. G. C. S. *Ber. Buns. Phys. Chem.*, **99**, 713–720 (1995).
26. Lobo, V. M. M., Ribeiro, A. C. F., and Andrade, S. G. C. S. *Port. Electrochim. Acta*, **14**, 45–124 (1996).
27. Lobo, V. M. M. and Ribeiro, A. C. F. *Port. Electrochim. Acta*, **12**, 29–41 (1994).
28. Lobo, V. M. M. and Ribeiro, A. C. F. *Port. Electrochim. Acta*, **13**, 41–62 (1995).

5 Structural Characteristics Changes in Erythrocyte Membranes of Mice Bearing Alzheimer's-like Disease Caused by the Olfactory Bulbectomy

N. Yu. Gerasimov, A. N. Goloshchapov, and E. B. Burlakova

CONTENTS

5.1 INTRODUCTION

We have shown that brain membrane fluidity alterations play important role in the development of the Alzheimer's-like disease [1], and these changes must be took into account during diagnostic and therapy of the dementia. The main problem is the impossibility to use pieces of human brain. In this case the most favorable body tissue is the blood, which contact with all other tissues. Therefore, it is important to study erythrocyte membrane structure alterations during Alzheimer's-like disease development.

Structural characteristics change in erythrocyte membranes of mice bearing Alzheimer's-like disease caused by the olfactory bulbectomy was studied. Erythrolysis, malondialdehyde (MDA) content, and lipid bilayer microviscosity was used as membrane structural characteristics. Erythrolysis and MDA content exhibited lipids peroxidation (LPO) level. Membranes microviscosity was measured by electron spin resonance of 2,2,6,6-tetramethyl-4-capryloyl-oxypiperidine-1-oxyl (lipidic label) and 5,6-Benzo-2,2,6,6-tetramethyl-1,2,3,4-tetrahydro-γ-carbolyn-3-oxide (proteinic label). Phasic changes were found in the structural characteristics of the erythrocyte membranes of mice involved in AD-like pathology. Those changes correlate with the stages of the forebrain membrane fluidity alterations. Liability of the erythrocyte and synaptosomal membranes are changing in opposite way for Sham operated (ShO) mice. Thus, structural state of the erythrocyte membranes can be used as measure of the forebrain membrane structural state. The obtained results testify failures in the system of LPO regulation.

5.2 EXPERIMENTAL

The samples of bulbectomized (BE) and ShO mice (NMRI strain, females) were kindly given to us by N. V. Bobkova (Institute of Cell Biophysics RAS). Every sample contained joint blood of 4–5 mice. Erythrocytes were isolated from blood by means of differential centrifugation at 1,000 g for 10 min. Membrane fluidity in two regions of lipid bilayer was estimated with the help of ESR technique. The 2,2,6,6-tetramethyl-4-capryloyl-oxypiperidin-1-oxyl (probe I) and 5,6-benzo-2,2,6,6-tetramethyl-1,2,3,4-tetrahydro-γ-carboline-3-oxyl (probe II) were used as a probes. Probes spin correlation time was calculated from obtained spectra [2, 3], which correspond to the period of radical's reorientation about π/2. A probe spin correlation time is proportional to the membrane microviscosity or inversely proportional to fluidity. It is known that probe II localizes mostly in near-protein areas, probe I in protein-free areas of lipid bilayer surface (2–4 Å) [4], thus the probes spin correlation time can show lipid-protein interactions in membranes. In addition, erythrocytes lysis, and MDA content changes were determined as a LPO rate index.

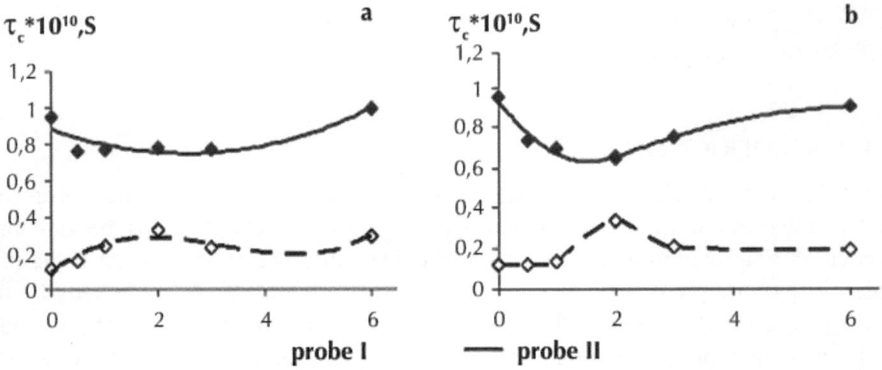

FIGURE 1 The probes spin correlation time in the erythrocyte membranes, (a) ShO mice and (b) BE mice.

The MDA content and hemolysis were measured before (mechanical hemolysis) and after (peroxile hemolysis, PH) incubation at 37°C by the reaction with thiobarbituric acid.

5.3 DISCUSSION AND RESULTS

Spin correlation time alterations of the spin probes in the erythrocytes membrane of the BE and ShO mice are shown in Figure 1(a). Inverse relation between probe I and probe II spin correlation time changes were found in the erythrocyte membranes of ShO mice for the first 4 months after operation. Usually, such antibate relation is observed in health [5]. After 4 months there are symbatic alterations in probe I and probe II spin correlation time. Apparently lipid composition is changed due to aging processes in this case. It is worthy of note that microviscocities of both region, lipidic and near-protein, of the erythrocyte, and the synaptosomal membranes, shown at [1], are changing in opposite way for ShO mice.

One can see more complicated alterations for BE mice (Figure 1(b)). Relative membranes microviscosity changes are shown in Figure 2 (ShO mice considered as control group).

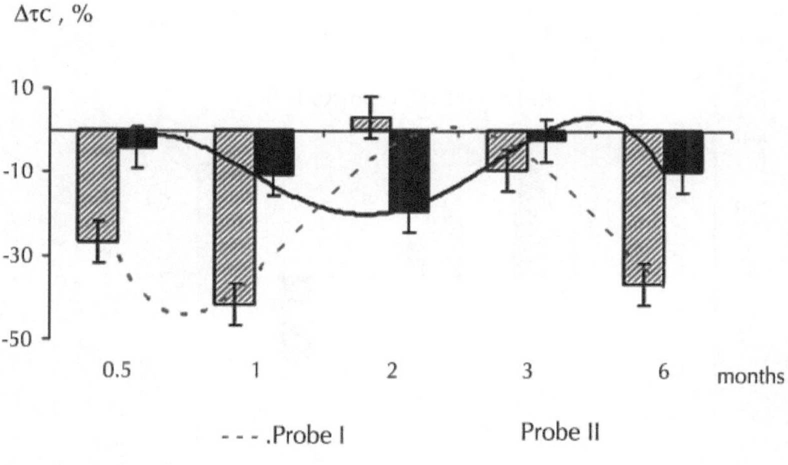

FIGURE 2 The BE mice erythrocyte membrane relative microviscocity alterations.

Microviscosity of the erythrocyte as well as forebrain [1] membranes lipid phase were deeply decreased at the early stage of the dementia development. Enhanced fluidity of the lipid phase leads to increase the liability of the near-protein region of erythrocyte membranes. At the 3rd month after olfactory bulbectomy microviscosity of the BE mice erythrocyte membranes returned to normal value. Thereafter fluidity of the both lipidic and near-protein membranes region was symbatic growing.

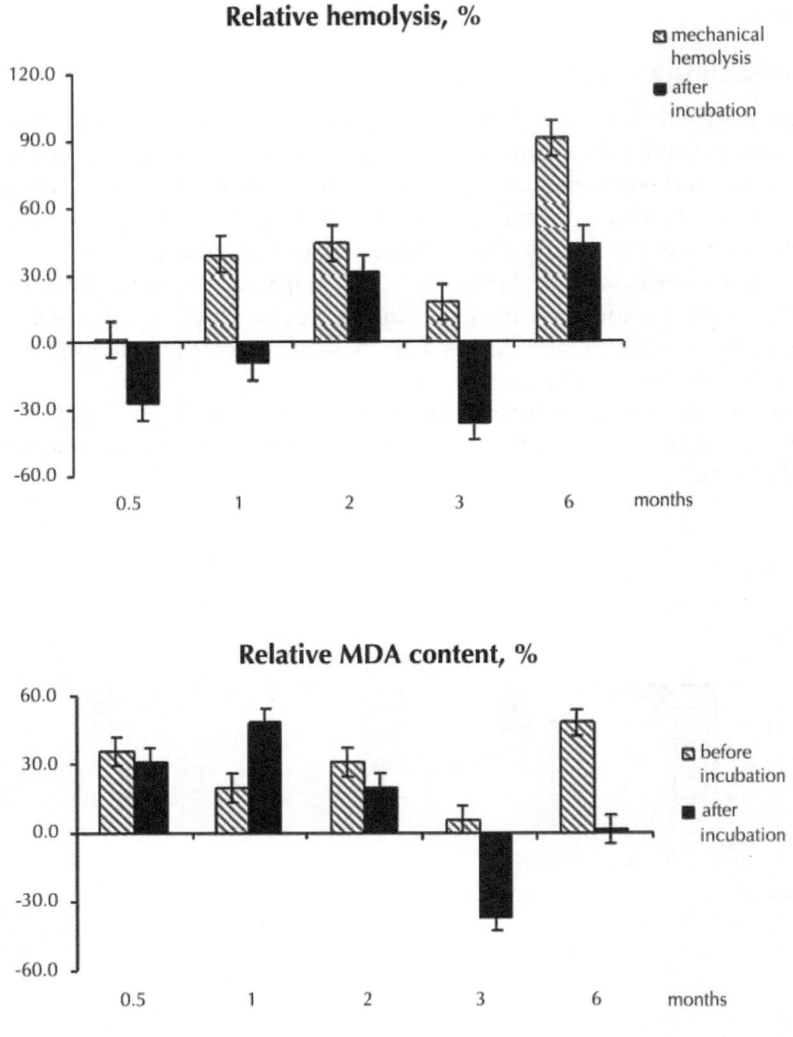

FIGURE 3 The BE mice erythrocyte relative LPO characteristics changes.

Hemolysis and MDA content before incubation in erythrocytes of the BE mice were increased in relation to ShO mice (Figure 3), which were growing during incubation (Table 1), except for 3rd month after operation. As we can see, there were phasic changes of LPO rate: which increases in the early stage of the pathology development, returns to normal in the 3rd month, and thereafter it increases again.

TABLE 1 The LPO characteristics of the erythrocytes.

ShO mice				BE mice			
Hemolysis, %		MDA, mmol/10^6eryth.		Hemolysis, %		MDA, mmol/10^6eryth.	
before incubation	after incubation	before incubation	after incubation	before incubation	after incubation	before incubation	after incubation
11	22	3,1	3,9	11	17	4,1	5,5
12	18	3,7	4,7	17	16	4,5	9,1
12	12	3,5	3,5	18	17	4,5	4,4
10	23	3,5	3,9	12	17	3,7	2,9
10	16	3,7	3,7	19	28	5,5	3,8

Enhanced LPO rate against high liability of the membrane leads to further increasing of the membrane fluidity, which testifies failures in the system of LPO regulation [6, 7].

Thus, phasic changes were found in the structural state of the BE mice erythrocyte membranes. The phases of these changes correlate with the stages of the experimental pathology development [8, 9] and with the phases of the forebrain membrane fluidity alterations [1]. In addition, microviscosity of the erythrocyte, and synaptosomal membranes are changing in opposite way for ShO mice. Consequently, structural state of the erythrocyte membranes can be used as measure of the forebrain membrane structural state.

5.4 CONCLUSION

We found phasic changes in the structural characteristics of the erythrocyte membranes of mice involved in AD-like pathology. Those changes correlate with the stages of "clinical" features and the phases of the forebrain membrane fluidity alterations. Liability of the erythrocyte and synaptosomal membranes are changing in opposite way for ShO mice. Thus, structural state of the erythrocyte membranes can be used as measure of the forebrain membrane structural state. The obtained results testify failures in the system of LPO regulation.

KEYWORDS

- **Alzheimer's disease**
- **Lipid-protein interactions**
- **Membrane fluidity**
- **Membrane structure**
- **Spin labeling**

REFERENCES

1. Gerasimov, N., Goloshchapov, A., and Burlakova, E. The fluidity changes of the membranes isolated from forberain of mice bearing Alzheimer's-like disease caused by the olfactory bulbectomy. *In Modern Problems in Biochemical Physics. New Horizons*. By Sergei D. Varfolomeev, Elena B. Burlakova, Anatoly A. Popov, and Gennady E. Zaikov (Eds.). Nova Science Publishers, New York (2011).
2. Wasserman, A., Buchachenko, A., Kovarskii, A., and Nejman, M. Investigation of Molecular Motions in Polymers with Paramagnetic Probe method. *Visokomol. Soed*., (in Russian), **10A(8)**, 1930 (1968).
3. Sukhorukov, B., Wasserman, A., Kozlova, A., and Buchachenko, A. *Bulletin of the Academy of Sciences of the USSR*, (in Russian), **177**, 454 (1967).
4. Biniukov, V., Borunova, S., Goldfeld, M. et al. Investigation of Structural Transition in Biological Membranes with Spin Probe Method. *Biokhimiya*, (in Russian), **43**(6), 1149 (1971).
5. Goloshchapov, A. and Burlakova, E. Study of Thermoinduced Structural Transitions in Membranes of Animal Organs during Injection of Antioxidants and Malignant Growth. *Byofizika*, (in Russian), **25**(1), 97 (1980).
6. Aristarkhova, S., Arkhipova, G., and Burlakova, E. et al. *Bulletin of the Academy of Sciences of the USSR*, (in Russian), **228**, 215 (1976).
7. Burlakova, E. and Khrapova, N. *Membranes Peroxidation and Natural Antioxidants Uspekhi Khimii*, (in Russian), **54**(9), 540 (1985).
8. Aleksandrova, I., Kuvichkin, V., Bobkova, N. et al. Increased Level of Beta-amiloid in the Brain of Bulbectomized Mice. *Biokhimiya*, **69**(26), 76 (2004).
9. Nesterova, I., Bobkova, N., Medvinskaya, N. et al. Morphofunctional State of Neurons in the Temporal Cortex and Hippocampus in Relation to the Level of Spatial Memory in Rats after Ablation of the Olfactory Bulbs. *Neurosci. Behav Physiol*., **38**(4), 349 (2008).

6 Metal Soap Greases

Esen Arkış and Devrim Balköse

CONTENTS

6.1 INTRODUCTION

Metal soaps are transition metal salts of the fatty acids and the alkaline earth. Although, the alkali salts of the fatty acids such as sodium and potassium are water soluble, metal soap is water insoluble but more soluble in nonpolar organic solvents [1]. Some commercially important metal soap include those of Alluminum, barium, cadmium, calcium, cobalt, copper, iron, lead, lithium, magnesium, manganese, potassium, nickel, zinc, and zirconium. Significant application areas for metal soaps include lubricants and heat stabilizers in plastics as well as driers in paint, varnishes, and printing inks. Other uses are as processing aids in rubber, fuel and lubricant additives, catalysts, gel thickeners, emulsifiers, water repellents, and fungicides [2].

Metallic soap properties are determined by the nature of the organic acid, the type of metal and its concentration, presence of solvent, additives, and the method of production. Fatty acids which can be used to produce metallic soaps include higher monocarboxylic acids having from about 12 to about 22 carbon atoms. Saturated or unsaturated, substituted, or unsubstituted fatty acid is useful [3]. Metallic soaps are prepared commercially by three general processes: Precipitation from aqueous solutions of metal salts and alkali soaps, fusion of metal oxides, hydroxides or salts with organic acids or esters and direct solutions of finely divided metals in heated organic acids [4].

A lubricant is a substance (usually a liquid) introduced between two moving surfaces to reduce the friction and wear between them. A lubricant provides a protective film which allows for two touching surfaces to be separated, thus lessening the friction between them. Lubricants are an essential part of modern machinery. The two moving surfaces could be metal-metal or polymer-metal. Metal soaps are used to regulate the viscosity of lubricating oils and molten polymers.

From the view point of colloid chemistry, the greases can be considered as two component system consisting of a dispersion medium and the dispersed phase. The dispersed phase (5–25%) can consist of salts of high molecular weight carboxylic acids known as soaps.

The dispersed phase forms in the course of grease production forms a three-dimensional (3D) skeleton penetrating the dispersion medium throughout its volume. Skeleton elements (dispersed phase particles) have colloidal size in two (often more than one) measuring directions. 60–80% of the dispersion medium is held in the 3D grease skeleton by adsorption bonds, and the remaining part, mechanically. Soap thickeners are polymorphic thickeners, whose interaction with dispersion medium strengthens as they change over into high temperature mesomorphic phases that are thickeners that do not contact with oils at ordinary temperatures and colloidally disperse at elevated temperatures. The dispersion is additionally facilitated and accelerated by a vigorous stirring or other physical means [5].

Solid films on surfaces can reduce the coefficient of friction and wear. Thick film formation was noted with saturated carboxylic acids such as stearic acid but not with oleic acid. Copper oleate can generate boundary films tens to hundreds of nanometers thick in slow speed rolling contact. Free carboxylic acids are rarely used as organic friction modifiers: however, their salts such as metal oleates are all employed as additives in lubricating oils. It is difficult to relate their structure to film formation and resultant film friction performance since they are relatively impure therefore, it should be synthesized to measure their boundary film forming behavior. Oleates of metals above iron in the electrochemical series form thin boundary layers whereas thick boundary layer occurs only for metals lower than iron in the electrochemical series and is due to a redox reaction involving from the steel surface and the metal oleate [6].

A good monolayer lubricant is not only characterized by a low shear strength for the bonds it forms with an asperity but also that the shear strength should not significantly increase (ideally should decrease) under a compressive load. A solid film lubricant should have high compressive strength and it should form an interface of low shear strength at an asperity contact [7].

This review has intended to summarize the studies on metal soap grease for easy understanding. Metal polymer and metal-metal contact lubrication in use are discussed, and classification, properties, colloidal, and morphological behaviors of metal soaps were introduced. The composition researches of grease regarding base oil additive and metal soaps were reviewed from literature with respect to tribological, rheological, and morphological point of view with some examples. Finally CaF_2 and antimony dithiocarbamide additives were reviewed as additives for the metal soap grease.

6.2 LUBRICATION OF POLYMER-METAL CONTACTS

6.2.1 Polyolefins

One of the underlying rationales for incorporating comonomers into the backbone of polyethylene is to modify the long and short chain branching distribution and alter the polymer morphology. Copolymerization can result in softer resins with better optical properties, improved tensile and impact strengths, superior low temperature characteristics, and enhanced heat sealability. Predictably, it also increases inherent surface friction and tack, making these materials more difficult to process and convert [8]. Lubricants can tie up catalyst residues, usually calcium stearate, stearate, and ethylene-bis-stearamide waxes are sometimes used in processing of fine powdered polefines, erucamides preferred for films and mouldings, floropolymer 'alloy' give better use of polymer [9].

6.2.2 Polyvinyl Chloride (PVC)

The PVC Lubricants are classified as "internal" and "external" according to their chemical structure. They are internal if they reduce melt viscosity, are soluble in PVC, provide little metal release and/or have little effect on fusion or promote fusion. They are classified if they provide metal release and are not soluble in PVC [10]. The stearates are used broadly as viscosity stabilizers particularly where low temperature profiles are required. Metallic stearates are inexpensive and able to modify melt viscosity without delaying fusion and are among the widely used internal lubricants in polymers. Among the disadvantages are that they are supplied as fine powders and so can cause a dust hazard and nuisance in processing. They also tend to exude from the polymer compound, leaving a surface residue and are not generally suitable for mouldings that will require secondary finishing such as printing, painting, bonding, or electroplating [9].

Practical experience shows that external lubrication using oil and grease is more effective than internal lubrication in reducing friction and wear of plastics. It was found that they act as surfactants and slip layers but do not fit the "internal and external" model by synergy of lubricants with fatty acid salts [11]. Later it was shown that flow properties are mainly affected by the concentration of paraffin wax, while the characteristics of fusion may be greatly influenced by calcium stearate [12]. Slip occurs when the lubricant layer forms a continuous layer which modifies the polymer-metal interface character with the presence of metal carboxylate experiment with PVC [13].

Hartitz delineated two groups of lubricants one of which was composed primarily of metal soaps, while the other consisted of less polar components, for example paraffin waxes low molecular weight polyethylenes and oils. Metal soaps alone or in combination with additives of the second group can delay gelation of the PVC [14].

6.3 LUBRICATION OF METAL-METAL CONTACTS

Metal-metal contacts are lubricated with lubricants and greases. Greases mainly contain a mineral and vegetable oil and a thickener. Thickener types are metallic soap thickeners, complex metallic soap thickeners, and non-soap thickeners. The non-soap thickeners are organoclay, polyurea, and Poly Tetra Fluoro Ethylene (PTFE). The metallic soap thickeners are alluminum soap, hydrated calcium soap, sodium soap, and lithium soap. Complex metallic soap thickeners give the grease a higher temperature boost with better oxidation and high drop points. Complex metallic soap thickeners are alluminum complex, calcium complex, barium complex, and lithium complex [15]. This is by far the most widely used category of grease in industry today. This type of grease varies by the additive that forms the soap in the lubricants chemical makeup.

FIGURE 1 The TEM picture of Li stearate grease [16].

Greases have gel structure as seen in Figure 1. If soap based grease is heated, its penetration increases only very slowly until a certain critical temperature is reached. At this point the gel structure breaks down, and the whole grease becomes liquid. This critical temperature is called the DROP POINT.

6.4 GREASE TYPES

6.4.1 Powder Form Greases

Dry lubricants contain no liquid and are used where dripping or spilling cannot happen or oil and grease are not recommended. They have a wide temperature range and may act as both a lubricant and a sealant. A dry powder lubricant composition was formulated comprising 48% by weight sodium acetate, 13% by weight lead acetate, 21% by weight sodium oleate, 12% by weight powdered graphite, and 6% by weight sodium chloride. This dry powder lubricant was sprayed on the hot cavity dies of a forging press used to produce motor vehicle wheels. The temperature of the dies was about 370°C. The results, from a lubrication standpoint, were found to be equal to those obtained using conventional oil base lubricants. However, the emission of smoke was considerably reduced. Furthermore, no flame up was noted. Similar results can be obtained using other 1-10 carbon fatty acid salts of the group IV-A metal as well as using other alkali metal soaps in place of the respective lead acetate salt and the sodium oleate soap [17].

6.4.2 Single Soap Greases

Sodium-base Greases

Sodium-base greases are also general purpose greases. Because they have a higher dropping point (approximately 148 to 176°C), they are often used to lubricate machine parts operating near heat. Sodium greases made with lighter oils are used for ball and roller bearing lubrications, as are combinations (mixed base) of calcium and sodium grease. Sodium soap greases have a spongy or fibrous texture and are yellow or green in color. Because of their working stability and intermediate melting point, they are used for lubricating wheel bearings (other than disc brakes) and for general purpose industrial applications. Typical examples include rough, heavy bearings operating at low speeds, as well as skids, track curves, heavy duty conveyors, universal joints, and antifriction bearings. Sodium soaps with higher melting points (100°C) were the first generation of high temperature greases but their solubility limited their use in many applications. Due to the large fiber size of traditional sodium soaps, they do not contribute lubricity to the grease. Also, they suffer from hardening in the storage. The sodium soap base is usually processed at high temperatures to drive off the water, after which the mineral oil is worked in the base. Sodium soap greases are thus usually anhydrous and water is not needed as a stabilizer. Sodium base greases are preferred where operating temperatures are likely to be high and where wet operating conditions are not involved [18].

The type of fat used to make the sodium soap base is an important factor in the fiber structure or fiber length of the finished grease. Low titre, unsaturated fatty oils and acids produce finished soda greases of the long fiber structure tending to be tough and elastic. High titre, saturated fats, and acids tend to produce greases of the short fiber type. The texture or structure of soda base greases is affected materially by details of processing and manufacturing variations are numerous.

Lithium Soap Greases

Lithium soap grease handles extremes of temperature quite well, which makes them highly suitable for both high and low temperature application. They have a dropping point of approximately 176°C, and can be used in continuous temperatures of 135°C. One reason for their successful low temperature performance is that they are made with oil having a low pour point. In fact, lithium soap greases have been used successfully at temperatures of –51°C. Use of lithium soap grease at higher temperatures requires a different formula however: the same grease cannot be used at both extremes of high and low temperatures because the change would be in the viscosity of the oil used in the grease.

Barium Soap Greases

Barium soap greases are general purpose types, valued for their ability to work over a wide temperature range. Their dropping point is approximately 176°C or higher, although they are not intended to be used in continuous operation at temperatures above 135°C.

Calcium Soaphydrate Greases

Calcium soaphydrate greases, also called lime soap greases, are probably the best known and most often used of all greases. They are smooth and buttery in texture, have excellent water resistance, a fair mechanical stability, are easy to apply and have melting points just under the boiling point of water. Depending on the method of manufacture they are usually relatively inexpensive. Uses include axle grease, water pump grease, and general machinery applications.

The major advantage of calcium soap grease is that they do not dissolve in water. However, it is not suited to use in lubricating high speed bearings. Ordinary general purpose calcium soap greases have a dropping point of approximately 79 to 93°C. Because its water content begins to dry out, and the soap and oil separate, calcium-soap grease is not suited to applications where the temperature will get above 71°C while they may survive 93°C for short periods of time.

Calcium soap is produced with a small residual water content which acts as stabilizer for the matrix and thus provides the required structure of the thickener. They are made by chemically reacting hydrated lime with tallow fatty acid in presence of mineral oil. Saponification under pressure of 15–100 lbs/inch^2 is frequently used to insure complete reaction in minimum time. After saponification is complete, more mineral oil is added, along with water as structure modifier, to obtain the desired consistency. Water in concentrations of from a fraction up to several percent in finished greases since the calcium soap oil system is essentially an emulsion wherein the water acts as a stabilizer or coupling agent.

Anhydrous Calcium Soap Greases

Anhydrous calcium greases offer advantages of traditional calcium soap greases, excellent adhesion, low temperature adhesion, and water resistance. Anhydrous calcium thickeners are based on 12-hydroxy stearic acid like their lithium counterpart and resulting grease offers excellent mechanical property [9].

Alluminum soap Greases

Alluminum soap greases are special purpose lubricants. These greases are made a process called double decomposition. The alluminum soap is prepared by the action of water soluble alkaline soap on a solution of alluminum sulfate or chloride. The precipitated alluminum soap is washed free of impurities and dried. It is then dissolved in mineral oil and thickened to the required nature. Due to the tri-valency of alluminum, three forms of alluminum stearate are possible. Grease making alluminum stearates are usually mixtures, with the mono or di stearates dominating, although pure forms of the mono, di, and tri stearates are available.

Their particular advantage is that they are very sticky making them perfect for applications requiring surface lubrication. Alluminum soaps of unsaturated acids such as oleic have poor gelling properties and tend to form soft, unstable greases. Alluminum stearate provided tacky substance suitable for the chassis but the complete lack of stability since they are sensitive to shear excluded its use in rolling elements. They have significant texture changes with the temperature. They can be used only in low temperatures below 76°C because they tend to soften rapidly. They have excellent water resistance but very poor mechanical stability.

6.4.3 Mixed Soap Greases

These greases are made with two or more metallic soap in combination to produce a lubricant that contains some of the desired properties of both. The most successful of these has been 16% sodium soap combined with 2% calcium soap with highly inhibited oil base plus additives. This has been very effective in packing for the socalled pre lubricated antifriction bearing [19].

Greases based on mixed soaps are classified into a separate subgroup of soap greases. They contain a thickener that is a mixture of soaps differing by the cation such as calcium-sodium, lithium-potassium, lithium-calcium, and so on soaps. Greases with such mixed soaps are as well called calcium-sodium (CaNa), lithium-calcium (LiCa), lithium-lead (LiPb),and so on greases, that soap cation whose share in the thickener is greater being the first indicated one. However, when carrying out various technical and economical analyses of grease production and application and selecting the grease production method, such greases are included in the group of the predominant soap cation [5].

Other combinations have been used, such as alluminum-sodium, calcium-zinc, lithium-calcium-sodium, and lithium-sodium, however, most of these have been developed for highly specialized use, and some are still in the experimental stage [5].

6.4.4 Complex Soap Greases

Sodium Complex Grease

The primary intent of mixed base greases is to secure a very short fiber structure, approaching as closely as possible a smooth and buttery texture while retaining the high melting point properties of the soda base. In most instances, soda soap predominates in the mixture and the resultant greases accordingly have properties substantially similar to straight soda lubricants [8].

Lithium Complex Soap Grease

Lithium complex grease performance is like that of lithium greases except dropping points is about 50°C higher. Lithium complex is a misnomer used to describe high dropping point greases. Because lithium is monovalent (meaning it can only react with one acid per ions), it cannot form a traditional soap where two or more acids reacted with one basic ion. However there are several components that can be used to enhance the molecular interactions of the soap molecules and increase the dropping point enough to call the resulting grease a "lithium complex". The most common method is by forming a lithium salt of a dibasic acid (usually azelaic or sebacic) *in situ* with lithium 12-hydroxy stearate soap. Lithium complex greases provide good low temperature performance and excellent high temperature life performance in tapered roller bearings. It is the most popular of the complex greases and has wide spread application [20].

Barium Complex Grease

Barium complexes were one of the first multipurpose greases. They are made by reacting barium hydroxide in a crystalline form with a fatty acid, complexing the soap with stabilizing substances and then blending with the desired amount of oil. Textures can vary from buttery to fibrous depending on the complexing agent used. The fibrous is the most common. The dropping points range from 193–251°C, and is fairly stable to shear and working. They are water resistant act as fair rust preventatives. They are not very pumpable at cool temperatures, but can be made so by adjusting the base oil. Barium complex is fairly good multipurpose grease, but is expensive and difficult to manufacture. These greases work very well in wheel bearings, water pumps, chassis, and universal joints. They also work well as an outside gear lubricant because of the water resistance and have excellent adhesive properties [21].

Calcium Complex Grease

Calcium complex grease has unusually high heat resistance making it of considerable value in extreme pressure applications. The dropping point of this type of grease is 260°C or even higher. This means that this type of lubricant will maintain its stability while running continuously at high temperatures. However, this type of grease has not replaced lithium soap greases because they are not as mechanically stable [21]. Calcium complexes should not be considered as multipurpose greases. They are very useful, but should be considered carefully beforehand [17].

Alluminum Complex Grease

The soap used for alluminum complex grease is alluminum benzoyl stearoyl hydroxide. These greases provide a wider range of applications than multipurpose types. They are made from two dissimilar acids reacted with alluminum isopropoxide to form a complex soap molecule, or this soap is formed by reacting alluminum isopropylate with stearic acid, benzoic acid and water. They have high dropping points, excellent water resistance, good shear stability and good pump ability depending on the mineral oil used. They respond well to additive treatments which fortifies the grease for high loads. They are oxidation and rust inhibited [13].

6.5 AQUEOUS LUBRICANTS AND PHASE DIAGRAM OF SOAPS IN WATER

Emulsions of soaps are also used as metal-metal lubricants. Hollinger et al., 2000 investigated high pressure lubrication with lamellar structures in aqueous lubricant [22]. Two solutions of dispersed organic fatty particles in water were essentially compared: a vesicle solution (called vesicles) and a lamellar crystallites solution, (called lamellar). The two systems had the same chemical basis. The water phase contained primary and secondary amines, as well as mineral phosphate esters. The solution was buffered at pH = 8.5. The oily phase was composed of fatty phosphates and fatty acids, which were surfactants, since they had polar heads. The lamellar solution was obtained by adding copper and zinc complex ions to the vesicle solution and was characterized by a change in color from a white milky emulsion to a blue one. The lubrication mechanisms of water-based fluids were not yet well understood, especially when extreme load conditions was applied. It was shown that the presence of such nanostructures in the sliding interface could provide extreme pressure lubrication. This water-based lubricant was particularly efficient when one of the two solid surfaces was made of brass. In this case, a multilayer was created in the contact. It was formed by a brass layer transferred to the counter face, the brass surface itself and a lamellar film adherent to the former layers [22].

Phase Diagram of Soap Species in Water

The typical phase diagram of a soap species, in the presence of a solvent system has the general form depicted in Figure 2. The diagram is characterized by the T_k line corresponds to Krafft boundary, which essentially gives the temperature above which the soap molecules exist in a molten but aggregated form [23]. The Krafft point is defined as the point of intersection on the temperature axis between the curve of molecular solubility for the solid soap, depicted by the dashed line extrapolation, and the curve of critical micella concentration (CMC), depicted by the dotted line. As the soap molecules are introduced into water, at very low concentration (to the left of the dotted line), they exist as individual molecules in solution, exhibiting preferential adsorption at surfaces or excess concentrations in the interfacial regions. Upon an increase in concentration beyond CMC curve, the molecules start to form spherical micelles. The individual micelles start packing together as the charge repulsion diminishes in the presence of higher counterion concentrations. This leads the formation of liquid crystalline phases in the molten state. Hexagonal phases (H_1) are composed of rodlike micelles packed in a hexagonal pattern, whereas maximum packing of molecules is achieved through formation of lamellar structures (L_α), wherein flattened disc micelles come together and form large stacks of double layered sheets, as shown in Figure 2. The Krafft boundary the soap molecules can form chain frozen solid crystals or gel phases, which consist of rigid lamellae of soap double layers with a disordered liquid solution sandwiched between them. The chain frozen solid soap crystals exhibit a number of forms such as fine fibers, ribbons, twisted fibers, rods, and flat platelets depending on crystallization conditions and the type and mixture of soap molecules involved.

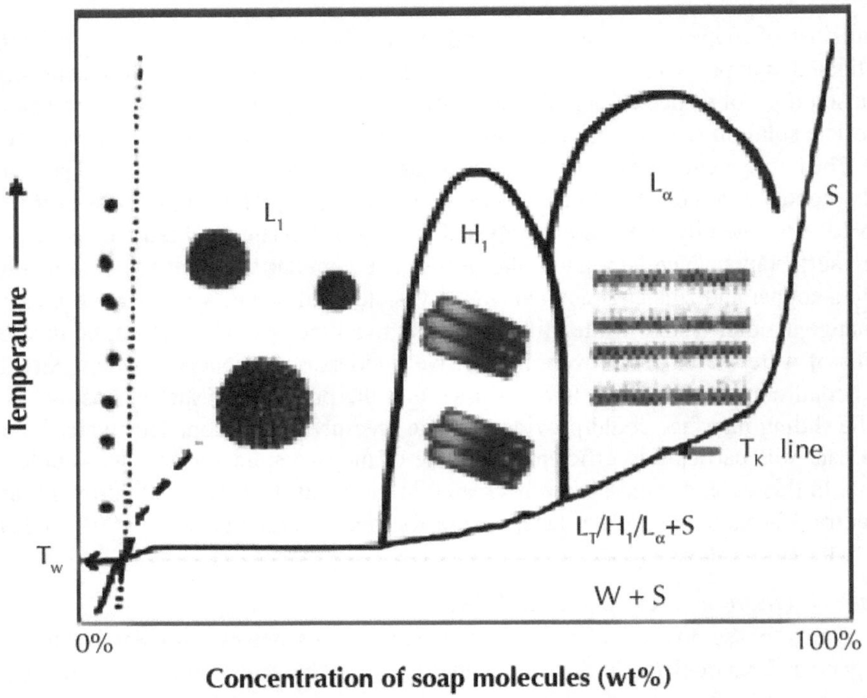

FIGURE 2 Simplified schematic T-x phase diagram of the aqueous soap system. L_1, micellar solution, H_1, hexagonal phase, L_α, lamellar phase, S, solid phase, T_K, Krafft boundary, dotted line, CMC curve, dashed line, and molecular solubility curve for solid phase [23].

6.6 GREASES WITH DIFFERENT BASE OILS: MINERAL OIL AND VEGETABLE SOYBEAN OIL

Recent environmental awareness has forced consideration of the use of biodegradable fluids such as vegetable oils and certain synthetic fluids in grease formulations. The higher cost of synthetic greases limits their use only in aerospace, computer, and medical applications. The soybean oil as lubricating base oil is rapidly biodegradable and the thickeners such as lithium soaps are not harmful in the environment, however vegetable oil based greases have poor thermo-oxidative stability resulting from the highly unsaturated nature of vegetable oils and thus cannot be used at high temperatures. Typical grease formulation is outlined in Figure 3 and typical grease recipies are shown in Table 1.

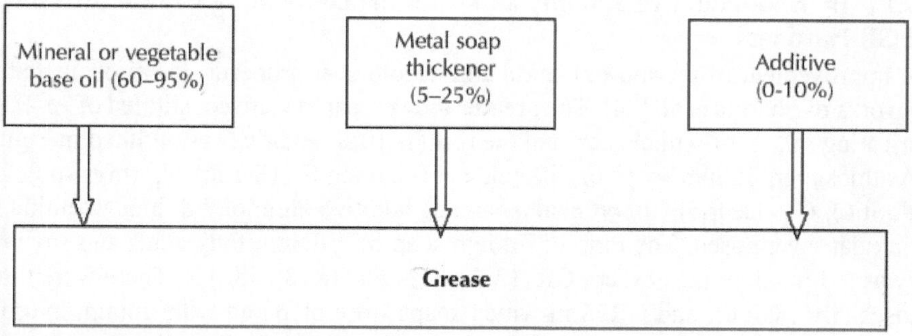

FIGURE 3 A typical grease formulation.

TABLE 1 Typical grease recipes.

Fatty Acid Type of Li-soap or Grease type	C number	Li Soap, %	Oil visc mm^2s^{-1}	Soybean Oil, %	Additive, %	NLIG Grade	Referans
Lauric	(C_{12})	25		75		00	
Myristic	(C_{14})	25				0	
Palmitic	C_{16})	25				2	
Stearic	(C_{18})	25		75		2–3	
Linoleic	(C_{18})	25				1	
Oleic	(C_{16})	34		66		4	
Oleic	(C_{16})	25		75		2–3	
Oleic	(C_{16})	20		80		2	
Stearic	(C_{18})	24.5		73.5	2	2	
Stearic	(C_{18})	24		72	4	2–3	
Li Greases		7	200	93		0–1	
Li Greases		7	30	93		0	
Li Greases		14	200	86		2–3	
Li Greases		14	30	86		2	
Li Complex grease		7	200	93		0	
Li Complex grease	7	30		93		0 00	

6.6.1 Thermo-oxidative Stability and National Lubricating Grease Institute (NLGI) Hardness

The improvement of thermo-oxidation and tribological properties were evaluated in the works of Sharma et al. [24]. The greases they prepared were constituted of 75–95% lubricating oil, 5–20% thickener, and the rest (1–10%) additives as outlined in Figure 3. As thickener, lithium soaps of different acids having C_{12} (lauric), C_{14} (myristic), C_{16} (palmitic), C_{18} (stearic) carbon chains, and as additive antimony dithiocarbamide as antioxidant were used. The ratio of lithium soap of different fatty acids and soybean oil was 1:3 in all greases except GR11 with 1:2 and GR28 with 1:4. There were three greases GR5, GR12, and GR28 in which soaps were prepared with lithium hydroxide monohydrate and fatty acids with different degrees of unsaturation, 0 (stearic), 1 (oleic), and 2 (linoleic), respectively. The concentration of antimony dithiocarbamide was between 2 and 4% to study the additive effect.

Weight loss in vegetable oil greases at 150°C for 24 hr was approximately 10% which is lower compared to weight loss in mineral oil based greases (approximately 50%) under similar conditions. Figure 4(a) showed that weight loss decreased with increasing oil ratio or decreasing soap content. The rate of weight loss was constant from 100 to 150°C and increased rapidly as the temperature was increased from 150 to 225°C (Figure 4(b)). This indicated the breakdown of molecules in soybean oil-based greases as a result of oxidative degradation which is more rapid after 150°C. Increasing the unsaturation to 2 yields a reduction in the weight loss (Figure 4(c)).The presence of more double bonds in the fatty acid structure of the soap are attractive sites for reaction with primary oxidation products, which results in more polymerization (leading to grease hardening) and less volatile oxidation product formation. Figure 4(d) showed that the oxidation onset temperature for addivate samples GR13 and GR25 is much higher (150°C) compared to GR5 (<100°C) sample without any additive.

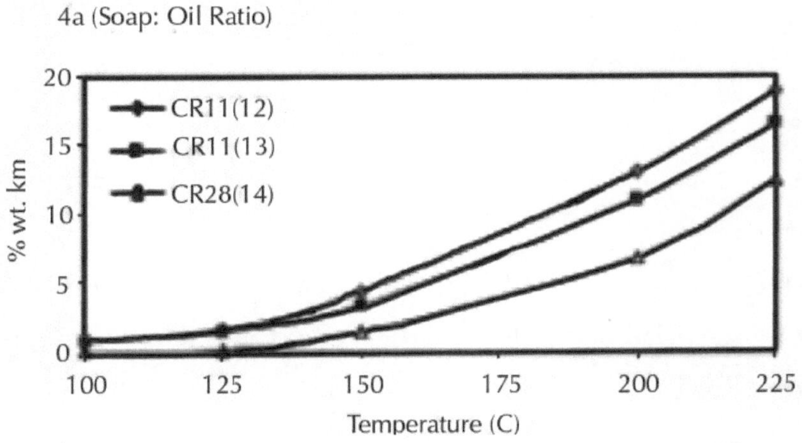

FIGURE 4 *(Continued)*

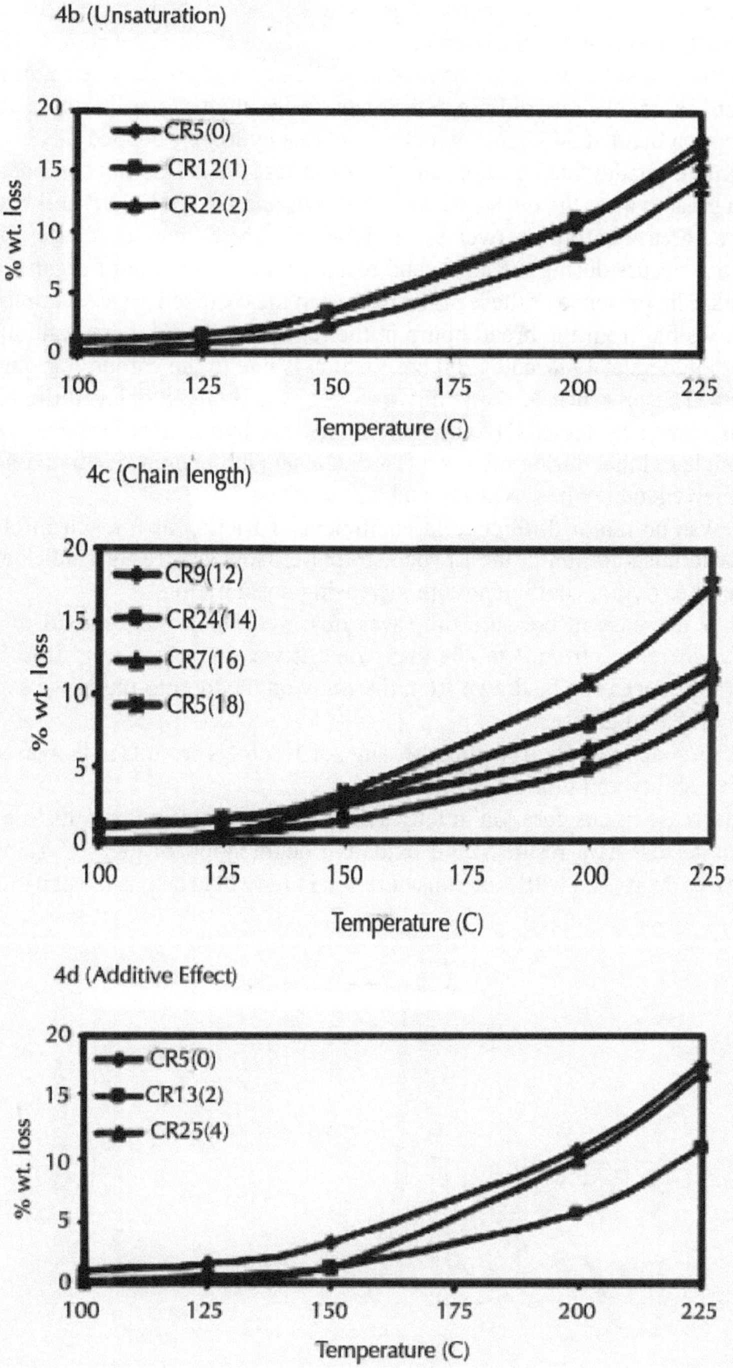

FIGURE 4 Compositional effects of grease at different temperatures using TFMO (a) for soap ratios 1.2, 1:3, and 1;4, (b) for unsaturation 0, 1 and 2, (c) for chain lengths 12 C, 14 C, 16 C, and 18 C, and (d) for additive percentage 0, 1 and 2 [24].

Figure 5 shows the FTIR spectra of grease GR5 before the test and greases GR13 and GR25 after oxidizing in the oven at 100°C for 123 hr. The absorption at 1579 cm^{-1} caused by the thickener was not changed in addivated and unaddivated greases. This shows that lithium stearate did not deteriorate during the test and is stable at 100°C. The absorption band at 3473 cm^{-1} which is due to a hydrogen-bonded OH group, was stronger in the unaddivated grease after the oven test (GR5-123 hr) compared to the addivated greases after the oven test (GR13-123 hr and GR25-123 hr) and least in the unaddivated grease before the oven test (GR5-0 hr). The oxidation of base oil present in the grease occurs during the aging and results in the formation of organic acids as byproducts. The presence of these organic acids in the oxidized grease samples (GR5-123 hr) is visible from the broad hump in the region 3100–2500 cm^{-1} and also from the peak shoulders at 1717 and 1700 cm^{-1} which is due to the carbonyl group of acid byproducts. The presence of the additive in GR13 and GR25 delayed the oxidation process as shown by lesser OH absorption and acid byproducts in these greases. It maybe concluded that during the oven test thickeners in the grease remain unchanged whereas the vegetable oil is oxidized and polymerized.

There was no major difference in coefficient of friction as a result of change in the fatty acid unsaturation in the Li soap structure. However, results indicate a small increase in the friction coefficient with increasing unsaturation.

A larger decrease in pressure drop was observed when the antioxidant concentration was increased from 2 to 4% than when it was increased from 0 to 2%. The important inferences can be drawn from the study on the thermo-oxidative stability of vegetable oil based greases.

- A high soap to oil ratio (within the range of 1:2 to 1:4) results in increased oxidation stability and lubricity of greases.
- A decrease in unsaturation in fatty acid chains of molecules results in increased volatile loss as a result of the oxidative degradation of greases. Lubricity of greases decreased with more unsaturation in fatty acid chains of soap molecules.

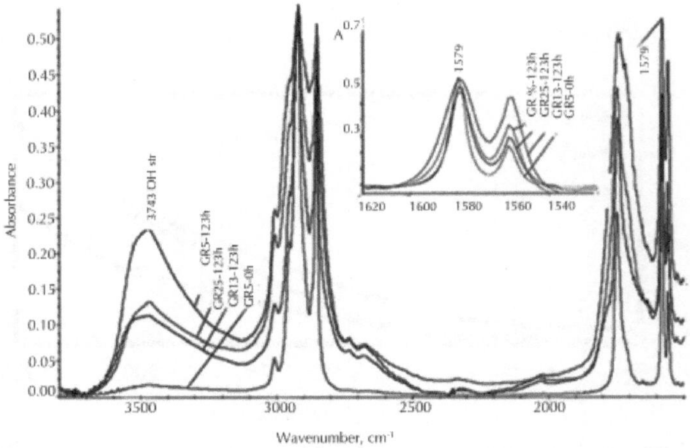

FIGURE 5 The FTIR spectra of grease GR5 before test (GR5–0 hr) and GR5, GR13, and GR25 after 123 hr in oven test at 100°C [24].

- The oxidation stability and lubricity of grease decrease with increasing fatty acid chain length (within the range of 12–18) in soap molecules.
- The thermo-oxidative stability and lubricity of soybean oil based greases can be improved by using suitable additives [18].

In another test [25], lithium hydroxide monohydrate and lauric, myristic, palmitic, stearic, oleic, and linoleic fatty acids with antimony dithiocarbamide and sulfurized olefin antioxidant additive were used in soybean oil. Penetration test was used to find NLGI Hardness with cone penetration (Table 2) and thermal analysis was conducted at pressurized differential scanning calorimetry (PDSC) to find onset temperature that is the temperature when a rapid increase in the rate of oxidation is observed in the test sample. Using 1:1 and 1:0.75 equivalent ratio of lithium to fatty acid in the thickener system (Figure 6), the weight ratio of soybean oil to lithium soap is varied from 65:35 to 80:20 to obtain grease with NLGI No.2 hardness and the higher is metal to fatty acid ratio: the higher was the thermal stability. (Figure 6) With the 6 wt% of additive and equivalent weight ratio of 1:1 maximum observed onset temperature of oxidation was achieved as 165°C at 3450 kPa constant air pressure.

TABLE 2 The NLGI grease classification.

Grade	ASTM Penetration (10^{-4} m) (ASTM D 217)
000	445–475
00	400–430
0	355–385
1	310–340
2	265–295
3	220–250
4	175–205
5	130–160
6	85–115

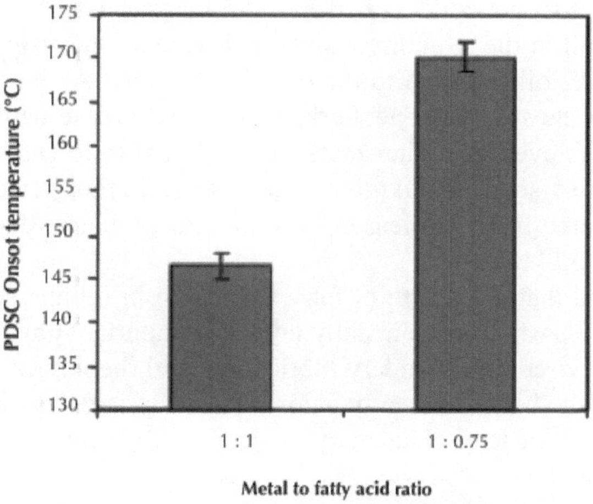

FIGURE 6 Effect of lithium hydroxide to fatty acid ratio on the oxidative stability of soy grease 1:3 wt% ratio of soap to base oil, NLGI no.2 and antioxidant level of 2 wt% [25].

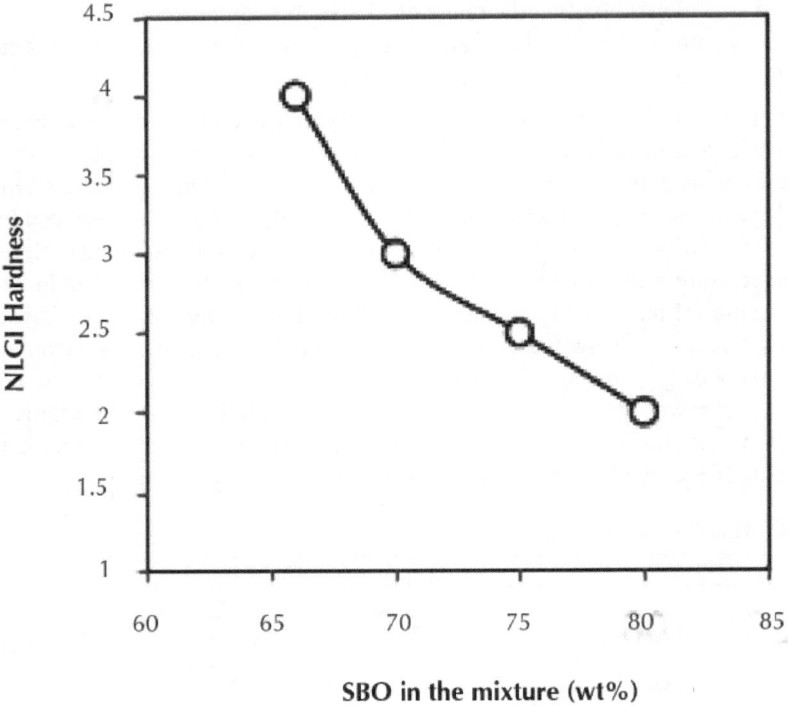

FIGURE 7 Variation of grease NLGI hardness with weight% soybean oil in the reaction mixture using 1:1 equivalent ratio of lithium to fatty acid in the thickener system [25].

The NLGI hardness decreases as the wt% of soybean oil increases as seen in Figure 7. The metal soap in the thickener is capable of holding a certain amount of base fluid within its network. This is a two-stage process. First, absorption and adhesion of base oil in the structure results, and second, the soap structure swells when the remaining oil is added to the reaction mixture. As long as the base oil is confined within the soap fiber network, NLGI No.2 grease with good oxidative stability can be achieved. A higher metal to fatty acid ratio can result in greases with better oxidative stability. Oxidative stability and other performance properties were deteriorated if oil is released from the grease matrix due to overloading of soap with base oil [1].

It was observed that the length of fatty acid chain in lithium soap structure affects grease hardness which subsequently influence important physical performance properties such as viscosity, boundary lubrication, and rheological behavior. Metal soaps prepared with short chain fatty acids resulted in softer grease. Grease consistency increased with long chain fatty acids used for synthesis of lithium soap thickener (Figure 8).

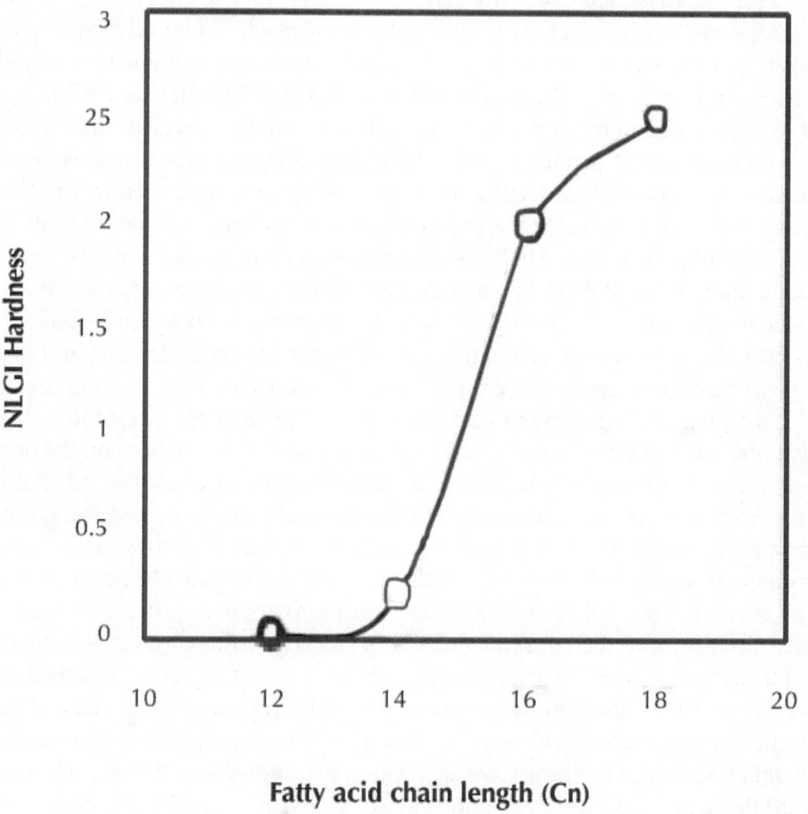

FIGURE 8 Lithium soap fatty acid chain length effect on soy grease NLGI hardness (1:1 equivalent ratio of lithium to fatty acid in the thickener; 1:3 wt% ratio of soap to base oil) [25].

6.7 THE EFFECT OF COMPOSITION ON RHEOLOGY IN ROLLING CONTACT WITH LITHIUM AND ALLUMINUM COMPLEX GREASES

The effect of composition on rheology was investigated with two objectives by Couronne et al [26, 27]. Firstly, to analyze the influence of the interactions between grease components on the film thickness in a rolling contact. Secondly, to try to find how the ability of grease to give fully flooded lubrication is related to its intrinsic properties. The varying parameters were the base oil type (mineral or ester), the soap type (lithium or alluminum complex and the presence of two macromolecular additives (V and T, respectively viscosity and adherence improvers). The soap concentration was (12% w/w) were the same for all the samples. When present V and T concentrations were respectively 5.5 and 0.03% w/w. Film thickness was measured by optical interferometry. Microstructure and physico-chemical structure were observed by Transmission Electron Microscopy (TEM). Stress sweep was conducted to evaluate the elastic modulus (G'). G' is the quantitaive estimation of the grease stiffnes in the low strain range.

The film thickness increases with speed and reaches a relatively high thickness with a slope of 0.67, characteristic of a fully flooded elastohydrodynamic contact. Composition has effect on lubricating ability and determining parameter is soap-base oil interaction. Li-mineral oil-additive and Al complex soap-ester base oil and additive have large capacity to form a lubricating film. Rheology has effect on lubricating ability with determining parameter is G'. Ester-base oil which has high solvency with lithium soap and polymer molecules, gives a strongly interconnected microstructure and thus a high value of G'. At the opposite, the mineral base oil that presents lower solvency with lithium soap and polymers leads to an heterogeneous network and lower G' (elastic modulus). All these factors have an influence on the grease's elasticity and the elastic modulus seems to be a good indicator of grease tribological behavior [26].

Another rheological characterization, including continuous, transient, and dynamic tests was carried on eleven different lubricants with Li greases, Li-complex greases and Urea greases. The varying parameters were the soap nature (lithium, lithium complex, diurea, and tetraurea), the soap concentration (7 and 14%), and the base oil viscosity (30 and 200 mm^2s^{-1}). It should be noted that greases contained no additive.

Stress sweeps were imposed to determine the yield stress, σ_y and the stationary flow properties (apparent viscosity μ_{ap}) at very short and very long time intervals. The instantenous elastic modulus G_i and the instantenous critical elastic strain γ_c were deduced from the first milliseconds after the stress was applied. γ_c was observed when the creep stress was applied or removed. G_m the mean elastic modulus was measured during the sample recovery which typically takes 5–10 min. G_m is defined as the ratio of the creep test over the total recovered strain. With regard to yield stress determination, the stress sweep was chosen to measure σ_y, It consisted of plotting the shear stress/strain rate values in linear scale sand extrapolating linearly the measured stresses toward the stress axis: the intersection gave σ_y yield stress. At similar concentrations lithium soap showed a larger yield stress than lithium complex and tetraurea lubricants.14% weight lubricants had higher σ_y yield stresses than 7% weight lubricants at low and medium strain rates. At higher strains the base oil viscosity effect appears.

Dynamic tests were obtained at 0.5 and 5 Hz frequencies. For G',dynamic elastic modulus, the largest strain in the linear reached 1% and with respect to dynamic loss modulus the maximum strain reached 5.8%. For given thickener nature and concentrations the samples made with 30 mm^2s^{-1} base oil gave larger linear domains than those made with 200 mm^2s^{-1} base oil. From TEM observations greases made from 200 mm^2s^{-1} base oil appear more fragile due to heterogeneities. It has also observed that network for Lithium complex greases are more open and thus unable to sustain large strains. Dynamic viscosities obtained at 0.5 Hz are typically six times higher than those measured at 5 Hz: this confirms the strong shear-thinning behavior already seen during stress sweeps. The authors suggest using rheometry as an alternative technique to standardized methods. Yield stress varies in the same way as penetration does, but is further influenced by the sample microstructure [27].

6.8 ADDITIVES IN GREASES

Antioxidants and viscosity regulators are the main additives in greases. The improvement of thermo-oxidation of soybean oil was obtained when antimony dithiocarbamide was used 2–4% as antioxidant additive. [24]. The Low-density polyethylene (LDPE) is also an example of additives to grease. Recycled LDPE as additive may improve the rheological properties of common lithium lubricating greases. Recycled LDPE act as a filler in the soap matrix increasing particles interaction resulting with low mechanical stability but high value of viscosity and viscoelastic modulus. [28]. Teflon is added to some grease to improve their lubricating properties. Gear greases consist of rosin oil, thickened with lime and mixed with mineral oil, with some percentage of water. Special purpose greases contain glycerol and sorbitan esters. They are used, for example, in low temperature conditions. Some grease is labeled "EP", which indicates Extreme pressure greases contain solid lubricants, usually graphite and/or molybdenum disulfide, to provide protection under heavy loadings. The solid lubricants bond to the surface of the metal, and prevent metal-to-metal contact and the resulting friction and wear when the lubricant film gets too thin. Copper is added to some grease for high pressure applications, or where corrosion could prevent disassembly of components later in their service life.

6.8.1 Tribological Investigation of CaF_2 Nanocrystals as Grease Additive

Since, heavy metals involve sulfur ion, they are potential threat to environment. Alkali metals could be used since they have low shear strength and stable thermophysical and thermochemical properties at elevated temperatures. Due to small scale effect, nano-sized materials dramatically lowered melting points compared with their bulk counterparts. Antifriction and wear properties of CaF_2 nano crystals prepared by ion exchange method were evaluated on a four ball tester. The friction and wear tests were conducted at a rotating speed of 196, 294, 392, 490, and 588 N for test duration of 30 min. The wear scar diameter on three lower balls were measured by using an optical microscope and friction coefficients were recorded automatically with a strain gauge equipped with four ball tester.

The TEM images showed that CaF_2 particles in hydroxystearate soap fiber when added in grease. From EDS and XPS analysis, disproportion of Ca and F in boundary lubricating film indicated that tribological reaction of CaF_2 nanocrystals occurred on the worn steel surface at severe trbological conditions. Upon addition of 0.5%, friction coefficient and wear scar decreased. After addition of 1% friction coefficient and wear scar increased. Therefore, optimum concentration is about 1 wt% compared to lithium grease alone and lithium grease containing 1% CaF_2 as a function of applied loads, 1% CaF_2 is lower and stable from friction coefficient and wear scar diameter tests [29].

6.9 CONCLUSION

A phase diagram of an aqueous soap shows that fiber structure property of soap can be manipulated through processing. A typical composition of grease includes base oil, metal soap, and additive. Soybean oil used as vegetable base oil is better than mineral base oil for environmental concerns. Lithium soap is widely used among other types of metal soaps and lithium alone and lithium soap complex mixed lithium soaps were studied with respect to their rheology, chain length, oxidation stability, soap to oil ratio, and saturation. Additive as antimony dithiocarbamide has good oxidative stability and CaF_2 nanocrystal can lower melting points compared with their bulk counterparts.

This study is focused primarily on stearate based metal soaps. Optimal selection of grease constituents from different points of view in this work may trigger the insight of readers for their future work.

KEYWORDS

- **Critical micella concentration**
- **Metal soap grease**
- **Metal-metal contacts**
- **Polymer-Metal contacts**
- **Transmission electron microscopy**

ACKNOWLEDGMENT

We acknowledge the permissions from publishers Elsevier for Figure 1 in reference 16 and American Chemical Society for the Figures 2–8 published in references 23, 24, and 25.

REFERENCES

1. Gönen, M., Balkose, D., Inal, F., and Ulku, S. *Industrial and Engineering Chemistry*, **44**, 1627–1633 (2005).
2. Öztürk, S. *Preparation and Characterization of Metal Soap Nanofilms*, Master Thesis, Graduate School of Engineering Sciences of Izmir Institute of Technology (2005).
3. Balkose, D., Ulku, S., and Atakul, S. Synthesis of Zinc Borate by Inverse Emulsion Technique. *Journal of Thermal Analysis and Calorimeter*, doi: s10973-010-1159-0 (2010).
4. Gönen, M. *Process Development of Metal Soap.* Master Thesis, Graduate School of Engineering Sciences of Izmir Institute of Technology (2003).
5. Ischuk, Yu. L. Overview of Lubricating Greases. In *Lubricating Grease Manufacturing Technology*. New Age, pp. 1–10 (2005).
6. Ratoi, M., Bovington, C., and Spikes, H. In situ study of metal oleate friction modifier additives. *Tribology Letters*, **14**(1), 33–40 (2003).
7. Nieminen, J. A., Sutton, A. P., Pethica, J. B., and Kaski, K. Mechanism of lubrication by a thin solid film on a metal surface. *Modeling Simul. Mater. Sci. Eng.*, **1**, 83–90 (1992).
8. tsikot.yehey.com/forums/showthread.php?t=8987 – Filipinler, Mola Kula Philipinnes (2006).
9. Murphy, J. Modifying Processing Characteristics. In *Additives for Plastics Handbook.* Elsevier Advanced Technology, pp. 205–218 (2001).
10. Summers, J. W. Lubrication mechanism of poly(vinyl chloride) compounds: Changes upon fusion (Gelation). *J. Vinyl. Technol.*, **11**, 57–62 (2005).
11. Rabinovitch, E., Lacatus, E., and Summers, J. W. The lubrication mechanism of calcium stearate/paraffin wax system in PVC compounds. *J. Vinyl. Technol.*, **6**, 98 (1984).
12. Collins, E. A., Fahey, T. E., and Hopfinger, A. J. The effects of lubricants on the extrusion of poly vinyl chloride. *Polym. Science Tech.*, **26**, (Polym. Addit.), 351–370 (1984).
13. Fras, I., Cassagnau, P., and Michel, A. Lubrication and slip flow during extrusion of plasticized PVC compounds in the presence of lead stabilizer. *Polymer*, **40**, 1261–1269 (1999).
14. Hartitz, J. E. The effect of lubricants on the fusion of rigid poly(vinyl chloride). *J. E. Polym Engng. Sci.*, **14**(5), 392–398 (1974).
15. *www. ampacet.com. (2011)*
16. Adhvaryu, A., Sung, C., and Erhan, S. Z. Fatty Acids and Antioxidant Effects on grease microstructures. *Industrial Crops and Products*, **21**, 285–291 (2005).

17. Johnston, W. G. and Milz, W. C. *Dry Powder Lubricant US Patent* 3962103 (1976).
18. Gow, G. "Grease Thickeners" *Chemistry and Technology of Lubricants,* Third Edition FM. Mortier, MF. Fox, and ST. Orszulik (Eds.). Springer, **420**, (2010).
19. www.usaallamerican.com (2011).
20. Georgi, G. W. and Stucker, J. R. Fats and Fatty Acids for Lubricating Grease Manufacture. *Journal of the American Oil Chemists Society*, **24**(1), 15–19 (1947).
21. Ron Hughes from ReliabilityWeb.com (2010).
22. Hollinger, S., Georges, J. M., Mazuyer, D., Lorentz, D. G., Aguerre, O., and Dua, N. High-pressure lubrication with lamellar structures in aqueous lubricant. *Tribology Letters,* **9**(3–4), 143 (2000).
23. Raut, J. S. and Naik, V. M. Soap the Polymorphic Genie of Hierarchically Structured Soft Condensed-Matter Products. *Ind. Eng. Chem. Res*., **47**, 6347–6353 (2008).
24. Sharma, B. K., Adhvaryu, A., Perez, J. M., and Erhan, S. Z. Soybean Oil Based Greases: Influence of Composition on Thermo-oxidative and Tribochemical Behavior. *Journal of Agricultural and Food Chemistry,* **53**, 2961–2968 (2005).
25. Adhvaryu, A., Erhan, S. Z., and Perez, J. M. Preparation of Soybean Oil Based Greases: Effects of Composition and Structure on Physical Properties. *Journal of Agricultural and Food Chemistry*, **52**, 6456–6459 (2004).
26. Couronne, I., Vergne, P., Mazuyer, D., Truong-Dinh, N., and Girodin, D. Effects of Grease composition and Structure on Film Thickness in Rolling Contact. *Tribology Transactions*, **46**, 31-36 (2003).
27. Couronne, I., Blettner, G., and Vergne, P. Rheological Behavior of Greases:Part I-The Effect of Composition and Structure. *Tribology Transactions*, **43**, 619–626 (2000).
28. Martin-Alfonso, J. E., Valencia, C., Sanchez, M. C., Franco, J. M., Gallegos, C. Development of new lubricating grease formulations using recycled LDPE as rheology modifier additive, *Eur, Polym, J,* **43** 139–149 (2007)..
29. Wang, L., Wang, Bo., Wang, Xiabo, Wang, Liu.. Tribological investigation of CaF_2 nanocrystals as grease additives. *Tribology International,* **40**, 1179–1185 (2007).

7 Effects of Cyclic Loading and Binding to Extracellular Matrix on Macromolecular Transport in Soft Biological Tissue

M. A. Akhmanova and S. P. Domogatsky

CONTENTS

7.1 INTRODUCTION

The solute transport is especially important in avascular or damaged tissues, where it has to provide a sufficient supply of nutrients to cells and modulate biochemical environment of the cells [1]. Mass transfer can be accelerated by convection induced by cyclic loading of the tissue that has boundary with a fluid reservoir and has the ability to squeeze fluid out of the pores due to deformation. There is experimental evidence that biosynthetic activity of the cells is altered by mechanical stimulation [1, 2, 4, 10], and that the most important stimuli are fluid velocity relative to cells and shear strain [1]. Studies of articular cartilage repair suggest that the activation of protein synthesis can partly be triggered by altered molecular transport due to convection [2, 9]. In support of this idea, insulin-like growth factors (IGF) and dynamic compression applied together showed a synergistic effect on biosynthetic response of chondrocytes [9, 10]. However, it is poorly understood how the binding of macromolecules to extracellular matrix influences their transport induce by loading. The factor of binding seems important because most proteins and signaling molecules have the ability to interact with matrix components [8].

Recently a number of transport models that consider dynamic loading of biological tissue have been reported [2, 9-15]. These models are devoted to molecular transport specifically in artificial and native cartilage [2, 10-12], in intervertebral disk [13, 14], or more generally in soft gels [2, 9, 15]. Soft biological tissue has the structure of a gel: it consists of polymer matrix and interstitial fluid inside the pores. Therefore, the predictions of the models for different tissues and gels can be compared. Although some of these predictions were successfully confirmed by experiments in vitro, many questions arise. First of all, the authors used continuum mechanics theory of small deformations to analyze large deformations (up to 20%). Also, many parameters has been taken into account (such as dependence of diffusivity [11, 15], permeability [2, 9, 11, 14, 15], and porosity [10, 12, 13] on deformation) that made it impossible to understand which of them are minor, and which are of major importance. Finally, there is no general model even for small deformations that considers solute binding to the matrix, although it is known that many of the morphogenic proteins, growth factors, and other regulatory molecules have the ability to specifically interact with extracellular matrix components. To date only the molecular transport model by Zhang et al. (2007) [12] has taken into account both solute binding and cyclic loading, but this model was developed specifically to describe transport of IGF in cartilage explants. The results of this study suggest that the greater increase in solute uptake inside the gel caused by cyclic loading could be achieved for the free (unbound) IGF and for lower bath concentrations. This model was modified for the application to intervertebral disks by Travascio (2009) [13].

The main goal of the current study was to develop a mathematical model that adequately describes the response of a gel that has a boundary with bathing solution, to small strain cyclic deformation. The model predicts the value and time course of fluid flow, induced by deformation in the physiological range of frequencies. To explore the efficiency of solute transport from the bath to the gel the solute transport equation was solved. This equation includes diffusion, binding of the solute to the matrix, and convective flow of fluid due to cyclic deformation. The aim of the current computational model is to predict an optimal combination of external deformation and material parameters of the tissue that would promote solute transport.

During their physiologic function biological tissues are commonly deformed by external cyclic loading. The concept of tissue cell activity dependence on mechanical stimulation is accepted in biomechanics, but there is a lack of evidence about the mechanisms underlying the capacity of cells to sense mechanical stimuli. One of the hypotheses is that cyclic loading can enhance molecular transport by cyclic convective flow. This idea was supported by theoretical and experimental studies on neutral non-reactive solute transport in avascular tissues such as cartilage. However, the effect of solute binding to the extracellular matrix has not yet been explored. In the present study, we develop a transport model for a matrix binding solute using a biphasic model of biological tissue deformation. A set of non-dimensional parameters was derived. These parameters characterize the correlation of reaction, material and loading parameters and govern the efficiency of transport. Our results suggest that there is an optimal loading frequency for a particular matrix and a solute that leads to maximal transport acceleration. The load-induced uptake of a solute relative to pure diffusion is higher if this solute can bind to the matrix. Most dramatic effect is seen on uptake of unbound solute: it can be several times higher as compared to the uptake of non-binding solute under the same conditions.

Artificial bone implants are usually connected to soft biological tissue. External mechanical loading induces fluid flow in tissue. It is an important stimulus for cells, which governs the rate of their biosynthetic activity and even differentiation that in turn promote osseointegration of the implant [1]. It is hypothesized that cyclic loading can enhance solute transport in biological tissues [2, 3], and through this mechanism improve tissue nutrition or response to regeneration factors. The existing experimental data that support this statement are rather limited, and were mainly obtained on cartilage samples [3-7]. Recently, several theoretical models for solute transport in loaded gel and tissue were developed [2, 9-15]. These models predict enhancement of free solute transport induced by cyclic fluid flow for different ranges of tissue parameters. Only two studies have taken into consideration the effect of solute binding to the matrix [12, 13], but their results are limited to protein transport in cartilage and intervertebral disk.

The objective of this study is to examine the effect of cyclic loading theoretically and to investigate at what conditions convective transport induced by dynamic loading might significantly alter solute accumulation, taking into account reversible binding of solute to polymer matrix. Our results could be used to choose the appropriate characteristics of implant coatings, help in selection of effective growth factors for tissue treatment, and optimize the design and stimulation protocols of controlled release devices.

7.2 MODEL SYSTEM

The biological tissue undergoing cyclic deformation is modeled by the following system (Figure 1): the rectangular piece of a gel of length h is placed between two impermeable plates. The spacial coordinates (x-axis) are associated with the fixed bottom plate. The upper plate is cyclically loaded. At a point $x = h$ the gel has a free surface, which is the boundary with a bath solution and is perpendicular to the plates. At the distance h from the surface (at point $x = 0$) there is a symmetry plane, or impermeable wall. There is no displacement of the gel at this point. Perpendicular to the cross section, presented on Figure 1, the gel is considered of an infinite width. This assumption means that there is no deformation in this dimension.

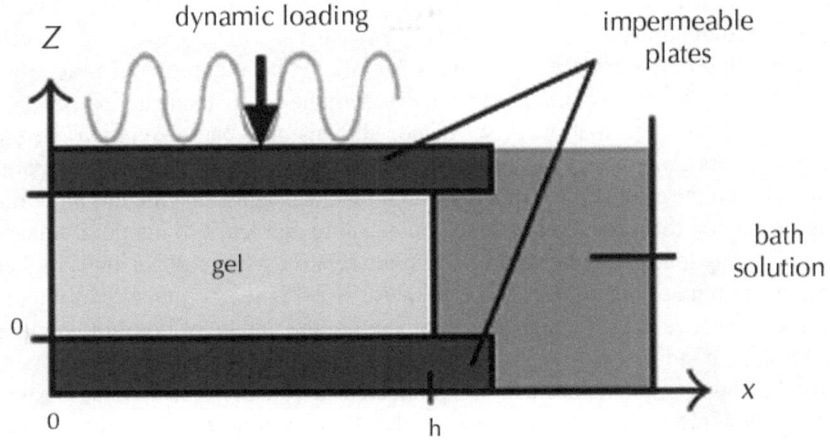

FIGURE 1 The geometry of the model system.

Let us assume that the friction between the gel and the adjacent plates is zero and Poisson ratio is also zero. Consider the deformation of the gel while the upper plate moves down by Δd. Then the process of this deformation can be seen as superposition of the two deformations:

(1) The deformation with no volume change (which means that the volume of the gel redistributes in such a way that the free surface moves to the right by $\Delta x = h\frac{\Delta d}{d}$).

(2) The compression of the layer adjacent to the surface to compensate to this movement.

As a result, the boundary remains at the same point $x = h$, but the solid phase near this surface is compressed. The volume of fluid that flows out to the bath is equal to the volume by which the plate moved. Additionally, stresses and strains do not depend on z because of any friction assumption [2, 16]. Hence, as we will see, all functions in the governing equation depend only on x.

Let us introduce the virtual displacement $\hat{u}$ of solid phase that at each time moment means the displacement relative to the state #1 in x-direction. It is called virtual because it is the displacement from the virtual state that has never existed. The displacement at the state #1 relative to the undeformed state is $-\varepsilon_{\perp} \cdot x$, where $\varepsilon_{\perp}$ is the applied strain in z-direction. Hence, the real displacement $u_x(x,t)$ in x-direction is a sum of the two described virtual displacements: $u_x = \hat{u} - \varepsilon_{\perp} \cdot x$.

7.3 MATHEMATICAL MODEL

7.3.1 The Governing Equation for Fluid and Matrix Motion

The mathematical model, that describes fluid motion, solid phase deformation, and solute transport in the model system, is presented. The model is based on biphasic theory, developed by Mow et al. in 1980 for mechanics of cartilage [16]. The gel is considered as porous media, consisting of fluid phase (interstitial fluid), solid phase (polymeric network), and solute.

Porosity of the gel is defined as follows:

$$\phi = \frac{V^f}{V},$$

where V^f is the volume of fluid phase, and V is the total volume of the gel. The volume of the solute phase can be assumed to be zero. Consequently, the volume fraction of solid phase is equal to $1-\phi=\frac{V^s}{V}$.

The governing equations for mechanics of the gel will be derived under the following assumptions:

(1) Solid and fluid phases are incompressible.
(2) Solid phase is linearly elastic (this assumption is valid for polymeric gels if strains are not more than 5% [17], and this will be maximum strain amplitude in the present study) and elastic modulus is independent of strain.
(3) Body forces are not considered.
(4) Poisson ratio of the gel is zero that is the whole volume change during deformation is due to fluid outflow from the pores or reverses inflow. There is no friction between the gel and the adjacent impermeable plates.
(5) Permeability k is constant (it has been shown (data not presented) using theoretical formula [15] that k differs negligibly for strain <=5%).
(6) Diffusivity D is constant (it has been shown (data not presented) using theoretical formula [15] that D differs negligibly for strain <=5%).

According to the biphasic theory [16, 18] under the assumption (4) the balance of momentum for the whole system reads [2]:

$$\frac{\partial \sigma}{\partial x} = 0 \tag{1}$$

where $\sigma(x,t)$ is the total stress of the gel, that, taking into account assumption (1), can be expressed as the sum of stresses in fluid and solid phases[16, 18]:

$$\begin{gathered} \sigma^s = -(1-\phi)\cdot p + \sigma^e \\ \sigma^f = -\phi \cdot p \\ \sigma = \sigma^f + \sigma^s = -p + \sigma^e \end{gathered} \tag{2}$$

where p is the hydrostatic pore pressure, σ^e is an effective stress of the solid matrix, that describes elastic stress of polymeric network:

$$\sigma^e = H_A \cdot \varepsilon \tag{3}$$

where H_A is the aggregate equilibrium modulus, ε is strain, that is expressed through the solid phase displacement $u(x,z,t)$ as follows:

$$\varepsilon = \frac{\partial u_x}{\partial x} + \frac{\partial u_z}{\partial z} \tag{4}$$

where u_x and u_z are components of $u(x,z,t)$.

The continuity equations for the fluid and solid phases, assuming (1), yield [10]:

$$\nabla\left(\phi \cdot v^f + (1-\phi)\cdot v^s\right) = 0$$

where v^f and $v^s = \frac{\partial u}{\partial t}$ are the velocities of fluid and solid relative to a fixed representative element volume (REV) in the frame of reference associated with the fixed lower plate [10]. Note, that $\varepsilon_z = \frac{\partial u_z}{\partial z}$ is equal to the strain applied to the upper plate, $\varepsilon_\perp(t)$. Strains in the matrix do not depend on z. Hence $u_z = z \cdot \varepsilon_\perp(t)$. Also the z-components of solid and fluid velocities are equal: $v_z^f = v_z^s = \frac{\partial u_z}{\partial t}$. Combining the expressions in this paragraph yields:

$$\frac{\partial v_x^f}{x} + \frac{\partial v_z^f}{z} + \frac{(1-\phi)}{\phi}\left(\frac{\partial^2 u_x}{\partial x \partial t} + \frac{\partial^2 u_z}{\partial z \partial t}\right) = 0$$

that may be written as:

$$\frac{\partial v_x^f}{x} + \frac{\partial \varepsilon_\perp}{\partial t} + \frac{(1-\phi)}{\phi}\frac{\partial}{\partial t}\left(\frac{\partial u_x}{\partial x} + \varepsilon_\perp\right) = 0 \cdot$$

By integrating this equation with respect to x with the condition that $v_x^f = \frac{\partial u_x}{\partial t} = 0$ at $x = 0$ we get:

$$v_x^f = -\frac{(1-\phi)}{\phi}\frac{\partial u_x}{\partial t} + \frac{1}{\phi}x\frac{\partial \varepsilon_\perp}{\partial t} \tag{5}$$

The apparent fluid velocity relative to the solid matrix, v, is proportional to the gradient of pore pressure p as given by the Darcy's law[10]:

$$v = \phi \cdot (v_x^f - v_x^s) = -k \cdot \frac{\partial p}{\partial x} \tag{6}$$

Making use of Equation (2–6), we can rewrite Equation (1) as differential equation for solid displacement:

$$\frac{\partial^2 u_x}{\partial x^2} - \frac{1}{H_A \cdot k}\left(\frac{\partial u_x}{\partial t} + x\frac{\partial \varepsilon_\perp}{\partial t}\right) = 0. \tag{7}$$

Using new variable $\widehat{u}(x,t) = u_x(x,t) + \varepsilon_\perp(t) \cdot x$, this equation may be transformed into:

$$\frac{\partial^2 \widehat{u}}{\partial x^2} - \frac{1}{H_A \cdot k}\frac{\partial \widehat{u}}{\partial t} = 0 \tag{8}$$

Note, that the apparent fluid velocity relative to the solid can be calculated using the $\widehat{u}(x,t)$ function, because

$$v = -H_A k \frac{\partial^2 \widehat{u}}{\partial x^2},$$

as may be deduced from left part of the Equation (6).

Equation (8) is the first equation that describes porous gel mechanics under load. Interestingly this equation coincides with the equation derived for confined compression of articular cartilage [18].

7.3.2 The Governing Equation for Molecular Transport

Total solute concentration is the sum of c_F– free solute, dissolved in the fluid, and c_B– solute that is bound to the solid phase:

$$c_w = c_F + c_B,$$

where c_F and c_B are defined as moles of substance contained in a liter of fluid phase.

Note, that all concentrations should be calculated relative to coordinate system that is tied with solid phase. The purpose of the model is to calculate $c_i(x,t)$–concentration dependence on time t and special variable x of a bound ($i = B$) or free ($i = F$) or total ($i = w$) solute within the tissue. The volume filled by matrix is considered as "tissue", but not the fluid, that flows out of the pores and becomes the bath solution. As the tissue undergoes deformation, each particular point on matrix is displaced relative to the fixed frame of reference used. That is why the concentration of the solute should be considered relative to a fixed REV. To put it another way, the solid phase in the current model represents extracellular matrix with cells, and the aim of the study is to estimate concentrations relative to cells.

The equations for mass balance for each type of solute in each phase can be introduced. As concentrations in the bath do not depend on z, the continuity equation for a free solute reads:

$$\frac{\partial \phi \cdot c_F}{\partial t} + \frac{\partial}{\partial x}\left(\phi \cdot c_F \cdot (v^{C_F} - v_x^s)\right) = q \tag{9}$$

where v^{C_F} is the free solute velocity relative to the fixed frame of reference and q is the solute mass sink due to binding to the matrix.

The solute velocity relative to the fluid is described by Fick's law of diffusion. Making use of Equation (6), the mass flux of the solute relative to the solid phase is given by:

$$c_F \cdot (v^{C_F} - v^s) = c_F \cdot (v^{C_F} - v_x^f) + c_F \cdot (v_x^f - v_x^s) = -D\frac{\partial c_F}{\partial x} - c_F \cdot \frac{v}{\phi} \tag{10}$$

The continuity equation for a bound solute reads:

$$\frac{\partial \phi \cdot c_B}{\partial t} + \frac{\partial}{\partial x}\left(\phi \cdot c_B \cdot (v^{C_B} - v^s)\right) = -q \tag{11}$$

where v^{C_B} is the bound solute velocity relative to the fixed frame of reference, the sink term $-q$ equals the sink term of free solute with the opposite sign, because binding and unbinding are the only mechanisms of solute exchange between the two phases. By neglecting the diffusion of a bound solute in a solid phase the mass flux term in Equation (11) can be omitted.

Porosity of the gel generally depends on strain [10]:

$$\phi = \frac{\phi_0 + \varepsilon}{1 + \varepsilon}.$$

From this expression follows that for $\phi_0 \approx 1$ and small deformations ≤ 0.05, the derivatives of ϕ with respect to x and t are practically zeros.

Combining the Equation (7, 9–11) and assuming that porosity $\phi_0 \approx 1$, yields:

$$\frac{\partial c_F}{\partial t} + \frac{\partial c_B}{\partial t} = D\frac{\partial^2 c_F}{\partial x^2} + \frac{\partial}{\partial x}(v \cdot c_F) \tag{12}$$

Equation (12) is the governing equation for the free solute transport. To relate unknowns c_F and c_B to each other, the equation for a binding reaction must be proposed.

7.3.3 Bimolecular Reaction

It is assumed that the free solute (F) can bind to the binding site on matrix (N). And the bounded form of the solute (B) can dissociate from the matrix:

$$F + N \underset{k_r}{\overset{k_f}{\longleftrightarrow}} B \tag{13}$$

where k_f is the rate constant for the forward reaction of association and k_r is the rate constant for the reverse reaction. This reaction could be described in terms of bimolecular reaction [12, 19]: the rate of increase of bound solute concentration is proportional to the free solute and free binding sites concentrations, and the complex dissociates proportionally to the bound solute concentration. This relationship is expressed in the following Equation [19]:

$$\frac{dc_B^s}{dt} = k_f \cdot c_F(N^s - c_B^s) - k_r \cdot c_B^s \tag{14}$$

where c_B^s is the concentration of the bound solute with respect to the solid phase volume, N^s is the concentration of the binding sites with respect to the solid phase volume. (The values of parameters k_f and k_r are taken from literature considering this notation).

Let us transform Equation (14) so that all variables attribute to the same volume that is volume of the fluid phase. The bound solute and the binding sites contain in the solid phase, so their concentrations with respect to the solid phase volume can be presented as:

$$c_B^s = \frac{\phi}{1-\phi} c_B$$

And

$$N^s = \frac{\phi}{1-\phi} N$$

where c_B and N are concentrations with respect to the fluid phase volume. Then, Equation (14) takes the form:

$$\frac{\phi}{1-\phi}\frac{dc_B}{dt} = \frac{\phi}{1-\phi} k_f \cdot c_F(N - c_B) - k_r \frac{\phi}{1-\phi} \cdot c_B$$

that is equal to

$$\frac{\partial c_B}{\partial t} = k_f \cdot c_F(N - c_B) - k_r \cdot c_B \tag{15}$$

which is the expression for the net rate of change of bound concentration.

7.3.4 Boundary Conditions

The dynamic loading to the upper plate (see Figure 1) is applied under displacement control. The deformation is set to be sinusoidal and comprise compression and tension phases that is:

$$\varepsilon_{\perp} = \varepsilon_0 \cdot \sin(2\pi \cdot f \cdot t) \tag{16}$$

where ε_0 is a deformation amplitude (relative to the width of the gel), which will be not more than 0.05, f is the loading frequency. The argument could be made against this choice of applied strain expression, that the real loading is often only compression with strain changing from zero to maximum negative value and back. But after the relaxation time this load can be described as a superposition of a prestressed state (in which strain equals half maximum value) and compression-tension sinusoidal load. By defining applied deformation by Equation (16) we make the following simplifications: skip the transition state during relaxation time and do not account for a prestress value.

As the solute concentration changes only in x-direction, and the problem considered is one-dimensional, the boundary conditions for the displacement $u(x,t)$ are:

$$\begin{cases} u_x(0,t) = 0 \\ u_x(h,t) = 0 \end{cases}$$

and yield the boundary conditions for $\widehat{u}(x,t) = u_x(x,t) + \varepsilon_{\perp}(t) \cdot x$:

$$\begin{cases} \widehat{u}(0,t) = 0 \\ \widehat{u}(h,t) = h \cdot \varepsilon_0 \cdot \sin(2\pi f t) \end{cases} \tag{17}$$

The boundary conditions for the solute concentration are derived from the assumption that the bath solution is well mixed and therefore the concentration of free solute on the surface of the gel is constant. Impermeable wall at $x = 0$ yields zero flux condition:

$$\begin{cases} c_F(h,t) = c_0 \\ \left. \dfrac{\partial c_F}{\partial x} \right|_{x=0} = 0 \end{cases} \tag{18}$$

7.3.5 Initial Conditions

The process of solute accumulation will be analyzed by the present model that is why the initial conditions reflect zero concentration of the solute in the gel and no displacement:

$$\begin{cases} c_F(x,0) = 0 \\ c_B(x,0) = 0 \end{cases} \tag{19}$$

$$\widehat{u}(x,0) = 0 \tag{20}$$

7.3.6 Nondimensionalization of Governing Equations

In order to find general solution and to determine governing parameters of the model system, all the variables for the set of equations (Equation (8), (12), (15) with conditions

(17), (20)) are transformed to dimensionless variables, as has been proposed by Mauck et al. [2]:

$$x' = \frac{x}{h},\ \widehat{u}' = \frac{\widehat{u}}{h},\ t' = \frac{D}{h^2}t,\ c'_{F,B} = \frac{c_{F,B}}{c_0},\ N' = \frac{N}{c_0} \tag{21}$$

After substitution of these variables, the governing system of equations reduces to:

$$\begin{cases} \dfrac{\partial \widehat{u}'}{\partial t'} - R_g \cdot \dfrac{\partial^2 \widehat{u}'}{\partial x'^2} = 0 \\ \dfrac{\partial c'_F}{\partial t'} + \dfrac{\partial c'_B}{\partial t'} = \dfrac{\partial^2 c'_F}{\partial x'^2} + \dfrac{\partial \widehat{u}'}{\partial t'}\dfrac{\partial c'_F}{\partial x'} \\ \dfrac{\partial c'_B}{\partial t'} = k_1 \cdot c'_F (N' - c'_B) - k_2 \cdot c'_B \end{cases} \tag{22}$$

where $R_g = \frac{H_A \cdot k}{D}$, $k_1 = \frac{k_f \cdot c_0 h^2}{D}$, $k_2 = \frac{k_r \cdot h^2}{D}$ are three non-dimensional parameters. Following Mauck et al. [2], it is useful to note, that R_g represents the ratio of characteristic velocity of fluid in the gel under load $\frac{H_A \cdot k}{h}$ to the characteristic diffusive velocity of solute relative to the fluid $\frac{D}{h}$. k_1 is the ratio of the characteristic diffusion time $\frac{h^2}{D}$ to the characteristic binding time $\frac{1}{k_f \cdot c_0}$ on the edge of the gel. k_2 is the ratio of the characteristic diffusion time $\frac{h^2}{D}$ to the characteristic unbinding time $\frac{1}{k_r}$, which is independent of concentration.

The boundary and initial conditions respectively transform to:

$$\begin{cases} \widehat{u}'(0,t') = 0 \\ \widehat{u}'(1,t') = \varepsilon_0 \cdot \sin(2\pi \cdot f' \cdot R_g \cdot t') \\ c'_F(1,t') = 1 \\ \left.\dfrac{\partial c'_F}{\partial x'}\right|_{x'=0} = 0 \end{cases} \tag{23}$$

and

$$\begin{cases} c'_F(x',0) = 0 \\ c'_B(x',0) = 0 \\ \widehat{u}'(x',0) = 0 \end{cases} \tag{24}$$

where the non-dimensional frequency

$$f' = \frac{f}{f_g} \tag{25}$$

is normalized by $f_g = \frac{H_A \cdot k}{h^2}$–the characteristic frequency of the gel (or "gel diffusion" frequency [3]), that represents the frequency of loading at which the matrix of the gel is deformed throughout the full thickness h.

7.3.7 Numerical Solution

The matrix laboratory (MATLAB) software is employed to solve numerically the governing set of partial differential equations (PDE). The ordinary differential equations (ODEs) resulting from descritization in space are integrated to obtain approximate solutions at times specified. The spacial domain $0 \le x' \le 1$ is descritized into 1,000 equally sized increments. The time steps during PDE solving are chosen by solver, and for output the time span is defined by the user so that there are no less than thirty time increments per loading cycle. The displacement $\tilde{u}'(x',t')$ is calculated separately by solving the first PDE in the system (22), which is independent of other equations. Then the set of two PDEs for solute concentration is solved with implementation of the available $\tilde{u}'(x',t')$ function.

To get the solution for unknowns $\tilde{u}'(x',t')$, $c_F'(x',t')$, $c_B'(x',t')$, it is required to specify seven non-dimensional parameters: R_g, k_1, k_2, N', ϕ, ε_0, and f'. To define the appropriate values for these parameters the ranges of values of H_A, k, k_f, k_r, D, and N for different types of tissues, biological gels and solutes found in the literature are summarized (see Table.1).

From these data it may be estimated that the value of R_g lies in the range 10–1,000, and will be varied during our analysis by the order of 10. It is assumed that porosity $\phi = 1$. Although for cartilage this approximation is not reasonable, for other biological gels its difference from unity can be neglected. ε_0 range from 0.05 to 0.005. The frequency considered is between 10 Hz and 0.01 Hz as this range is physiological: blood pulse frequency is 1 Hz the mean frequency of muscles contraction is about 0.01–10 Hz depending on type of action (walking, standing up, chewing, etc.).

7.4 DISCUSSION

Using the present model, the solute transport enhancement by cyclic deformation of the tissue has been investigated. It is useful to compare our predictions for two cases of solute transport: without binding to the matrix and transport influenced by solute binding.

As stated in results, for the case with no binding the gain does not depend on frequency in the sense that the same gain is achieved at different times for different frequencies. The gain is roughly proportional to the deformation amplitude. Binding of the solute to the matrix can increase the gain (relative to the gain for non binging solute) by few orders of magnitude, but this effect depends on parameters of reaction, frequency, and their ratio. The model predicts that there are optimal reaction constants, ratio of the binding sites concentration to the bath concentration and frequency, for which maximum gain in solute uptake can be achieved.

TABLE 1 The material parameters for native, engineered biogels and tissues.

Matrix	H_A, kPa	$k, \times 10^{-6}$ $mm^2/kPa \cdot s$	ϕ	Solute	$D, \times 10^{-6}$ mm^2/s	$k_f, 1/\mu M \cdot s$	$k_r, 1/s$	N
Fibrin gel 1-2%, plasma clot	0.1-1 [20,21]	100000-10000 [19, 20, 21]	0.99 [20]	plasmin	50 [19, 22]	0.1-0.01 [19]	0.0001 [19]	60 uM [19]
Collagen gel 1%[]	5[23]	1000[24]	0.99 [23]	fibronectin	25 [24]	0.1 [25]	0.001 [25]	30 nM [25]
Cartilage	200-700 [2, 18, 12]	2-0.2 [2, 18, 12]	0.8 [2, 18]	IGF-1	5 [5, 12]	0.4 [12]	0.001 [12]	50nM [12]
				dextran 3 kDa	15[4]	-	-	-
				glucose	1000[3]			
Agarose gel 2-3%	10-20[15]	700-100 [15]	0.97 [15]	albumin	50 [15, 26]	-	-	-

The effect of frequency in the case of binding has been shown on the example of plasmin transport into the fibrin gel. Figure 12 shows that there is a most effective optimal frequency, at which the highest gain in average solute uptake can be achieved. For the higher or lower frequencies the value of average gain is somewhat lower. This result is hardly surprising and can be explained from the general principals of solute binding. The reaction of binding is considered as bimolecular, thus it has characteristic binding time, estimated by $\tau_f = \frac{1}{k_f(c_F + N - c_B)} \approx \frac{1}{k_f(c_0 + N)}$, and characteristic time of unbinding, $\tau_r = \frac{1}{k_r}$. During the first half of the loading cycle, when the fluid flows inside the gel, the solute has time $\tau = \frac{1}{2f}$ to bind to the matrix. All the molecules that did not bind are washed back during the second half of the cycle. The same time interval, $\tau = \frac{1}{2f}$, exists for the bound solute to unbind (assuming that all free floating solute is washed out). This qualitative discussion makes it possible to see that for frequencies $f >> k_f(c_0 + N)$ percent of successfully bound solute at each cycle will be low, and the beneficial effect of binding will be negotiated. On the contrary, for $f << k_r$ the solute has enough time to bind and unbind during the loading cycle, that also leads to reduction of the role of binding on solute transport inside the gel. If the solute has a tendency to adhere, that means that $\tau_f < \tau_r$ for this solute, and if the period of loading lies inside $[\tau_f; \tau_r]$, an increase in mass transport due to binding will be observed. Thus, the existence of the optimal frequency seems natural.

7.5 RESULTS

First the results in non-dimensional variables are presented. They show the trends of the obtained concentration profiles under load and allow to calculating the gain in solute uptake relative to pure diffusion. Then the example of model application to the fibrin gel mechanics and transport of plasmin-like protein is presented in real variables. This result illustrates the effect of dynamic loading and binding on solute uptake in an abundant biological system.

Following the approach used by Urciuolo et al. [15], the gain function has been introduced:

$$Gain(t) = 100 \cdot \frac{m_{dyn}(t) - m_{diff}(t)}{m_{diff}(t)} \ [\%] \tag{26}$$

which serves to evaluate the percent difference between the amount of solute, accumulated in the gel under loading condition $m_{dyn}(t)$, and that due to pure diffusion $m_{diff}(t)$. Amount of bound, unbound, or total solute uptake at time t is calculated as an integral over gel thickness.

$$\int_0^h c_i(t)dx$$

where $i = F, B, F + B$.

The way of presentation the results will become clear after short insight into mechanics of poroelastic materials, because the existence of convective transport in the gel during deformation and its effectiveness is closely tied with the ability of the gel to squeeze the fluid through its pores. The convective flow distribution inside the gel depends strictly on the frequency of loading. This can be shown by solving the first equation in the system (22) for three sample frequencies: $f' = 1,10,100$, $R_g = 100$, and $\varepsilon_0 = 0.05$. As a result, the fluid velocity amplitudes *versus* the normalized distance from the surface are presented on the Figure 1 for those frequencies. By analyzing the curves it may be concluded that there is a characteristic depth where the convective flow is significant. This distance depends on frequency. For higher frequencies the surface layer, which is able to increase mass transfer is narrower. At low frequencies the slope of velocity profile is more flat, but the amplitude is lower.

This relation has been pointed out by other authors [9, 15, 18]. The expression for the characteristic depth is $\widehat{h} = \sqrt{\dfrac{H_A \cdot k}{f}}$. The maximum convection velocity drops considerably by this depth and consequently, there is negligible mass transport acceleration deeper in the gel. The zero velocity depth can be estimated as $2 \cdot \widehat{h}$. When the concentration front reaches $x = 2\widehat{h}$, the gain in mass uptake starts to decrease. That is why we choose $2 \cdot \widehat{h}$ as the length of the gel ($h = 2\widehat{h}$) to solve the transport-reaction equations.

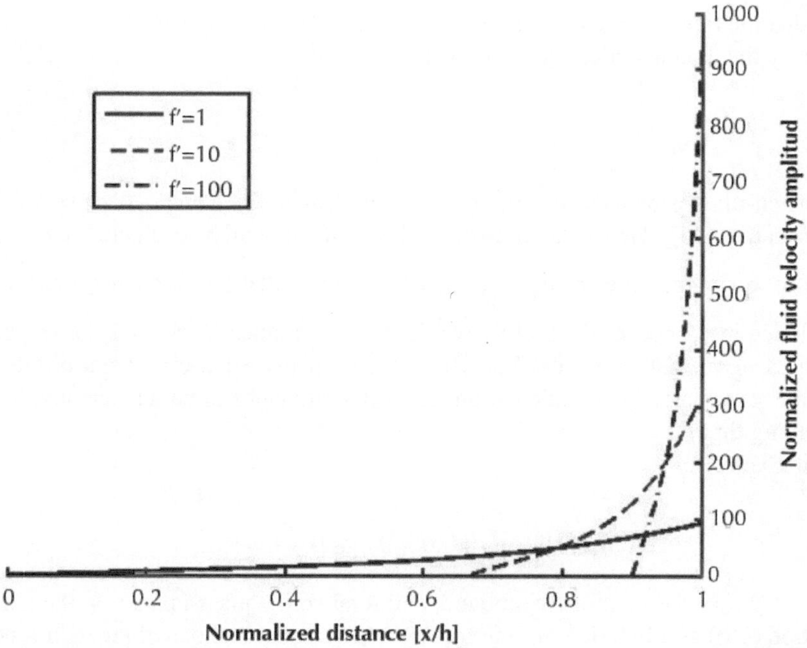

FIGURE 2 The maximum velocity of fluid flow during cyclic loading of the gel as a function of depth for the three frequencies.

A few remarks should be made. First, the constant value of deformation amplitude ε_0 corresponds to the same strain amplitude for any frequency, because we consider only deforming layer of the matrix. Strain amplitude depends on depth and it is higher at the surface (graph not shown), so that the maximum strain equals to ε_0. Also note that $\hat{h}$ depends on frequency, thus the characteristic time of reaching $x=2\hat{h}$ by the diffusing solute front will also depend on frequency.

Now, as we have determined the length of the gel to solve for, it is possible to determine the working range of other parameters. The value of the dissociation rate k_r can be of the order of magnitude about $10^{-2} \div 10^{-5}\frac{1}{s}$. The simplification $k_2 = \frac{k_r \cdot \hat{h}^2}{D} = k_r \frac{H_A k \cdot 4}{D \cdot f} = k_r \frac{4R_g}{f}$ yields $k_2 = 10^{-5} \div 10^4$. The value of the association rate k_f can be of the order of magnitude about $10^{-2} \div 1\frac{1}{\mu M \cdot s}$. To evaluate k_1 further assumption is needed, that will limit the value of c_0. If $N'=1$ is assumed, then, using the data from Table 1, $k_f \cdot c_0 = 10 \div 0.01\frac{1}{s}$. Hence,

$$k_1 = k_f \cdot c_0 \frac{4R_g}{f} = 10^{-2} \div 10^7 \tag{27}$$

Or, for any N',

$$k_1 = \frac{10^{-2}}{N'} \div \frac{10^7}{N'} \tag{28}$$

Note that the ratio $\frac{k_2}{k_1} < 1$, and it depends on c_0, but not on f and R_g. Similarly the dimensionless time can be represented as:

$$t' = t\frac{f}{4R_g} \tag{29}$$

In this non-dimensional consideration the dimensionless frequency will be set to $f' = 4$ (see Equation (25)). Hence, the time dependence of gain will be evaluated with respect to the following parameters: R_g, k_1, $\frac{k_2}{k_1}$, N', ε_0. Note, that the frequency dependence is included in k_1, not in f'. In other words, for any frequency we consider respective thickness of the gel, in which fluid flow is significant. Such consideration helps to keep the same precision of calculations for different values of parameters and to save computing time.

First let us set

$$R_g = 100,\ \varepsilon_0 = 0.05,\ N' = 1,\ \frac{k_2}{k_1} = 10^{-3}$$

Figure 3 shows the gain of free, bound, and total solute mass uptake by the gel (see Equation (26)) as a function of time for $k_1 = 10^3$. The oscillations of the gain function repeat the same behavior of the solute concentration at loading condition. Hence, its frequency equals to $R_g \cdot f'$. On Figure 4 the average of gain function for free solute is plotted over time for $k_1 = 10^3$ and $k_1 = 0$ (no binding case). The time interval is chosen to be $t' = [0, 0.2]$ because during this time the concentration front passes the characteristic depth. The graph shows that the average gain lines are slightly inclined downwards and do not intersect for different k_1. We see clear difference between the two cases. Therefore, the graph showing the average gain function at particular time will be representative for comparison of the effect of different values of k_1. The average gain at any time point is defined as the mean value between the two apexes of the gain function adjacent to this point.

On Figure 5 the average gain of free, bound, and total solute mass uptake by the gel at time $t' = 0.2$ is plotted as a function of k_1. For $k_1 \leq 1$ the graphs for free and total solute coincides because the concentration of bounded solute is orders of magnitude less than free solute concentration. Note, that different k_1 can represent different reaction rate or different frequency. In the last case the time variable t' will correspond to different real time. This must be taken into account when comparing the effect of frequency on mass transport at particular time moment. Also it is important to emphasize that for non binding case ($k_1 = 0$) the gain will be close to the gain for total solute uptake when $k_1 > 0$, which means that beneficial effect of binding is seen for free solute or for bound solute alone. The same tendency is seen when other parameters (R_g, N', ε_0) are varied.

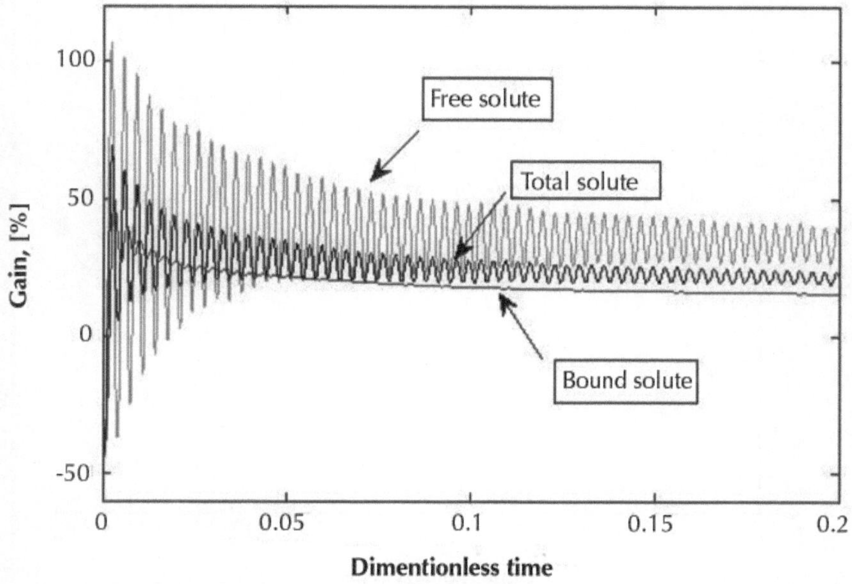

FIGURE 3 Percent gain of free, bound, and total solute mass uptake by the gel relative to the free diffusion case as a function of dimensionless time ($R_g = 100$, $\varepsilon_0 = 0.05$, $N' = 1$, $\frac{k_2}{k_1} = 10^{-3}$, $k_1 = 10^3$).

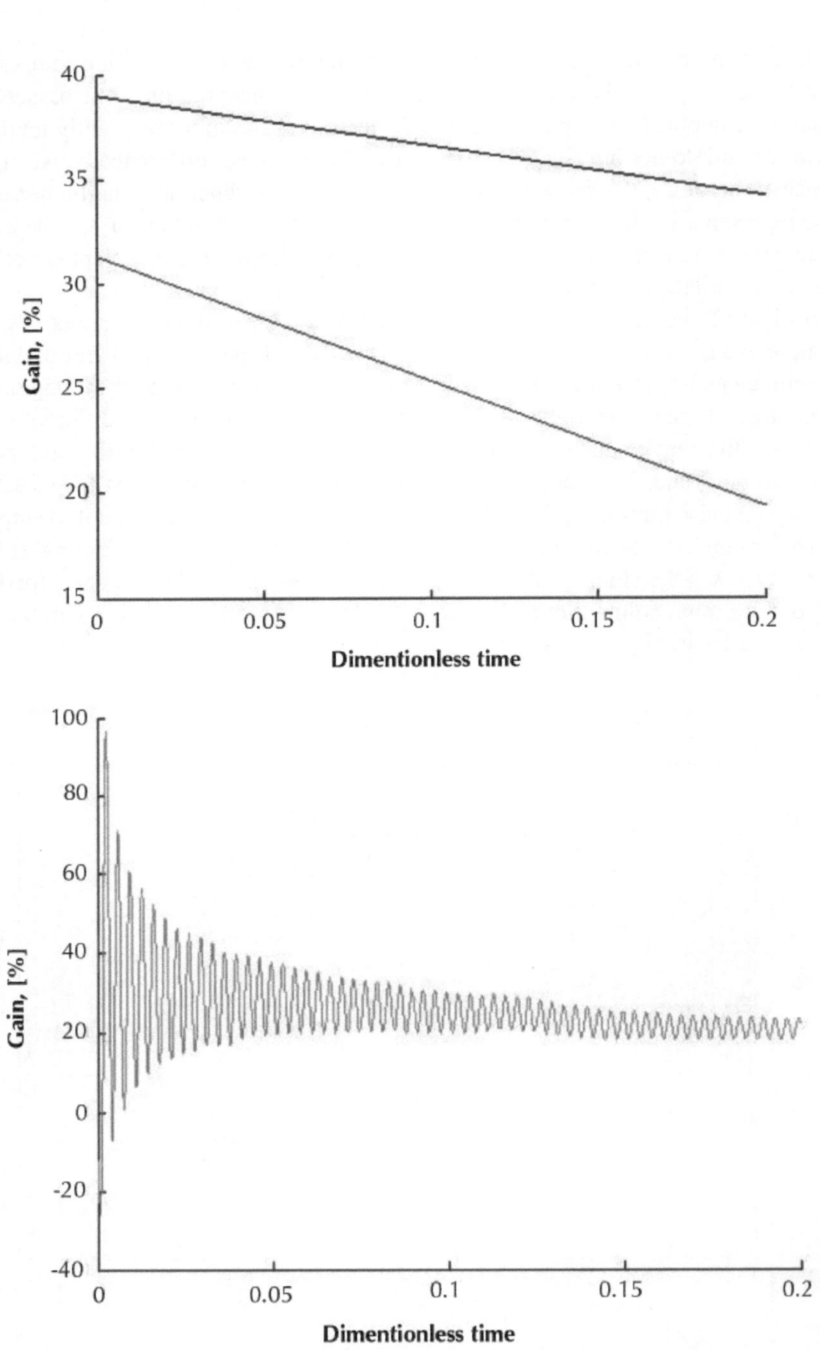

FIGURE 4 (A) The linear approximation of gain function for free solute uptake for $k_1 = 10^3$ (upper line) and $k_1 = 0$ – no binding case (bottom line) as a function of time ($R_g = 100$, $\varepsilon_0 = 0.05$, $N' = 1$, $\frac{k_2}{k_1} = 10^{-3}$). (B) The gain functions for $k_1 = 0$ no binding case. The value of average gain at $t' = 0.2$ is 21%.

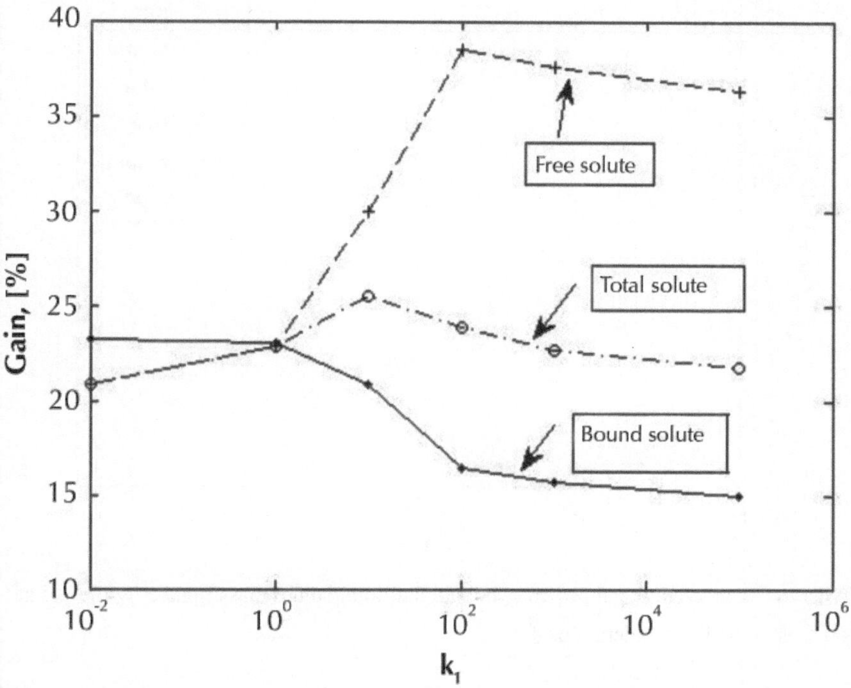

FIGURE 5 The average gain of the free, bound, and total solute mass uptake at time $t' = 0.2$ as a function of k_1. For $k_1 \leq 1$ the graphs for free and total solute coincides because the concentration of bounded solute is orders of magnitude less than free solute concentration.

On Figure 6 the average gain of free, bound, and total solute mass uptake by the gel at time $t' = 0.2$ is plotted as a function of $\frac{k_2}{k_1}$. For $\frac{k_2}{k_1} \geq 1$ the gain approaches its value for not binding solute, and the graphs for free and total solute coincide because the concentration of the bounded solute is orders of magnitude less than the free solute concentration.

Figure 7 shows the gain of free, bound, and total solute mass uptake by the gel as a function of time for $R_g = 1000$ and $k_1 = 10^3$. When compared with Figure 3, where the same graphs for $R_g = 100$ are plotted, the large increase in gain for bigger R_g can be seen. To generalize this result, Figure 8 shows the effect of R_g on the average gain of free, bound, and total solute mass uptake by the gel at time $t' = 0.2$ (for $k_1 = 10^3$, $\varepsilon_0 = 0.05$, $\frac{k_2}{k_1} = 10^{-3}$). Note that k_1 depends on R_g, but it was set at constant value in this plot because the effect of k_1 in the range $10^2 - 10^5$ can be neglected compared with the effect of R_g.

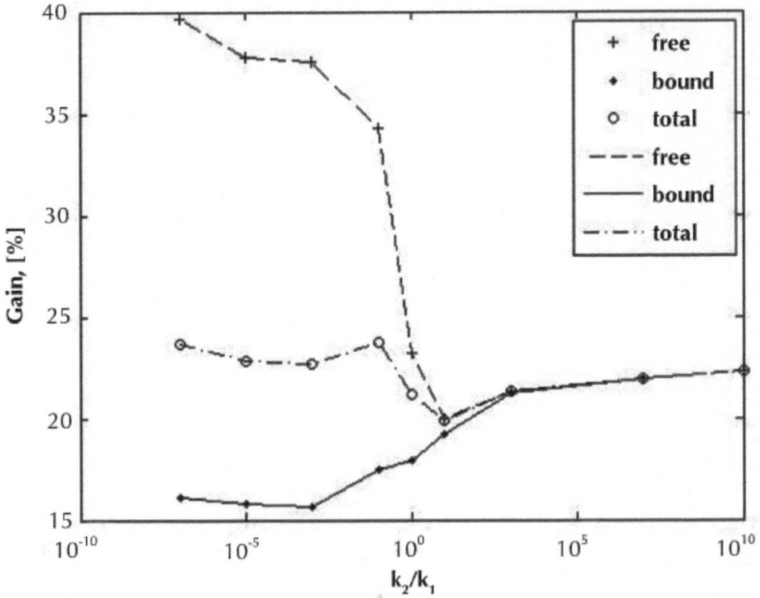

FIGURE 6 The average gain of free, bound, and total solute mass uptake by the gel at time $t' = 0.2$ is plotted as a function of $\frac{k_2}{k_1}$.

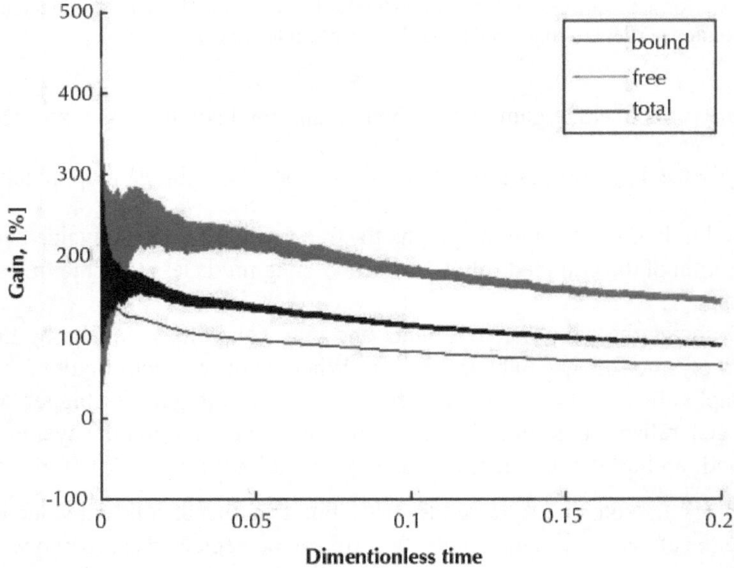

FIGURE 7 The gain function for free, bound and total solute $R_g = 1000, k_1 = 10^3$, $\varepsilon_0 = 0.05$, and $\frac{k_2}{k_1} = 10^{-3}$.

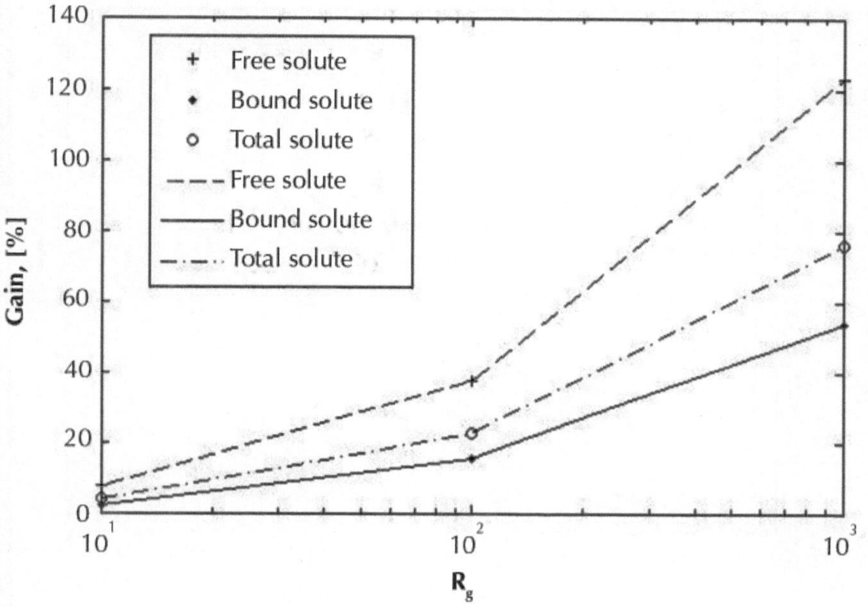

FIGURE 8 The average gain at time $t' = 0.2$ as a function of R_g.

Figure 9 shows the effect of ε_0 on the average gain of free, bound, and total solute mass uptake by the gel at time $t' = 0.2$ (for $k_1 = 10^3$, $R_g = 100$, $\frac{k_2}{k_1} = 10^{-3}$). As expected, the higher strain amplitude leads to greater solute uptake enhancement, while 0.5% amplitude produces near zero effect. The rise of enhancement of accumulation with amplitude appears to be most pronounced for unbound solute. Following studies [10, 15], it may be hypothesized that amplitudes larger than 5% will enhance solute transport even more. Unfortunately it is not possible to use the current model to investigate larger amplitudes, because it is derived from the assumptions (2), (5), (6) and uses continuum mechanics theory for infinitesimal strains.

Figure 10 shows the effect of N' that is the ratio of binding sites concentration to the bath solute concentration, on the average gain of free, bound, and total solute mass uptake by the gel.

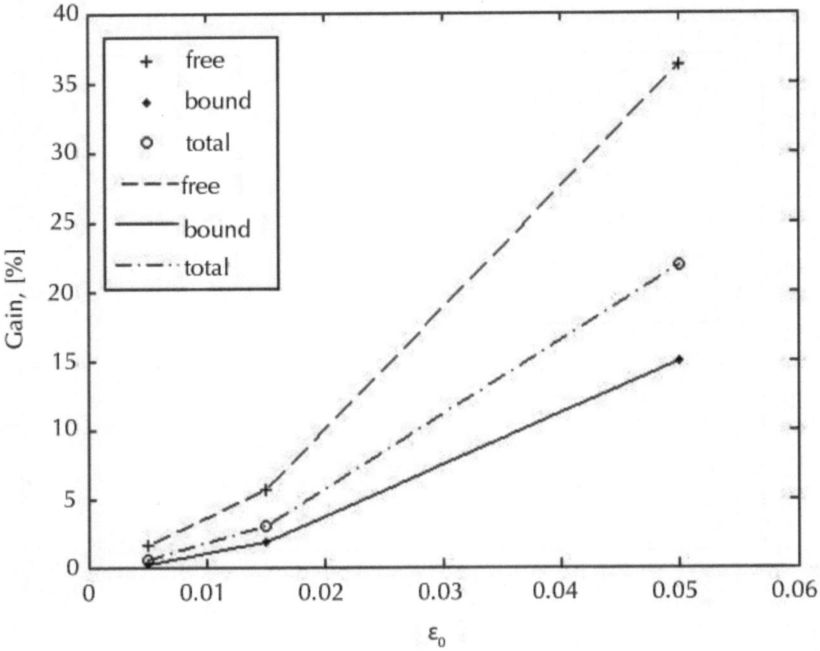

FIGURE 9 The average gain at time $t'=0.2$ as a function of ε_0 (for $k_1=10^3$, $R_g=100$, and $\frac{k_2}{k_1}=10^{-3}$).

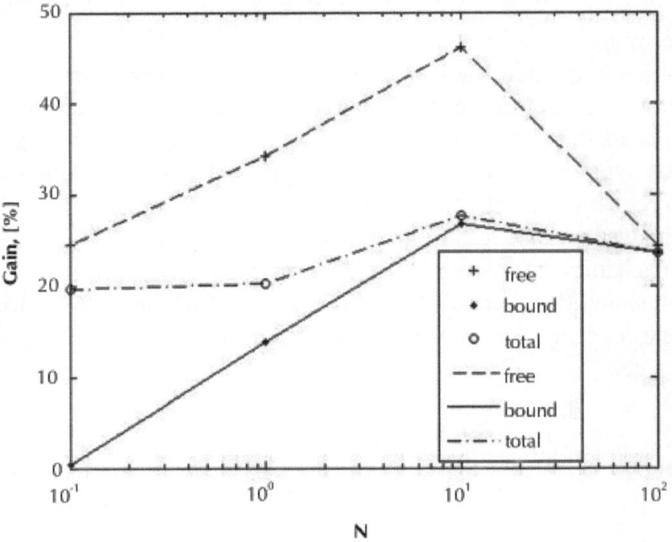

FIGURE 10 The average gain of free, bound, and total solute mass uptake by the gel at time $t'=0.2$ as a function of N' (for $k_1=10^3$, $\frac{k_2}{k_1}=10^{-3}$, $\varepsilon_0=0.05, R_g=100$).

From the presented results follows a qualitative conclusion, that the gain in mass uptake increases while increasing ε_0 or R_g, or decreasing $\frac{k_2}{k_1}$. As for parameters of binging intensity, k_1 and N', there exists an optimal value, that leads to the peak in gain. Roughly, these "peak" values are: $k_1 = 100$ and $N' = 10$ for $R_g = 100$. This finding is in contradiction with the results of Zhang et al. (2007), which states that the gain in free solute uptake gradually increases with the ratio of binding sites concentration to the bath concentration (N').

7.5.1 Transport of Matrix Binding Protein in Fibrin Gel

To show how the present model can be applied to a particular biological system, we present an example of fibrin gel and mass transport of a plasmin-like protein inside it. The parameters for this system are taken from the literature (see Table 1). The real parameters are set to:

$$c_0 = 1\mu M,\ k_f = 0.1\frac{1}{\mu M \cdot s},\ k_r = 0.0001\frac{1}{s},\ H_A = 1kPa,\ k = 10000\frac{\mu m^2}{kPa \cdot s},\ D = 50\frac{\mu m^2}{s}$$

which correspond to the following non-dimensional parameters:

$$R_g = 200,\ N' = 50,\ k_1 = 100,\ k_2 = 0.1 \text{ for } f = 1Hz$$

Figure 11 (A) represents the gain function for free, bound and total solute for these parameters as a function of time in seconds. Figure 11(B) shows the gain function for not binding solute for two frequencies: $f = 1Hz$ and $f = 0.1Hz$ as a function of time in seconds. For $f = 0.1Hz$ it is actually the same graph as for $f = 1Hz$, but stretched in time by the order of 10 (see non-dimensional analog of this graph on Figure 4(B) and use Equation (29) to transform normalized time into real). Since the average gain monotonically decreases, the concentration front at higher frequency will propagate further with respect to the deforming layer ("characteristic depth" $\hat{h} = \sqrt{\frac{H_A \cdot k}{f}}$). Therefore, if we measure the gain for different frequencies (but for the same strain amplitude) at the same time moment, we get less gain for higher frequency.

From Figure 11 (A) it is easy to see that if the solute adheres to matrix with high binding sites concentration, the shape of the gain curve is different compared with Figure 11(B) and Figure (3): it increases gradually to a peak value, then gradually decreases. Consequently, measuring the gain at the same time for different frequencies does not reflect the effect of frequency, because it is unknown at what part of the curve this time point falls. That is why, to investigate the effect of loading frequency on the transport enhancement, we calculate the average gain for frequencies $f = 10, 1, 0.1, 0.01Hz$ at particular time for each frequency. This time points t_f are chosen in such a way that the gains for non-binding solute at these times are equal (the value of the average gain is 35%): $t_{10} = 16s$, $t_1 = 160s$, $t_{0.1} = 1600s$, $t_{0.01} = 16000s$. By these times the concentration front passes the same part of the deforming layer for each frequency. Thus, the time points lie in corresponding parts of gain curves. The results are presented on

Figure 12. The peak average gain is achieved at $f = 1Hz$. At this frequency the gain in free solute uptake exceeds gain for no binding case by the order of 2.5.

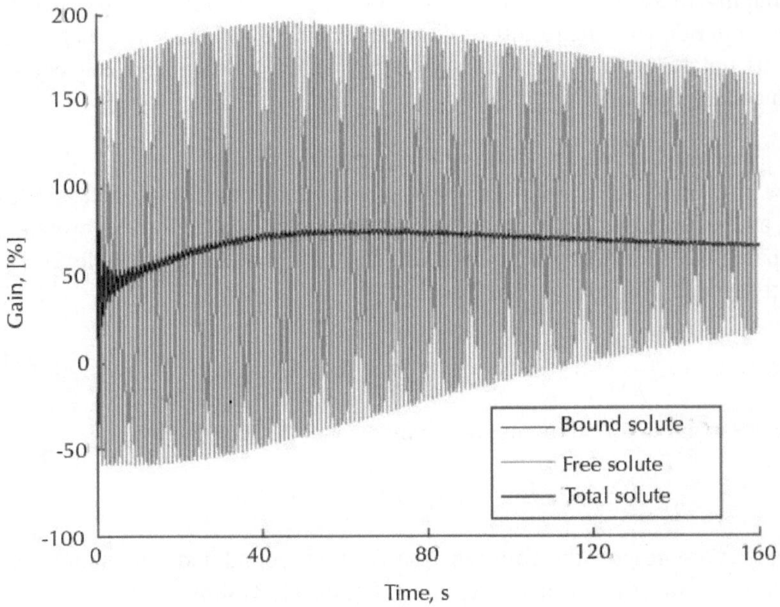

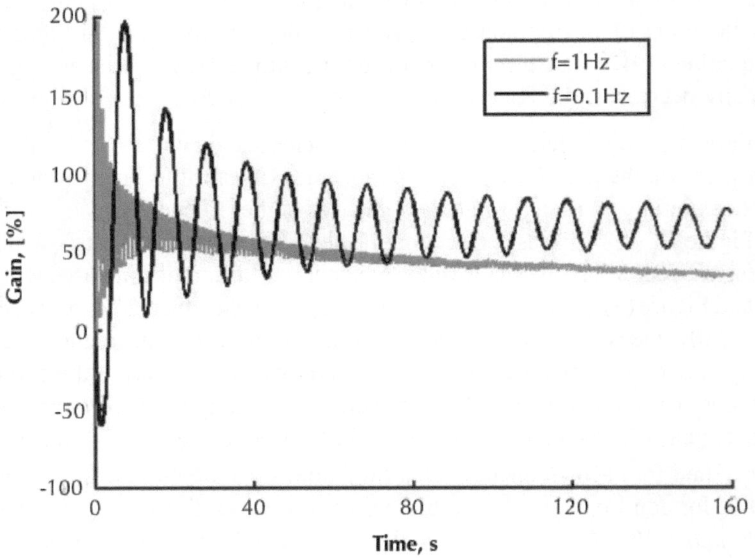

FIGURE 11 (A) The gain functions for free, bound and total plasmin uptake by the fibrin gel. Curves for bound solute and total solute nearly coincide.(B) The gain function for not binding solute for two frequencies: $f = 1Hz$ and $f = 0.1Hz$ as a function of real time.

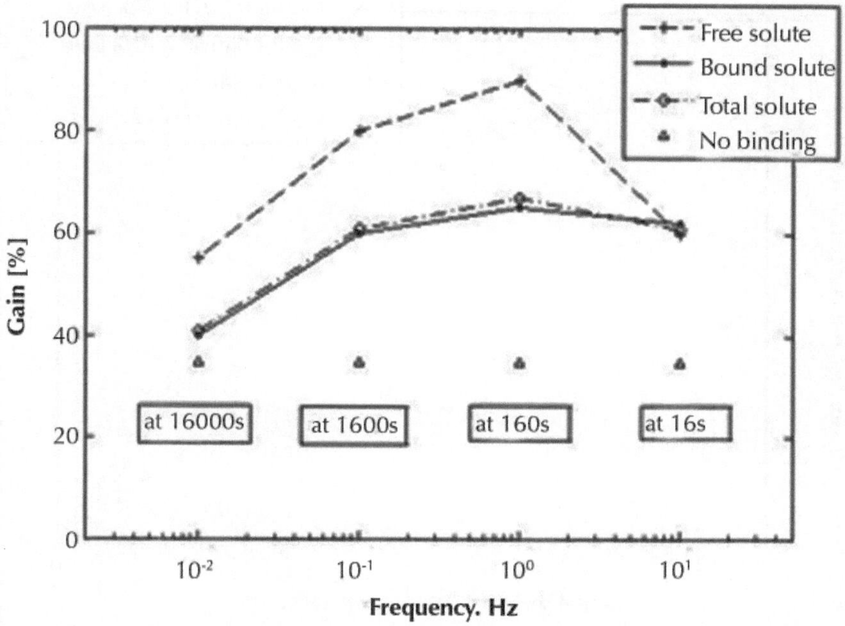

FIGURE 12 The average gains of free, bound, and total solute as a function of frequency. Note, that for higher frequency the time point of calculation is lower (times are shown in boxes under data). The average gain for the case of the solute that does not bind to matrix is shown by triangles.

7.5.2 Comparison with Experimental Data

Few experimental data on molecular transport in gels under cyclic loading is available to date, even for non-binding solutes. Experiments are performed mainly on cylindrical cartilage explants, subjected to high deformation amplitudes [3, 5, 7]. Desorption or absorption rates have been measured for solutes of different sizes. Their dependence on frequency and strain has been described. But it is questionable to use the experimental results for high strains (10–50%) to compare with the model predictions, because we use the assumption of small strains while deriving the equations.

To demonstrate the correlation of our model with experimental results, we have compared the theoretical predictions with the data obtained by Quinn et al. (2002) [4]. The authors investigated the radial desorption of 3kDa dextran from cylindrical cartilage explant under 0.001Hz, 5% dynamic loading. Using fluorescence, they have obtained the concentration profile of dextran as a function of radial distance from the lateral surface at 1 hr after the beginning of desorption from initial homogeneous state. Radius of cartilage disk was 1.5 mm. The corresponding concentration profile for diffusion without loading has also been obtained.

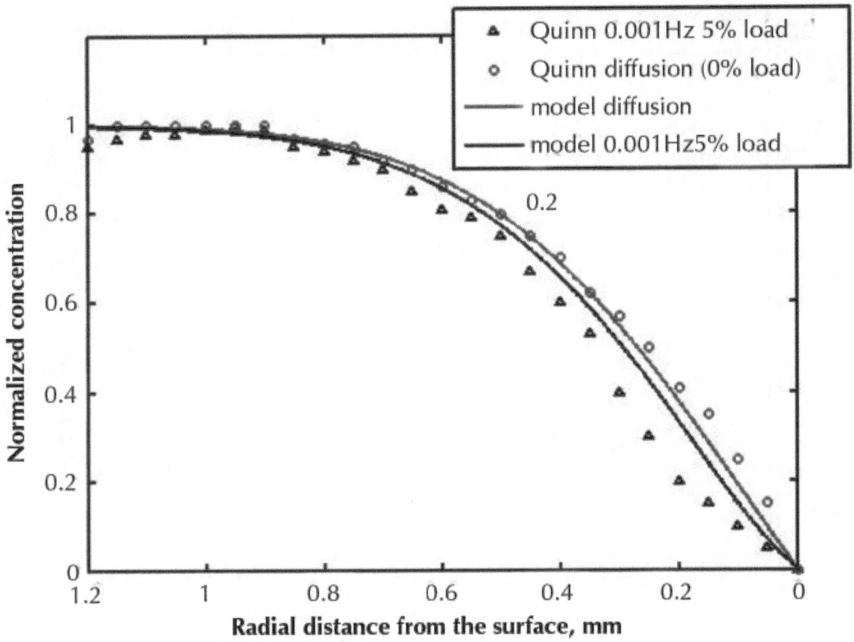

FIGURE 13 Comparison of the theoretical and experimental desorption profiles: concentration distribution of 3kDa dextran *versus* distance from the cartilage surface at time 3,600 s. Concentration value is normalized by initial concentration that was homogeneous through the thickness of explant at time 0.

To model the cartilage disk, we rewrite the governing equations in cylindrical coordinates (not shown). Unfortunately, some necessary parameters of the system are not mentioned in the text. Therefore, we take the value for aggregate modulus and cartilage permeability from other source (see Table 1). We solve the system of equations for the following parameters:

$$H_A = 250kPa,\ k = 2\cdot 10^{-6}\frac{mm^2}{kPa\cdot s},\ D = 15\cdot 10^{-6}\frac{mm^2}{c},\ f = 0.001Hz\cdot$$

Figure 13 shows the normalized dextran concentration profile inside the cartilage disk after 3,600 s of desorption, as calculated using the model. Symbols on Figure 13 denote the results from Quinn et al. There is no indication of experimental errors in this chapter. Nevertheless, it can be concluded that our model tends to underestimate the effect of cyclic loading on desorption from cartilage.

7.6 CONCLUSION

The present study describes theoretically the solute transport through biological tissue under cyclic loading with account of the solute binding to the extracellular matrix. The model is based on the theory of mechanics of poroelastic materials. This computational model was used to compare the process of the solute uptake from the bath solution due to free diffusion and due to advection-diffusion mechanism, induced by cyclic loading. The goal was to find the governing parameters for solute transport acceleration by the fluid flow.

The main results are the following:

(1) Dynamic deformation enhances both bound and unbound concentrations within the tissue in comparison to free diffusion. The beneficial effect for unbound solute is greater.
(2) The maximum gain for unbound solute uptake could be several times higher (2.5 times for parameters of fibrin gel and plasmin) for solutes that have an ability to bind the matrix, compared with the gain for non binding solutes.
(3) The effect of loading is considerable in the surface layer of the gel that is involved in the deformation, and approaches zero deeper in the gel. The thickness of this surface layer depends on loading frequency and material parameters of the gel.
(4) For the loadings with the same displacement amplitude the solute transport acceleration in the surface layer increases with frequency because strain amplitude will be higher for higher frequency.
(5) For the loadings with the same strain amplitude (or, equally, loading pressure amplitude) there exists an optimal frequency at which the solute transport acceleration reaches its maximum. This frequency depends on characteristics of the gel and on the parameters of the solute binding to extracellular matrix.

Finally, we conclude that solute binding to the matrix is an efficient mechanism that is synergistically acting along with cyclic convection on solute transport in dynamically loaded gels. The enhancement in transport of nutrients, morphogens or drug molecules can potentially trigger response of the cells, thus providing a mechanism of sensitivity of mechanical stimulus by the cells in healthy or regenerating tissue.

The applications of the model comprise predictions of optimal mechanical stimulation of the implant for better integration into bone. Also it can be used for deriving mechanical properties for implant coatings, designing of controlled drug delivery devices involving cyclic deformation of the carrier, and calculating drug distribution in the loaded tissues.

KEYWORDS

- **Biphasic theory**
- **Cyclic deformation**
- **Molecular transport**
- **Solute binding**
- **Tissue mechanics**

REFERENCES

1. Prendergast, P., Huiskes, R., and Søballe, K. Biophysical stimuli on cells during tissue differentiation at implant interfaces. *Journal of Biomechanics,* **30**(6), 539–548 (1997).
2. Mauck, R., Hung, C., and Ateshian, G. Modeling of neutral solute transport in a dynamically loaded porous permeable gel: implications for articular cartilage biosynthesis and tissue engineering. *Journal of biomechanical engineering,* **125**, 602 (2003).
3. Evans, R. and Quinn, T. Dynamic compression augments interstitial transport of a glucose-like solute in articular cartilage. *Biophysical journal,* **91**(4), 1541–1547 (2006).
4. Quinn, T., Studer, C., Grodzinsky, A., and Meister, J. Preservation and analysis of nonequilibrium solute concentration distributions within mechanically compressed cartilage explants. *Journal of biochemical and biophysical methods,* **52**(2), 83–95 (2002).
5. Bonassar, L., Grodzinsky, A., Frank, E. et al. The effect of dynamic compression on the response of articular cartilage to insulin-like growth factor-I. *Journal of orthopaedic research,* **19**(1), 11–17 (2001).
6. Garcia, A., Frank, E., Grimshaw, P., and Grodzinsky, A. Contributions of fluid convection and electrical migration to transport in cartilage: relevance to loading. *Archives of biochemistry and biophysics,* **333**(2), 317–325 (1996).
7. Chahine, N., Albro, M., Lima, E. et al. Effect of dynamic loading on the transport of solutes into agarose hydrogels. *Biophysical journal,* **97**(4), 968 (2009).
8. Swartz, M. and Fleury, M. Interstitial flow and its effects in soft tissues. *Biomedical Engineering,* **9**(1), 229 (2007).
9. Sengers, B., Oomens, C., and Baaijens, F. An integrated finite-element approach to mechanics, transport and biosynthesis in tissue engineering. *Journal of biomechanical engineering,* **126**, 82 (2004).
10. Gardiner, B., Smith, D., Pivonka, P. et al. Solute transport in cartilage undergoing cyclic deformation. *Computer Methods in Biomechanics and Biomedical Engineering,* **10**(4), 265–278 (2007).
11. Zhang, L. and Szeri, A. Transport of neutral solute in articular cartilage effects of loading and particle size. *Proceedings of the Royal Society A,* **461**(2059), 2021 (2005).
12. Zhang, L., Gardiner, B., Smith, D. et al. The effect of cyclic deformation and solute binding on solute transport in cartilage. *Archives of biochemistry and biophysics,* **457**(1), 47–56 (2007).
13. Travascio, F. *Modeling Molecular Transport and Binding Interactions in Intervertebral Disc.* Ph.D. thesis. (2009).
14. Ferguson, S. Ito, K., and Nolte, L. Fluid flow and convective transport of solutes within the intervertebral disc. *Journal of biomechanics,* **37**(2), 213–221 (2004).
15. Urciuolo, F., Imparato, G., and Netti, P. Effect of dynamic loading on solute transport in soft gels implication for drug delivery. *AIChE Journal,* **54**(3), 824–834 (2008).
16. Mow, V., Kuei, S., Lai, W., and Armstrong, C. Biphasic creep and stress relaxation of articular cartilage in compression: theory and experiments. *Journal of Biomechanical Engineering,* **102**, 73 (1980).
17. Grossberg, Yu. A. and Khokhlov, A. R. *Statistical Physics of Macromolecules.* Nauka Publishers, Moscow (1988).
18. Soltz, M. and Ateshian, G. Interstitial fluid pressurization during confined compression cyclical loading of articular cartilage. *Annals of Biomedical Engineering,* **28**(2), 150–159 (2000).
19. Diamond, S. and Anand, S. Inner clot diffusion and permeation during fibrinolysis. *Biophysical journal,* **65**(6), 2622–2643 (1993).
20. Weisel, J. The mechanical properties of fibrin for basic scientists and clinicians. *Biophysical chemistry,* **112**(2–3), 267–276 (2004).
21. Noailly, J., Van Oosterwyck, H., Wilson, W. et al. A poroviscoelastic description of fibrin gels. *Journal of biomechanics,* **41**(15), 3265–3269 (2008).
22. Matveyev, M. and Domogatsky, S. Penetration of macromolecules into contracted blood clot. *Biophysical journal,* **63**(3), 862 (1992).

23. Vader, D. et al. Strain-Induced Alignment in Collagen Gels. *PLoS One*, **4**(6), 5902 (2009).
24. Ramanujan, S., Pluen, A., McKee, T. et al. Diffusion and convection in collagen gels implications for transport in the tumor interstitium. *Biophysical journal*, **83**(3), 1650–1660 (2002).
25. Ingham, K., Brew, S., and Isaacs, B. Interaction of fibronectin and its gelatin-binding domains with fluorescent-labeled chains of type i collagen. *Journal of Biological Chemistry*, **263**(10), 4624 (1988).
26. Pluen, A., Netti, P., Jain, R., and Berk, D. Diffusion of macromolecules in agarose gels comparison of linear and globular configurations. *Biophysical journal*, 77(1), 542–552 (1999).

8 Structural State of Erythrocyte Membranes from Human with Alzheimer's Disease

N. Yu.Gerasimov, A. N. Goloshchapov, and E. B. Burlakova

CONTENTS

8.1 INTRODUCTION

Alzheimer's disease (AD) is the progressive neurodegenerative pathology accompanied with behavior disorders and failures of the cognitive function and memory [1]. It is assumed that the main pathological characteristics are the presence of the β-amilod patchers and the neurofibrillar glomerules [2], oxidative stress development, and cholinergic system deficiency [3]. The presence of the β-amilod connected with structural change in the membranes, which affect on the functioning of receptors and membrane proteins, including enzymes. Increasing fluidity of the membranes in the presence of the β-amilod were shown at [4, 5].

We have shown that brain membrane fluidity alterations play important role in the development of the AD, based on olfactory bulbectomy [6]. Find out the changes in the structural state of the erythrocyte membranes of mice after olfactory bulbectomy [7] correlated with the changes in the forebrain membranes. Thus, erythrocyte membrane structural alterations can exhibit the changes in the brain membranes. Therefore, it was interesting to study structural state of the erythrocyte membranes from human with AD.

Structural state of erythrocyte membranes from patients with AD was studied. Erythrolysis, Malondialdehyde (MDA) content, and lipid bilayer microviscosity was used as a membrane structural characteristic. Erythrolysis and MDA content exhibit Lipid Per Oxidation (LPO) level. Membranes microviscosity was measured by electron paramagnetic resonance spin labeling of 2,2,6,6-tetramethyl-4-capryloyl-oxypiperidine-1-oxyl (lipidic label) 5,6-Benzo-2,2,6,6-tetramethyl-1,2,3,4-tetrahydro-g-carbolyn-3-oxide (proteinic label). It was established in erythrocyte membrane both components (lipidic and proteinic) fluidity was increased of all patients with AD. Three groups were sorted out in terms of LPO. The LPO intensity was enhanced in the first group. This group was characterized by high hemolysis and MDA content before and after incubation. The MDA content and hemolysis were increased during incubation period. The LPO intensity was decreased in two other groups. Low erythrolysis and no changes during incubation period were observed in the second group. Ratio MDA/peroxide hemolysis was notably greater than unity. Third group was characterized by increased mechanical hemolysis by decreasing the erythrolysis and MDA content during incubation period.

We had already shown that erythrocyte membrane structural alterations can exhibit the changes in the brain membranes [6, 7]. Therefore, it was interesting to study structural state of the erythrocyte membranes from human with AD.

8.2 EXPERIMENTAL

Erythrocyte membrane structural state of the ten patients with different AD severity was studied. As they control the red blood cells of fifty one peoples over 40 years old was used. Erythrocytes were isolated from blood by means of differential centrifugation at 1,000 g for 10 min. Membrane fluidity in two regions of lipid bilayer was estimated with the help of Erythrocyte sedimentation rate (ESR) technique. 2,2,6,6-tetramethyl-4-capryloyl-oxypiperidin-1-oxyl (probe I) and 5,6-benzo-2,2,6,6-tetramethyl-1,2,3,4-tetrahydro-γ-carboline-3-oxyl (probe II) were used as a probes. Probes spin correlation time (τ_c) were calculated from obtained spectra [8, 9], which correspond to the period of radical's reorientation about $\pi/2$. A probe spin correlation time is proportional to the membrane microviscosity or inversely proportional to fluidity. It is known that probe II localizes mostly in near protein areas, probe I in protein free areas of lipid bilayer surface (2–4 Å) [10], thus the probe spin correlation time can show lipid-protein interaction in membranes. In addition, erythrocyte lysis and MDA content (measuring units was mmole per million of erythrocytes) are changes were determined as a LPO rate index. The MDA content and hemolysis before (mechanical hemolysis (MH)) and after (peroxile hemolysis (PH)) incubation at 37°C for 10 min were measured by the reaction with tiobarbituric acid.

8.3 DISCUSSION AND RESULTS

Obtained results shown that erythrocytes membrane both the regions (lipid and protein) fluidity were increased twice above normal (Figure 1). This fact testifies as a failure in the system of LPO homeostasis in these membranes [11, 12]. Burlakova E. B. has proposed the memory model with the membranes structure as the primary factor [13]. Based on this model memory impairments during AD can be explained. As a result of increased membranes lability (~twice above normal) membranes mesomorphic structure is altering insomuch rapidly, that both the lasting and recent memory are lost.

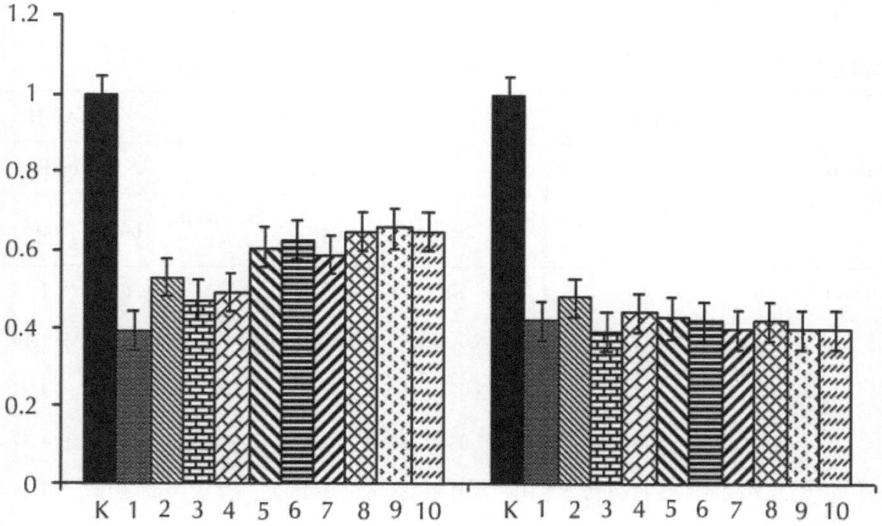

FIGURE 1 Relative changes of the erythrocyte membrane microviscosity. Control parameters (K) of τ_c: $1.10*10^{-10}$s for probe I (left side) $1.92*10^{-10}$s for probe II (right side)

Changes of LPO parameters are shown in Table 1. Mechanical hemolysis (MH) was increased for the first, second, and third patients. Erythrolysis were further growing during incubation period. At the same time MDA content were enhanced before incubation. These facts speak about enhanced LPO rate [14]. The MH were within normal limits for fourth patient, but extremely high after incubation. As for MDA content it was much higher than control before and after, incubation. Such parameters are specific for the stress situations [14]. In this case LPO rate were also increased.

Both hemolysis and MDA content were deeply decreased before incubation for the fifth, sixth, and seventh patients, with no growth of the hemolysis during incubation period. For all three patients MDA/PH relation were significantly increased. That means LPO rate suppressed because of substrate depletion. Such continuous that LPO rate decreasing can leads to remolding and disorganization of the membrane structure, enhancing percentage composition of phospholipids, disposed to reoxidation, and destruction of the cell due to lack of lipids. Perhaps, at this stage of the AD development organism tries to save erythrocytes from further lysis, which can lead to concentration of the unsaturated lipids in the membranes.

Hemolysis and MDA content were decreasing during incubation period for the eighth, ninth, and tenth patients. This fact indicates a substrate depletion and consequently,

LPO rate suppression in the erythrocytes. The MDA/PH relation after incubation was less then unity, whereas in most cases higher with absence of hemolysis growing [14]. This suggests failure in the system of LPO homeostasis in erythrocytes membrane. Apparently, further pathology development led to concentration of the unsaturated lipids in the membranes and accordingly, increasing its oxidability. Then there was burning out of this lipids and substrate depletion.

TABLE 1 The LPO rate parameters of the erythrocytes from patients with AD.

Patient, №	Hemolysis			MDA content			MDA/PH	
	MH, %	PH, %	PH/MH	Incubation			Incubation	
				before	after	increment, %	before	after
Control	1.69	2.9	1.72	2.8	3.7	32	0.97	1.28
1	2.35±0.09	4.67±0.09	1.99	5.01±0.15	5.55±0.19	11	1.07	1.19
2	2.57±0.09	3.63±0.09	1.41	3.47±0.15	6.33±0.25	82	0.96	1.74
3	2.89±0.15	4.15±0.16	1.44	7.03±0.25	7.36±0.25	5	1.69	1.77
4	1.71±0.07	5.19±0.19	3.04	5.91±0.25	6.62±0.25	12	1.14	1.28
5	0.60±0.05	0.70±0.05	1.17	2.99±0.15	1.69±0.15	-43	4.27	2.41
6	0.50±0.05	0.50±0.05	1.00	2.92±0.15	2.63±0.15	-10	5.84	5.26
7	0.30±0.05	0.30±0.05	1.00	1.80±0.15	3.09±0.15	72	6.00	10.30
8	2.00±0.08	1.00±0.08	0.50	3.46±0.15	3.12±0.15	-10	1.73	1.56
9	2.50±0.08	1.65±0.08	0.66	3.75±0.15	0.97±0.15	-74	1.50	0.39
10	2.00±0.08	1.75±0.08	0.88	2.27±0.15	1.27±0.15	-46	1.14	0.64

8.4 CONCLUSION

Erythrocytes membrane both the regions (lipidic and near protein) fluidity increased level were observed, which testify failures in the system of LPO homeostasis in these membranes [11, 12]. We have shown that LPO rate is enhanced not for all patients, as considered earlier. Three groups were marked out by LPO parameters. The LPO rates were increased for the first group. It was characterized by enhanced hemolysis and MDA content before and after incubation with increasing both hemolysis and MDA content. For the second and third groups LPO rates was decreased. The second group was characterized by low hemolysis without increment during incubation. Relations of the MDA content to the peroxide hemolysis were significantly greater than unity. Enhanced mechanical hemolysis and decreasing of MDA content and hemolysis during incubation period were observed for the third group. Probably, these groups were in a different AD severity.

KEYWORDS

- **Alzheimer's disease**
- **Lipid-protein interactions**
- **Membrane fluidity**
- **Membrane structure**
- **Spin labeling**

REFERENCES

1. Gavrilova, S. *Russ. Med.*, Zhurn., (in Russian) **5**, 1339 (1997).
2. Hardy, J. A. and Higgins, G. A. *Science*, **256**, 184 (1992).
3. Bartus, R. T., Dean, R. L., Beer, B., and Lippa, A. S. *Science*, **217**, 408 (1982).
4. Nagarajan, S., Ramalingam, K., Neelakanta Reddy, P. et al. *FEBS J.*, **275**(10), 2415 (2008).
5. Avdulov, N. A., Chochina, S. V., Igbavboa, U. et al. *J. Neurochem.*, **68(5),** 2086 (1997).
6. Gerasimov, N., Goloshchapov, A., and Burlakova, E. The fluidity changes of the membranes isolated from forberain of mice bearing Alzheimer's-like disease caused by the olfactory bulbectomy. *In Modern Problems in Biochemical Physics New Horizons.* Sergei D. Varfolomeev, Elena B. Burlakova, Anatoly A. Popov, and Gennady E. Zaikov, (Eds.). Nova Science Publishers, New York (2011).
7. Gerasimov, N., Golowapov, A., Molochkina, E., and Burlakova, E. Tezisy dokladov konferencii "Nejrohimija: fundamental'nye i prikladnye aspekty", M., (in Russian), p. 161 (2005).
8. Wasserman, A., Buchachenko, A., Kovarskii, A., and Nejman, M. Investigation of Molecular Motions in Polymers with Paramagnetic Probe method. *Visokomol Soed.*, (in Russian), **10**(8), 1930 (1968).
9. Sukhorukov, B., Wasserman, A., Kozlova, A., and Buchachenko, A. *Bulletin of the Academy of Sciences of the USSR,* (in Russian) **177**, 454 (1967).
10. Biniukov, V., Borunova, S., Goldfeld, M. et al. Investigation of Structural Transition in Biological Membranes with Spin Probe Method. *Biokhimiya*, (in Russian) **43**(6), 1149 (1971).
11. Aristarkhova, S., Arkhipova, G., Burlakova, E. et al. *Bulletin of the Academy of Sciences of the USSR*, (In Russian) **228**, 215 (1976).
12. Burlakova, E. and Khrapova, N. Membranes Peroxidation and Natural Antioxidants. *Uspekhi Khimii.* (in Russian), **54**(9), 540 (1985).
13. Burlakova, E. *Zhurn. Fiz. Khimii.,* (in Russian) **18**(5), 1311 (1989).
14. Bankova, V. Diss. *d-ra biol.*, (in Russian) nauk 18.02.91. M. (1990).

KEYWORDS

REFERENCES

9 Updates on Application of Silver Nanoparticles

N. I. Naumkina, O. V. Mikhailov, and T. Z. Lygina

CONTENTS

9.1 INTRODUCTION

Many medical and biological processes consist of photosensitive redox stages that occur with the participation of metal porphyrins and hydrogen peroxide. For instance photosynthesis consists of a large number of redox reactions, some of them are photocatalyzed by chlorophyll (Chl). During photosynthesis the interaction of electron donor and carbon dioxide leads to the formation of energy intensive organic compounds and generation of oxygen [1]. At present time it is proposed that water molecules are oxidized to O_2, and hydrogen peroxide is an intermediate of water oxidation [2, 3]. Moreover it is assumed that probably H_2O_2 is a source of photosynthetic oxygen [2]. It is important that H_2O_2 oxidation is less endothermic than water [4]. In photodynamic therapy metal porphyrins produce singlet oxygen, which is converted into hydrogen peroxide in water media [3]. Thus, the pharmacologically active porphyrin photosensitizers contact with the H_2O_2 also.

Creation of artificial photosensitive systems is useful for the study of coordination and photochemical interaction of metal complexes of porphyrins and reactive oxygen species. It is known that some metal complexes with porphyrins and phthlocyanines are effective catalysts of H_2O_2 decomposition in dark [5]. In this work, a photocatalytic activity of Chl and metal (Cr, Cu, Sn, Zn, Cd, Mg) are complexes of tetraphenylporphyrin (TPP) in the reaction of H_2O_2 decomposition was studied. Special attention is paid to the coordination of metal porphyrins and hydrogen peroxide. In addition a possibility of reduction of nicotinamidadenindinucleotid phosphate (NADP) and methyl red dye (MR) connected with H_2O_2 the decomposition of photocatalyzed by Chl was shown.

The coordination interaction between metal complexes of porphyrins, including Chl, and hydrogen peroxide was detected. The kinetic parameters of photocatalytic decomposition of H_2O_2 in the presence of Chl and metal porphyrins immobilized on silica were studied. Photocatalytic activity of a number of non-transition metal porphyrins is shown to correlate with their ability to generate photopotential. Processes of reduction of NADP and MR photocatalyzed by Chl in H_2O_2 solution are demonstrated.

Primary goal of our study is to show a possibility of reduction of electron acceptors under photocatalytic decomposition of H_2O_2.

9.2 EXPERIMENTAL

The Chl '*a*' were separated by known method [6]. Individuality and concentration of Chl were determined by UV-vis spectroscopy in quartz cells (1 cm) on hach dr/4000v spectrophotometer. Metal complexes of TPP were synthesized and purified in Ivanovo State University of Chemistry and Technology (Ivanovo, Russia). Hydrogen peroxide and sodium bicarbonate (Reakhim, Russia) were used without additional purification.

Immobilizations of metal complexes on silica L 40/100 (Chemapol) were realized by addition of silica (1 g) to the solutions of Chl in acetone and complexes of TPP in chloroform. Suspensions were kept in the darkness to evaporate the solvent. Samples were repeatedly washed out with distilled water and dried to constant weight in vacuum exicator over $CaCl_2$.

For kinetic experiments 10 ml of bicarbonate buffers (pH 8.5) containing 0.2 mol/l H_2O_2 and 200 mg of silica with one of the supported complexes were placed in the photochemical reactor. The obtained suspensions were irradiated by visible light using halogen lamp (150 W) with condenser and system of lenses at constant stirring. Concentration of H_2O_2 was determined by titration method in 0.2 mol/l H_2SO_4 medium using 0.01 N $KMnO_4$ solutions. All experiments were carried out at 20°C. In the experiments on reduction of electron acceptors the solutions of NADP ($0.7 \cdot 10^{-4}$ mol/l) or MR ($8.1 \cdot 10^{-4}$ mol/l) were added to water suspension of supported Chl. In the experiments with MR ethanol were added into reaction system. For determination of the amount of acceptor reduced forms of their spectral characteristics were used. UV-vis λ_{max} (water), nm (lgε): 320 (3.77) for reduced NADP and λ_{max} (ethanol), nm (lgε): 498 (2.68) for reduced MR [7].

9.3 DISCUSSION AND RESULTS

The Chl and metal complexes of TPP coordinate hydrogen peroxide according to the electron adsorption spectra (Figure 1). Solutions of H_2O_2 were sequentially added to the TPP metal complex solutions (10^{-5} mol/l) to concentrations from 10^{-7} to 10^{-3} mol/l. We can see that changes in the electronic spectra are concerned with only increase or decrease of the extinction in the bands and the isobestic point observation. All visible bands are correspond to π–π* transitions. At the addition of H_2O_2 energy parameters of the bands do not change. This means that the place of coordination of hydrogen peroxide molecule(s) is magnesium ion, but not the porphyrin macrocycle, since the frontier orbitals of d^0 and d^{10} metal complexes are localized on the ligand and they are very sensitive to outer coordination.

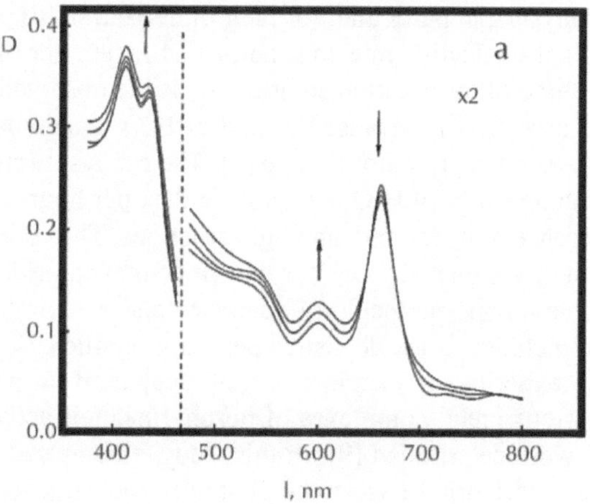

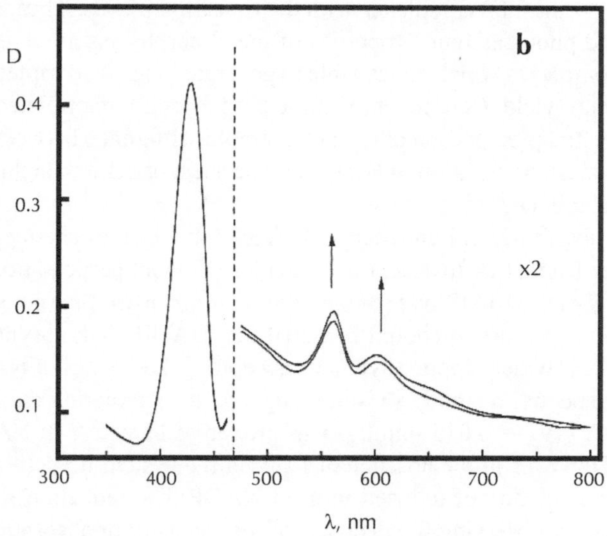

FIGURE 1 Electronic spectra of Chl in ethanol (50%) and ZnTPP in 0.05% surfactant Tween-20 at addition of H_2O_2.

Photocatalytic activity in the reaction of H_2O_2 the decomposition is shown to be demonstrated for Chl and for different metal complexes of TPP immobilized on silica. Kinetic parameters of their catalytic and photocatalytic activities are given in the Table 1. Under the visible light irradiation decomposition of H_2O_2 accelerates in the cases of all the metal complexes studied, except highly active complex with Fe^{III}. Complexes of TPP with magnesium, zinc, and Chl are the most active in the process of H_2O_2 decomposition.

It should be noted that the surface amount of metal complexes were much higher than a monolayer. Thus, the experimental kinetic data describe the collective activity of associated molecular ensembles of metal porphyrins. It is important that the coordination and the decomposition rate of hydrogen peroxide are effectively under aggregation of Chl and porphyrins, in particular, for their dimerization [8].

In the Table k_{ef} is the effective rate constant, N is the number of moles of (photo) catalyst, V is the volume of the reaction solution, k_{ob} is the observed rate constant that could be expressed as $k_{ef}(N/V)^n$ because the rate of H_2O_2 decompos $w = k_{ef}(N_{Cat}/V)^n [H_2O_2]^m$, where N/V, n and m=1 do not change [4]. The catalyst turnover number (TN) correspond to the number of mol H_2O_2 per mol catalyst per hour. Parameter η shows the percentage difference of k_{ef} for dark and light reactions. The coordination sphere of the metal ion also includes one (for Fe^{III}) or two (for Sn^{IV}) chloride anion. Ytterbium ion additionally contains one molecule of acetylacetone as extraligand on the third valence. Amount of metal complex deposited per 1 gm of silica.

Recently, in the experiments on the effect of Becquerel on photoelectrodes the modifiction with various metal complexes of porphyrins,their activity in the generation photopotential were determined [9]. In this study, a comparative analysis shows that there is a linear correlation between the photoelectrochemical and photocatalytic properties of d^0 and d^{10} metal complexes of TPP (Figure 2). No similar correlations were observed for the TPP complexes with the transition metals. Thus, the analogy in photovoltaic and photocatalytic properties of metal porphyrins are exists only in the case of metal complexes which are capable to generate long-lived triplet excited states with high quantum yield. Consequently, these properties are interrelated and could be used to estimate and predict each other. For example, difference between rates of dark and light induced decomposition of H_2O_2 could be a test parameter in the development of molecular solar energy converters.

To prove a possibility of light energy storage in the form of chemical potential an electron transfer from H_2O_2 to acceptor molecules with Chl participation was studied. Sodium salt of NADP and MR dye (p-dimetilaminoazobenzol-2-carboxylic acid) were used as electron acceptors. It should be noted that NADP is photosynthetic electron acceptor and MR is widely applied synthetic acceptor [8]. Kinetics of NADP reduction was studied by spectral method in visible range. At the irradiation during *ca.* 80 min the supported Chl (v_{im} = 0.015 μmol/g) was proved to be active in NADP reduction (Figure 3(a)). However in the absence of H_2O_2 in the system no activity of Chl was observed in accumulation of reduced form of NADP. The reduction reaction of MR was investigated by observing the decrease of the intensity of absorption band of the initial oxidized form. The obtained data confirm the photocatalytic activity of supported Chl (v_{im} = 0.063 μmol/g) in reduction of methyl red (Figure 3(b)). In the presences of hydrogen peroxide activity of Chl accelerates at conditions of photocatalysis. It was shown that after the photoexcitation the primary stage is electron transfer to the first molecule of hydrogen peroxide and the formation of the radical cation of Chl [3]. Further, the radical cation of Chl oxidizes the second molecule of hydrogen peroxide. Thus, in threefold system H_2O_2- Chl acceptor under the visibility of light irradiation the reduction of electron acceptors (NADP and MR) takes place what proves the ability of H_2O_2 to function as an electron donor in these conditions.

TABLE 1 Kinetic parameters of catalytic and photocatalytic activities of Chl and metal complexes of TPP immobilized on silica in H_2O_2 decomposition (pH 8.5)[a].

Metal complex	$\nu_{im}{}^{d}$, μmol/g	$(N/V)\cdot 10^5$, mol/l	$k_{ob}\cdot 10^5$, s^{-1}	k_{ef}, $l\ mol^{-1}\ s^{-1}$	TN, h^{-1}	$k_{ob}\cdot 10^5$, s^{-1}	k_{ef}, $l\ mol^{-1}\ s^{-1}$	TN, h^{-1}	η, %
			In dark			Under irradiation			
Chl	5.5	11.0	0.36	0.033	120	0.74	0.067	240	50
CrTPP	66	132	1.50	0.011	40	2.27	0.018	65	38
CuTPP	56	111	0.13	0.001	4	2.27	0.020	72	94
ZnTPP	55	110	3.87	0.035	130	4.35	0.040	145	10
CdTPP	52	103	0.39	0.004	14	0.60	0.006	22	36
SnTPP[b]	52	103	0.97	0.009	32	1.11	0.011	40	20
FeTPP[b]	21	41	17.9	0.438	1580	18.0	0.439	1580	0
MgTPP	49	98	–	–	–	5.0	0.051	185	–
PdTPP	55	110	7.79	0.071	255	–	–	–	–
YbTPP[c]	57	114	–	–	–	1.82	0.016	60	–

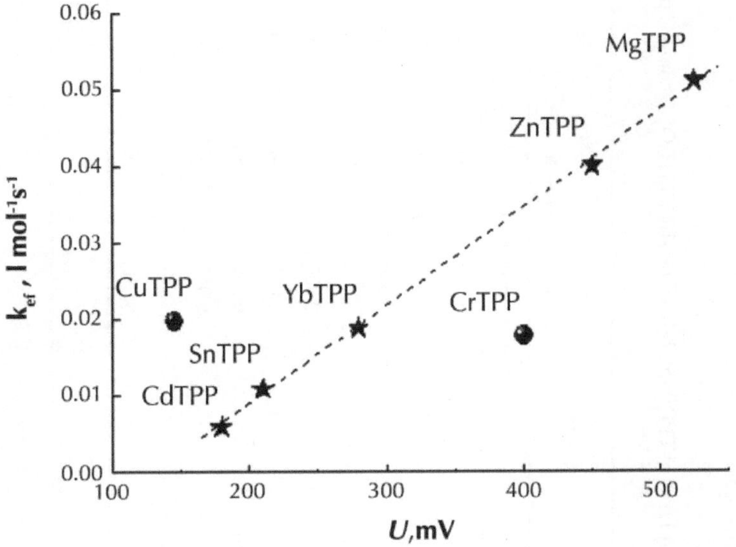

FIGURE 2 The linear correlation between the photocatalytic activity of non-transition metal, complexes of tetraphenylporphirin (shown by asterisks) and their ability to generate photopotential.

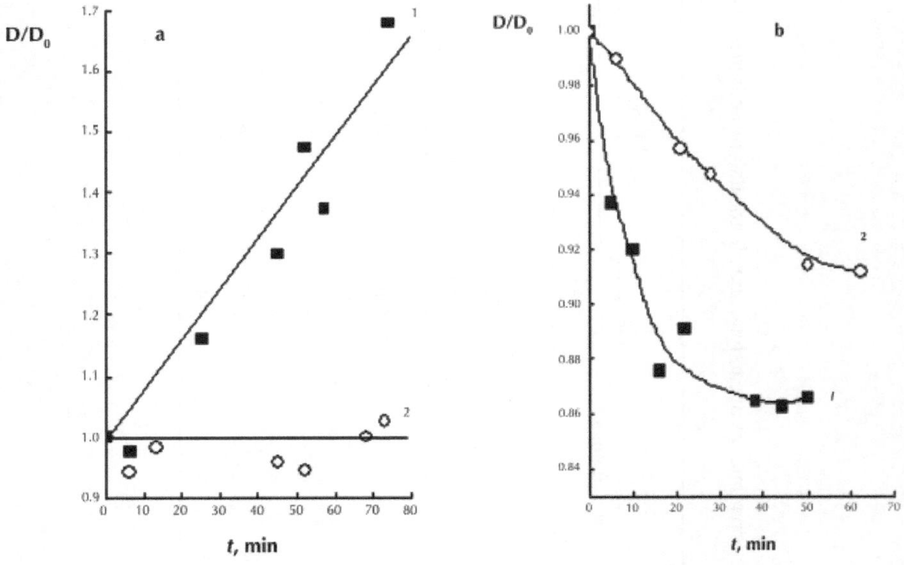

FIGURE 3 Reduction of NADP (*a*) and MR (*b*) under photocatalysis by Chl in the presence (*1*) and absence (*2*) of hydrogen peroxide.

These results show that interconversion of reactive oxygen species in photocatalysis of metal complexes of porphyrins in the certain cases is able to provide energy storage reaction with increase of the chemical potential. On the one hand, it is important for the development of processes of utilization of solar energy. On the other hand, it is of interest to study the energy of intracellular biochemical processes involving the pharmacologically active metal complexes.

9.4 CONCLUSION

The model photosensitive systems with hydrogen peroxide participation were prepared. Immobilized metal complexes of porphyrins and Chl shown to possess a catalytic activity in the reaction of H_2O_2 decomposition. Linear correlations between the photovoltaic and photocatalytic properties of non-transition metal complexes of TPP were found. It was demonstrated that electron transfer from H_2O_2 to acceptor molecules under visible light irradiation is possible.

KEYWORDS

- **Chlorophyll**
- **Electron transfer**
- **Hydrogen peroxide**
- **Metal porphyrins**
- **Photocatalysis**

ACKNOWLEDGMENT

The work was supported by Grant of President of the Russian Federation for State support of young Russian scientists - Candidates of sciences No MK-227.2011.3, RAS Presidium Programs No 3, No 25, Project of International Science and Technology Center No 3910 and Grant of support of Leading Scientific Schools No NSh-65059.2010.3.

REFERENCES

1. Hall, D. O.and Rao, K. K. *Photosynthesis*, Mir, Moscow p. 134 (in Russian) (1983).
2. Komissarov, G. G. *Photosynthesis physico-chemical approach*, (in Russian), Editorial URSS, Moscow p. 224 (2003).
3. Lobanov, A. V., Safina, Yu. A., Nevrova, O. V., and Komissarov, G. G. *Interaction of O_2 and H_2O_2 under irradiation of chlorophyll/silica/water suspensions saturated with air/in Problems of origin and evolution of biosphere*. (in Russian), E. M Galimov. (Eds.)., Librokom, Moscow p. 195 (2008).
4. Lobanov, A. V. *Photocatalytic processes involving hydrogen peroxide in natural and artificial photosynthetic systems*. (in Russian), Diss. Ph. D. Semenov Institute of Chemical Physics of RAS, Moscow, p. 117 (2004).
5. Berezin, B. D. *Coordination compounds of porphyrins and phthalocyanin*. (in Russian) Nauka, Moscow, p. 280 (1978).
6. Nevrova, O. V., Lobanov, A. V., and Komissarov, G. G. Decomposition of hydrogen peroxide photocatalyzed by chlorophyll, Cu^{II} and Cr^{III} porphyrins immobilized on silica. *Modern Problems in Biochemical Physics*, New Horizons. S. D. Varfolomeev, E. B. Burlakova, A. A. Popov, and G. E. Zaikov (Eds.). Nova Science Publishers, New York (2011).

7. Doson, R. and Elliot, D. *Biochemist's Handbook*. Mir, Moscow p. 544 (1991) (in Russian).
8. Lobanov, A. V., Vasiliev, S. M. and Komissarov, G. G. Interaction of Hemin and Hydrogen Peroxide Effect of Media. *Macroheterocycles*,. **2**(3–4), 268–270 (2009).
9. Rudakov, V. M., Ilatovskii, V. A., and Komissarov, G. G. Photoactivity of metal tetraphenylporphyrins. *Sov. J. Chem. Phys.*, **6**(4), 1021–1026 (1990).

10 Tetrapyrrolic Macrocycles with Magnesium, Alluminum, and Zinc in Hydrogen Peroxide Decomposition

O. V. Nevrova, A. V. Lobanov, and G. G. Komissarov

CONTENTS

10.1 INTODUCTION

Activity of tetrapyrrolic macrocycles such as chlorins and phthalocyanines with magnesium, alluminum, and zinc in hydrogen peroxide decomposition in dark and under visible light irradiation was studied. Using adsorption isotherms the conditions for obtaining photocatalytic systems with varying degrees of adsorption of phthalocyanines were revealed. In adsorption systems, the activity of metal complexes depends on their molecular association. The principles allow to obtain the most effective photocatalysts were found.

In this study catalytic and photocatalytic properties of metal tetrapyrrolic complexes were investigated. Special attention was paid to the molecular association of complexes.

Many metal complexes of tetrapyrrole macrocycles exhibit catalytic and photocatalytic activity in redox processes. It is known that high-spin Fe(III)-porphyrins and

Fe(III)-phthalocyanine with a distorted rhombic coordination structure are active in hydrogen peroxide decomposition that is also characteristic for catalase or peroxidase. It was found that Fe(III)-octacarboxyphthalocyanine supported on an amorphous concentrated fiber of artificial silk core element is a remarkably effective catalyst for the decomposition of hydrogen peroxide [1]. Both phthalocyanines in a solution of concentrated sulfuric acid and polymeric structure of phthalocyanines have the catalase activity. Phthalocyanines with metals of VIIIB group have the largest activity. Kinetic data for the H_2O_2 decay catalyzed by phthalocyanines, for a large number of metal complexes are summarized in [2, 3].

However, high rate constants of excited states radiationless deactivation are observed for complexes of tetrapyrrole macrocycles with open *d*-shell metals because of strong exchange interaction between metal unpaired electron and tetrapyrrole ligand molecular orbitals [4]. In contrast, complexes of porphyrins and phthalocyanines with non-transition metals (Mg, Al, and Zn) are capable to generate long-lived (up to 1 ms) triplet excited states with high quantum yield (60–90%) [5]. So, it is possible to realize light controlled processes in the presence of these metal complexes. Chlorophyll (Chl) also belongs to this group of tetrapyrrolic compounds, because its molecule contains a magnesium ion.

The study of the catalytic activity of adsorbed Chl and phthalocyanines-metal complexes in the hydrogen peroxide decay could be used for photodynamic therapy, artificial photosynthesis and molecular photonics [3, 6]. In this regard, the catalytic and photocatalytic activity of Chl and metal phthalocyanines adsorbed on silica in the H_2O_2 decomposition were investigated in this chapter. Alluminum (AlClPc) and zinc phthalocyanine (ZnPc) were used as an analogue of Chl, since their photophysical and photochemical properties are close to those of Chl, while their stability is much higher [3].

10.2 EXPERIMENTAL

The Chl was isolated and purified from dry nettle leaves by chromatography on a column with powdered sugar heated up at 100°C within 4 hr preliminary [7, 8]. The Chl mix was solved in hexane-ether system (3:7) and put on a column. Elution was realized by the same solvent system to distribute of pigments on the column fully. A fraction painted by Chl *a* was mechanically taken, and then it was washed off by ether. Individuality and concentration of Chl *a* were determined by UV-vis spectroscopy in quartz cells (1 cm) on spectrophotometer DR/4000V (Hach, USA). Found and literature [9] spectra of Chl were identical.

The Chl *a*. λ_{max} (ether) nm (lgε): 662 (4.96), 615 (4.14), 578 (3.88), 533.5 (3.57), 430 (5.08), and 410 (4.88).

The AlCl and Zn phthalocyanine complexes were synthesized and purified in Ivanovo State University of Chemistry and Technology (Ivanovo, Russia).

Equilibrium adsorption of Chl and MPc complexes on silica L 40/100 (Chemapol) was realized by addition of silica (1 g) to solutions of Chl in acetone, MPc complexes in N, N-dimethylformamide ("Fluka"). Concentration of solutions was varied for preparing of samples with different value of adsorption. Suspensions were in the dark for 24 hr with constant stirring. Recorded optical density of solutions at wavelength 674 nm before and after adding silica was used for determination of complexes quantity per one gram of silica. Samples were filtered on vacuum, repeatedly washed out with distilled water and dried to constant weight in vacuum-exicator over $CaCl_2$.

For kinetic experiments 10 ml of bicarbonate buffer (pH 8.5) containing H_2O_2 (0.2 mol/l) ("Reakhim") and 200 mg of silica with one of the supported Chl and MPc complexes were placed in the photochemical reactor. The obtained suspensions were irradiated by visible light using halogen lamp (150 W) with condenser and system of lenses at constant stirring. Concentration of H_2O_2 was determined by titration method in H_2SO_4 (0.2 mol/l) medium using $KMnO_4$. All experiments were carried out at 20°C.

10.3 DISCUSSION AND RESULTS

We considered the kinetics of H_2O_2 decomposition in the dark and under visible light irradiation in the presence of supported on silica gel Chl, AlCl and Zn phthalocyanine complexes with varying degrees of silica surface coverage. These metal tetrapyrroles were used either in equilibrium adsorbed or forcibly immobilized states. In this case, the complex association is controlled, the stability of the photocatalyst increases and the solid catalyst does not preclude the analysis of the concentration of H_2O_2.

The H_2O_2 catalytic decomposition rate can be expressed as: $w = k_{ef}(N_{Cat}/V)^n [H_2O_2]^m$, where k_{ef} is the effective rate constant, N_{Cat} is the number of catalyst moles, and V is volume of the reaction solution. If N_{Cat}/V = idem, we can identify $k_{ef}(N_{Cat}/V)^n$ as k_{ob}. Thus, we obtain $w = k_{ob} [H_2O_2]^m$. In all cases reaction kinetics are well linearized in the coordinates $\ln[H_2O_2]$-t (Figures 1, 4), that is m = 1, which led to assume the (pseudo) first order of H_2O_2 decomposition reaction.

Data on the catalytic activity of Chl, AlClPc, and ZnPc in the H_2O_2 decomposition are shown in Table 1. The presented values of catalyst turnover number (TN) correspond to the number of mol H_2O_2 per mol catalyst per hour. Parameter η shows the percentage difference k_{ef} of dark and light reactions.

Kinetic data show that the Chl photoactivity in the H_2O_2 decomposition decreased with increasing of silica grain surface filling degree with metal complexes (*a*) from 0.23 to 1.92 μmol/g (Table 1). The values of *a* is corresponding to kinks in the adsorption isotherm, which was specific for BET theory [11]. The hydrogen peroxide concentration changes in the experiment are shown in Figure 1. Both under light and in darkness the greatest activity was observed for Chl in the case of the substrate surface premonolayer coating.

Upon reaching the monolayer (*a* = 0.57 μmol/g) in this system, H_2O_2 and the magnesium ion coordination decreases because of a steric hindrance. As a result, during more growth of the Chl molecules number on silica surface inactive associates are formed and Chl activity decreases greater. In addition, there were the photochemical reactions and excitation energy dissipation competitions, which were to implement in the aggregates of dyes easily.

It should be noted that the catalytic activity of the samples used on the basis of Chl with filling degrees of 0.23 and 0.57 μmol/g practically coincided with their photoactivity (Table 1). Thus, deposited form of Chl is less sensitive to light than Chl in solutions because of photophysical properties changes of Chl due to possible coordination of Chl molecules with silica, where the central Mg ion is associated with OH-groups

TABLE 1 Catalytic properties of Chl, AlClPc, and ZnPc adsorbed on silica in H_2O_2 decomposition in bicarbonate buffer (pH 8.5) in dark and under the visible light irradiation.

Cat	a, μmol/g	$(N_{Cat}/V)\cdot10^5$, mol/l	$k_{ob}\cdot10^5$, s^{-1} (In dark)	k_{ef}, $l\ mol^{-1}\ s^{-1}$ (In dark)	TN, h^{-1} (In dark)	$k_{ob}\cdot10^5$, s^{-1} (Under irradiation)	k_{ef}, $l\ mol^{-1}\ s^{-1}$ (Under irradiation)	TN, h^{-1} (Under irradiation)	η, %
Chl	0.47	0.94	2.42	2.6	990	5.13	5.5	1700	50
	0.23*	0.46	3.4	7.6	2800	3.2	6.7	2700	~0
	0.57*	1.14	2.7	2.4	970	2.8	2.5	970	0
	1.92*	3.84	3.1	0.81	300	2.1	0.55	230	-30
AlClPc	0.45	0.90	0.33	0.37	150	1.17	1.3	620	75
	0.83	1.66	1.0	0.60	300	1.5	0.9	400	30
	0.95	1.90	0.83	0.44	190	1.33	0.7	310	40
	1.15	2.30	1.5	0.65	290	1.5	0.65	290	0
	2.34	4.68	1.0	0.21	100	3.33	0.71	250	70

*Chl was immobilized with non-equilibrium (forced) deposition by solvent evaporation [10].

of the support. This fact is confirmed by the dependence of the NT on the degree of application (Table 1). Number of H_2O_2 moles dissolved per mole of Chl per hour depends only on the substrate filling by the pigment, but not the presence of irradiation. In addition, Chl immobilized by non-equilibrium deposition on the surface of silica is associate in the system of a theoretical monolayer, and in the cases of smaller degrees of filling. These associates exhibit catalytic activity in H_2O_2 decomposition, reducing at integration associates. Special attention should be paid to the fact that for Chl with the degree of filling of 1.92 mmol/g photoactivity was lower than activity in darkness by 30%. Probably, in this case Chl excess on the carrier surface was photodestructed quickly. It explains the negative value of η. We showed that the Chl photodestruction was accelerated for a highly aggregated state [12]. Most likely, that for systems with $a = 1.92$ μmol/g the interaction of Chl-Chl predominates over the coordination of Chl with the OH-groups of silica. The results indicate the inefficiency of Chl coating large degrees using for photocatalytic reactions.

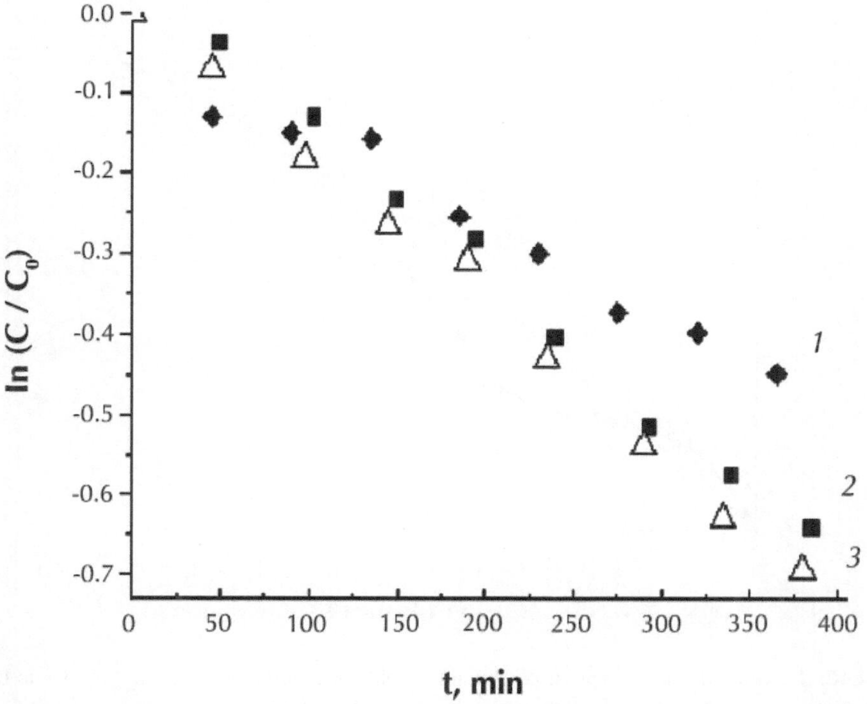

FIGURE 1 Semi-logarithmic linear anamorphoses of kinetic curves of H_2O_2 decomposition photocatalyzed by adsorbed Chl. The value of adsorption is 1.92 (*1*), 0.57 (*2*), and 0.23 μmol/g (3). Initial concentration of hydrogen peroxide is $[H_2O_2]_0 = 0.15$ mol/l.

In the case of equilibrium adsorption for system with $a = 0.47$ μmol/g, corresponding premonolayer substrate surface coating [13], the activity of Chl under light was twice higher than in the dark. Obviously, in this system packing of Chl molecules was different from the achieved for the samples with forced immobilization. As a result, coordination of Chl with the OH-groups of silica in this case has little effect on the photochemical process. It follows that in these systems use of the equilibrium adsorption is more effective.

Trend to photodecompose of H_2O_2, which is typical for natural Chl, was similar to one observed for synthetic analogue AlClPc. Table 1 show that AlClPc exhibits photocatalytic activity growing with decrease of phthalocyanine adsorption degree on silica (Figures 3 and 4). Adsorption isotherm for AlClPc is shown in Figure 2. Characteristic plateau corresponding to achievement of the tetrapyrrol monomolecular layer on the carrier surface is observed in the concentration range of 0.1–0.28 mmol/l with an average degree of adsorption a ≈ 0.95 μmol/ g.

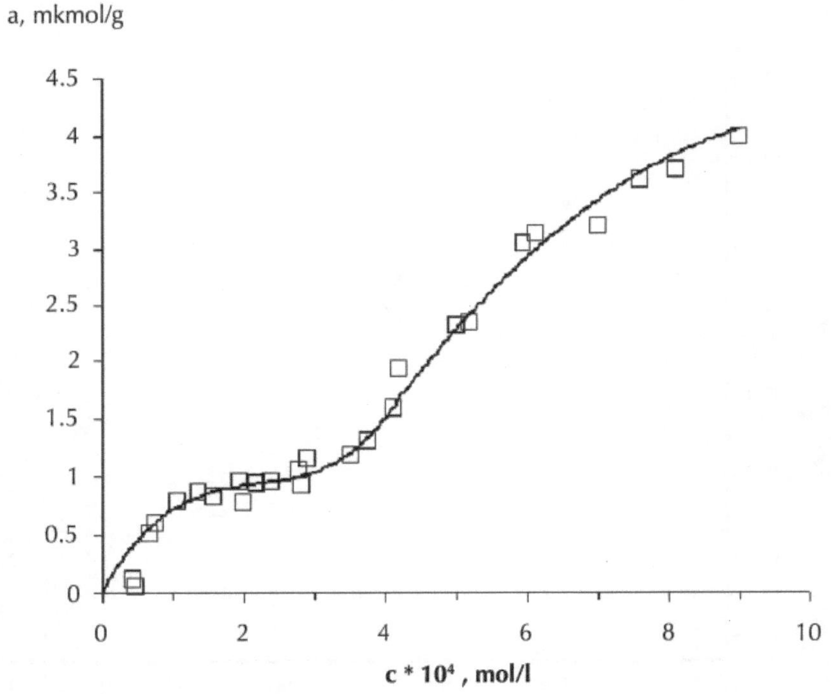

FIGURE 2 Isotherm of adsorption of alluminum phthalocyanine on silica, *a* - the value of adsorption, *c* - concentration of alluminum phthalocyanine in DMF.

The dark activity of AlClPc decreases to ~1.5 times in the case of monolayer adsorption and by ~70% in other cases (Table 1). In the absence of irradiation AlClPc activity in the hydrogen peroxide decomposition was the greatest for monolayer coating of the silica. In the case of AlClPc, as we have seen, the coordination of dye molecules with the silica does not influence on its activity significantly. Probably it is due to the nature of the metal complex. In alluminum phthalocyanine in a DMF solution metal central atom is covalently bonded with the ion Cl^-, which is replaced, apparently, by silica OH-groups during AlClPc adsorption process. Thus, the phthalocyanine molecular structure remains practically unchanged and an Al atom is not blocked. So the trend of Chl to decompose H_2O_2 under visible light was similar to the one that is shown for artificial dye AlClPc that is activity of the complex is decreasing with increase of pigments adsorption degree on silica for the investigated samples (Figure 5).

It has been found, that zinc phthalocyanine complex is active in the decomposition of hydrogen peroxide. Its activity increases under light in 1.5 times in comparison with the dark (Table 1). The activity of phthalocyanine zinc was higher than AlClPc, despite their very similar photophysical properties. It is possible that in the case of ZnPc coordination metal ion by molecules of H_2O_2 facilitated, as there is no extraligand in zinc coordination sphere.

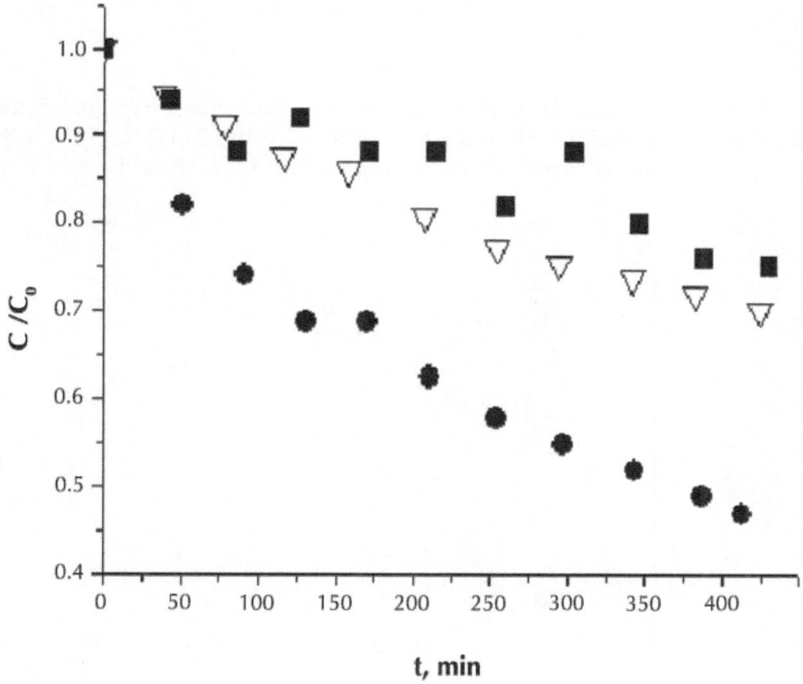

FIGURE 3 Kinetic curves of H_2O_2 decomposition photocatalyzed by adsorbed AlClPc. The value of adsorption is 2.34 (1), 0.95 (2), and 0.45 μmol/g (3). Initial concentration of hydrogen peroxide is $[H_2O_2]_0 = 0.15$ mol/l.

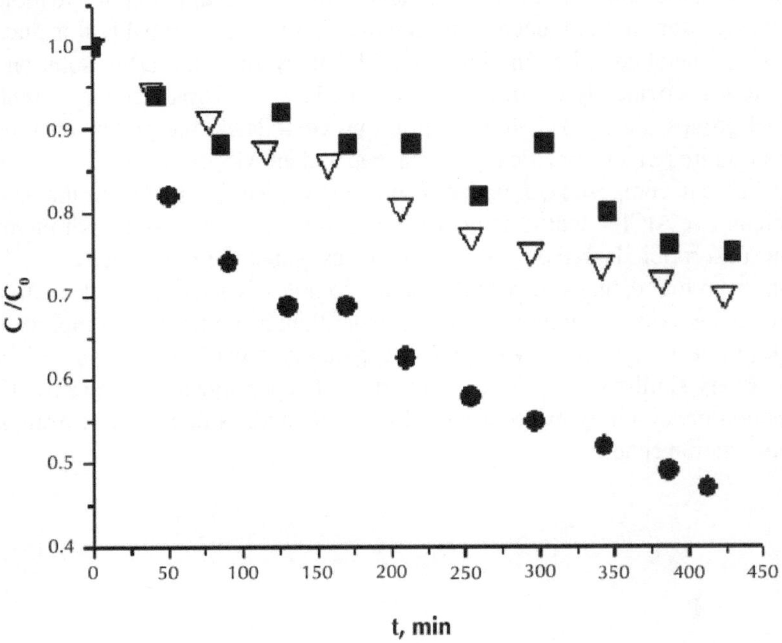

FIGURE 4 Semi-logarithmic linear anamorphoses of kinetic curves of H_2O_2 decomposition photocatalyzed by adsorbed AlClPc. The value of adsorption is 2.34 (1), 0.95 (2), and 0.45 μmol/g (3). Initial concentration of hydrogen peroxide is $[H_2O_2]_0 = 0.15$ mol/l.

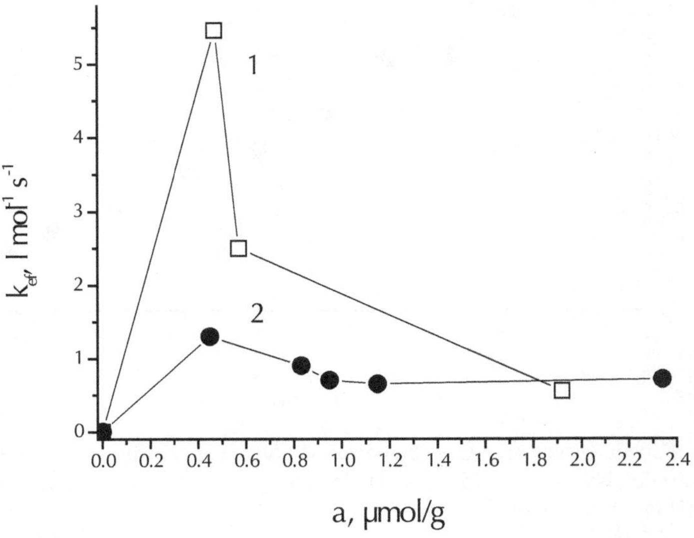

FIGURE 5 Dependence of effective rate constant of H_2O_2 decomposition photocatalyzed by Chl (1) and AlClPc (2) on their value of adsorption.

10.4 CONCLUSION

The study has shown that the use of metal complexes of tetrapyrrolic macrocycles adsorbed as premonolayer on silica at the equilibrium conditions are the most effective. Nature of the metal complex has a significant influence on the coordination of pigment to the substrate. Phthalocyanine complexes with Al, Zn, and Chl adsorbed on silica can be used in bioorganic chemistry, materials science, to create artificial models of photosynthesis and other chemical-technological processes as efficient catalysts for the decomposition of H_2O_2.

KEYWORDS

- **Adsorption layers**
- **Chlorophyll**
- **Hydrogen peroxide**
- **Metal phthalocyanines**
- **Tetrapyrrolic macrocycles**

ACKNOWLEDGMENT

The research was supported by RAS Presidium program P-24 Life origin and geobiological systems evolution and grant of support of leading scientific schools NSh-65059.2010.3.

REFERENCES

1. Hisanaga Tsuiki, Etsuko Masuda, Toshiki Koyama, Kenji Hanabusa, et.al. Functional metallomacrocycles and their polymers 34 Kinetics and mechanism of the biomimetic decomposition of hydrogen peroxide catalyzed by heterogeneous octacarboxyphtalocyaninato iron(III) supported on amorphous enriched rayon staple fibers. *Polymer*, **37**(16), 3637–3642 (1996).
2. Berezin, B. D. *Coordination compounds of porphyrins and phthalocyanin.* (in Russian), Nauka, Moscow p. 280 (1978).
3. Komissarov, G. G. *Fotosintez fiziko-khimicheskii podkhod Photosynthesis physicochemical approach.* (in Russian), Editorial URSS, Moscow (2003).
4. Askarov, K. A., Berezin, B. D., Bystritskaya, E. V. et al. *Porphyrins Spectroscopy, electrochemistry, application.* (in Russian), N. S. Enikolopyan (Ed.). Nauka, Moscow p. 384 (1987).
5. *Parmon, V. N. Molecular systems for water decomposition/in Photocatalytic transformation of solar enerlgy.* (In Russian), Part 2. Nauka, Novosibirsk p. 248 (1985).
6. Didier Astruc, Elodie Boisselier, and Catia Ornelas. Dendrimers Designed for Functions: From Physical, Photophysical, and Supramolecular Properties to Applications in Sensing, Catalysis, Molecular Electronics, Photonics, and Nanomedicine. *Chem. Rev.*, **110**(4), 1857–1959 (**2010**).
7. Komissarov, G. G. *Structural and functional modeling of natural photosynthetic systems. Diss.* (in Russian), ...dokt. khim. nauk. Semenov Institute of Chemical Physics of RAS, Moscow p. 379 (1973).
8. Iriyama, K., Ogura, N., and Takamiya, A. A simple method for extraction and partial purification of chlorophyll from plant material, using dioxane. *J. Biochem.*, **4**, 901–904 (1974).
9. Doson, R. and Elliot, D. *Spravochnik biokhimika.* (in Russian) Mir, Moscow, p. 544 (1991).
10. Nevrova, O. V., Lobanov, A. V., and Komissarov, G. G. Decomposition of hydrogen peroxide photocatalyzed by chlorophyll, Cu^{II}, and Cr^{III} porphyrins immobilized on silica. *Modern Problems in Biochemical Physics.* New Horizons. Sergei D. Varfolomeev, Elena B. Burlakova,

Anatoly A. Popov, and Gennady E. Zaikov. (Eds.). Nova Science Publishers, (in press) New York, (2011).

11. Brunauer, S., Emmett, P. H., and Teller, E. Adsorption of gases in multimolecular layers. *J. Am. Chem. Soc.*, 60(2), 309–319 (1938).
12. Lobanov, A. V., Nevrova, O. V., Vedeneeva, Yu. A., Golovina, G. V., and Komissarov, G. G. Photodestruction of Chlorophyll in Non-biological Systems In *Molecular and Nanoscale Systems for Energy Conversion*. S. Varfolomeev, L. Krylova et al. (Eds.). Inc. N., pp. 95–99 (2008).
13. Lobanov, A. V. *Photocatalytic processes involving hydrogen peroxide in natural and artificial photosynthetic systems*. (in Russian), Diss. Ph. D. Semenov Institute of Chemical Physics of RAS, Moscow p. 117 (2004).

11 Hemolysis of Erythrocytes under Exposure to Sulfur Nitrosyl Iron Complexes – Nitric Oxide Donors

E. M. Sokolova, T. N. Rudneva, N. I. Neshev, B. L. Psikha, N. A. Sanina, and S. V. Blokhina

CONTENTS

11.1 INTORDUCTION

Kinetics of hemolysis and time course of the intra-erythrocyte hemoglobin oxidation under action of synthetic sulfur-nitrosyl iron complexes are capable of releasing of nitric oxide (NO) in the course of spontaneous hydrolytic decomposition have been studied. The addition of given complexes to 0.2% mouse erythrocyte suspension resulted in hemolysis of erythrocytes which is went with characteristic for each complex induction time. It was shown that hemolysis have been preceded by hemoglobin oxidation under action of NO penetrating into cell. Oxidation of hemoglobin followed to the first order equation. The effective first order rate constants characterizing the NO-donating ability of each complex were determined. It was suggested that hemolytic effect of studied complexes is connected with formation of peroxynitrite: the cytotoxic product of interaction NO and superoxide anion radical.

Aim of this work to investigate kinetics and mechanism of erythrocyte hemolysis under action of NO-donors capable of spontaneous releasing of NO. It is known that the excessive production of NO in *vivo* result in cytotoxic effects which are considered as main pathogenetic factors for developing of inflammative and neurodegenerative diseases. In this connection the matter of special interest is the chemical and biochemical mechanisms of realization of NO cytotoxic potential in different types of cell and tissues.

The NO as a signal molecule executes significant biological function connected with the modulation of blood pressure level, inhibition of clot formation, neurotransmission, and nonspecific immune defense [1]. At the same time in many cases the excessive production of NO is observed. It cause cytotoxicity of NO directed against organism's own and come forward as main pathogenetic factor a lot of inflammative and neurodegenerative disease [2]. It is known that a cytotoxic effecter is not NO but the products of its complicated chemical transformation [3]. In this connection the investigation of chemical and biochemical mechanisms of realization of the NO cytotoxic potential in different types of cell and tissues acquires a special significance.

The purpose of this work is the investigation of kinetics and mechanism of erythrocyte hemolysis under action sulfur nitrosyl iron complexes NO-donors.

11.2 EXPERIMENTAL PART

11.2.1 Materials

Binuclear sulfur nitrosyl iron complexes (SNIC) bearing different ligands were used as NO-donors.[4]. We applied five such complexes: thiosulfate-contained tetranitrosyl iron complex (TNIC); pyrimidin-contained tetranitrosyl iron complex (Pym), cysteinamine-contained tetranitrosyl iron complex (Cys), methylimidazole tetranitrosyl iron complex (Mim), and benzothiazol tetranotrosyl iron complex (BTz). Structural base of SNICs is the typical tetragonal bridge structure including two iron atoms which carry two NO-group and two sulfur atoms which carry certain chemical ligands. Figure 1

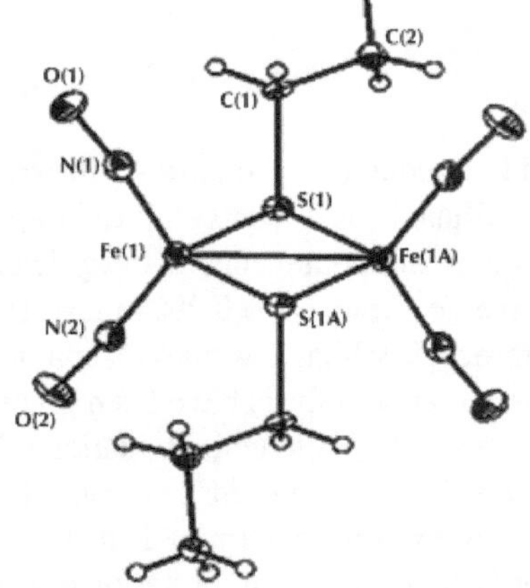

FIGURE 1 Crystal structure of Cys.

presents crystallic structure of cation of Cys obtained by X-ray diffraction [5]. Complexes were introduced into erythrocyte suspension from water solution (TNIC, Cys) or from dimethyl sulfoxide (DMSO) solution (BTz, Pym, and Mim). The solutions were prepared directly before experiment. The weight DMSO content in samples do not exceeded 3%.

11.2.2 Red Cell Preparation

Mouse blood was drawn into test tube containing 3. 8% sodium citrate solution. Blood was centrifugated for seven minute on 1,500 g. Plasma and white cells were removed and red cells were washed three times with phosphate buffered saline (0. 85% NaCl, 5 mM NaH_2PO_4/Na_2HPO_4, pH 7. 4). Red cell mass obtained after last centrifugation was kept at 4°C not more than 36 hr.

11.2.3 Erythrocyte Hemolysis

In experiments, 0.2% erythrocyte suspension was used. It was prepared through dilution of initial red cell mass by phosphate buffered saline on 450 times. Cell content in the suspension amounted 4.4×10^7 cell/ml. Hemolytic experiments were carried out at 37°C in plastic tubes at continuous slight stirring by magnetic stirrer. The course of hemolysis was registered spectrophotometrically on reducing the optical density (OD) of the erythrocyte suspension at 700 nm. At used level of the suspension dilution (0.2% of hematocrite) the OD of suspension is lineally depended on the concentration of undestroyed cell [6, 7]. The extent of hemolysis (γ) was determined by:

$$\gamma = \frac{A_0 - A}{A_0 - A_{H_2O}} \tag{1}$$

where A_0 and A are ODs of control and experimental samples, respectively; A_{H_2O} is OD of the completely hemolyzed sample.

11.2.4 Methemoglobin Determination

The 0.4 ml aliquots were removed from erythrocyte suspension and diluted 1.4 ml distilled water. After 1 min. incubation the OD at 630 nm was measured. The methemoglobin content was determined by:

$$[HbFe^{3+}] = \frac{\Delta A_{630}}{\varepsilon_{met} - \varepsilon_{oxy}} \times d \tag{2}$$

where ΔA_{630} = increasing of OD at 630 nm; ε_{met} = 3.8 мM^{-1} sm^{-1} и ε_{oxy} = 0.11 Mm^{-1}sm^{-1} molar extinction coefficients of methemoglobin and oxyhemoglobin at 630 nm, respectively [8]; d = factor of dilution.

11.3 DISCUSSION AND RESULTS

The influence of SNICs on the erythrocyte hemolysis was studied in the wide concentration range. All of the studied complexes displayed considerable concentration dependent on hemolytic effect. Figure 2 presents time course of erythrocyte hemolysis (γ = extent of hemolysis), which were obtained at equimolar concentration (4×10^{-5} M) of studied complexes. As individual kinetic characteristic of hemolytic activity of each complex the duration of induction period of hemolysis was used. The duration of induction period was defined as a time of reaching of the extent of hemolysis of 0.1 (dash line). The induction time corresponding each of studied complexes are showed on inset of Figure 2.

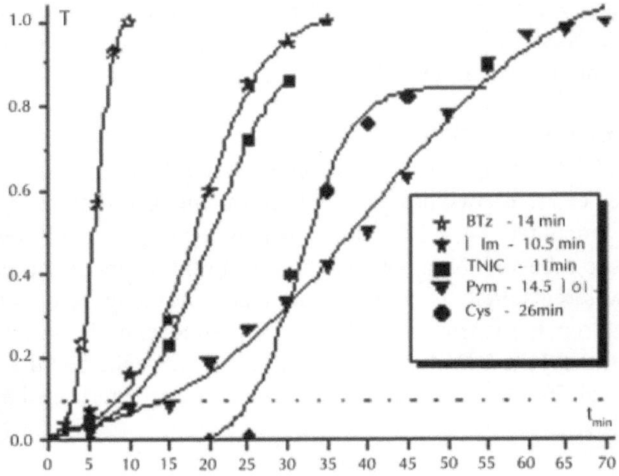

FIGURE 2 Hemolysis of erythrocytes under of SNICs.

Having carried out the hemolytic experiments we often visually observed the color change preceding to beginning of hemolysis. It might indicate to the chemical processes occurred inside of cell. Spectrophotometric investigation of hemolysates revealed the typical pattern of the hemoglobin spectral change suggesting on oxidation of oxyhemoglobin to methemoglobin (inset, Figure 3).

Obviously, it arises from interaction of NO penetrating into cell with oxyhemoglobin. To investigate the time course of methemoglobin formation we took out the aliquots of erythrocyte suspension then quickly destroyed the cells by distilled water and determined the methemoglobin content by the absorbance measurements at 630 nm. Figure 3 shows time course of methemoglobin formation inside of erythrocytes under exposure to Cys and Pym. It is clear that the methemoglobin formation in presence of each SNIC proceed with different rate. In addition it was established that kinetics of methemoglobin formation followed to the first order equation:

$$[HbFe^{3+}] = [HbFe^{3+}]_{\infty} \cdot (1 - e^{-kt}) \tag{3}$$

Therefore, the first order rate constants were determined (Figure 3, inset). It is clear that obtained rate constants differ almost in order of quantity.

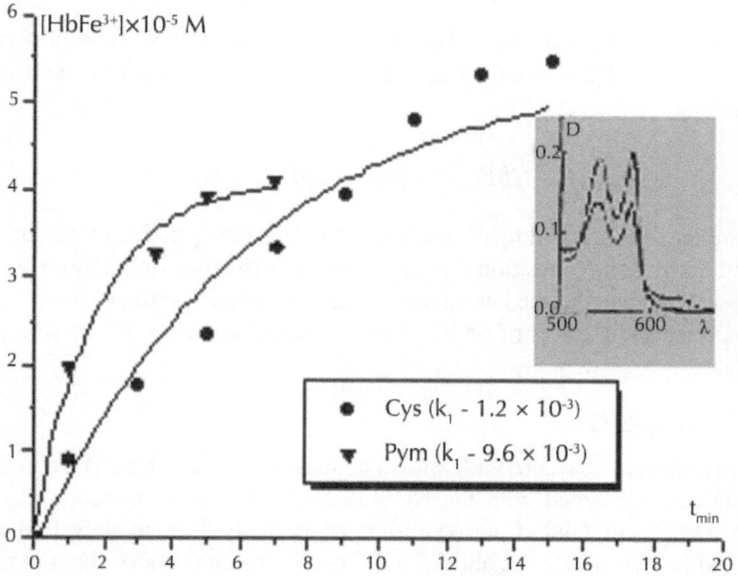

FIGURE 3 Formation of methemoglobin inside of erythrocytes under action of SNICs.

The releasing of NO from SNIC may be represents by following scheme:

$$[Fe_2(SR)_2(NO)_4] \xrightarrow{k_1} 4NO + [decomposition\ products] \qquad (4)$$

NO penetrating into erythrocytes reacts with oxyhemoglobin [9]:

$$NO + HbFe^{2+} \cdots O_2 \xrightarrow{k_2} HbFe^{3+} + NO_3^- \qquad (5)$$

Obtained kinetic dependence (Figure 3) and determined the effective first order rate constants characterize the rate of reaction (5) in presence of studied SNICs. Same as the rate constants characterize NO-donating ability of SNICs. In accordance with this NO-donating ability of Cys must be about 10 times lower than one of Pym. Comparing the hemolytic activities given complexes (induction periods, Figure 2) we see that they dependent from their NO-donating abilities. Namely, the slighter donor, Cys have larger induction period of hemolysis in comparison with Pym. Thus, obtained result allows interpreting presence certain interrelation between the level of NO formation and subsequent hemolysis of erythrocytes.

Earlier results evidence that the main origin of cytotoxicity is not NO itself, but the product of bimolecular interaction NO and superoxide anion radical is peroxynitrite [10]:

$$^{\bullet}NO + {^{\bullet}O_2^-} \rightarrow ONOO^- \quad (6)$$

Distinctive feature of erythrocyte is maintaining the relatively constant level of superoxide because of it is continuously formation in the reaction of oxyhemoglobin autoxidation:

$$HbFe^{2+} \cdots O_2 \rightarrow HbFe^{3+} + O_2^- \quad (7)$$

Thus, simultaneous proceeding of reaction (4) and (7) may provide formation of initial products for reaction (6) resulting in peroxynitrite formation. Taking into account the earlier results evidencing the hemolytic action of synthetic peroxynitrite [11] we supposed that hemolytic effect of SNICs may be caused by peroxynitrite formed inside erythrocytes.

11.4 CONCLUSION

Obtained results are in accord with known opinion on peroxynitrite as the main source of cytotoxicity concerned with the NO metabolism in living systems. The realization of cytotoxic potential of peroxynitrite occurs likely through activation oxidative process inside erythrocyte. Establishing of specific mechanism of that process is the matter of our further investigations.

KEYWORDS

- **Erythrocytes**
- **Hemoglobin**
- **Hemolysis**
- **Nitric oxide and peroxynitrite**
- **Sulfur-nitrosyl iron complexes**

REFERENCES

1. Bryan, N. S., Bian, K., and Murad, F. Discovery of the nitric oxide signaling pathway and targets for drug development. *Frontiers in Bioscience*, **14**, 1–18 (2009).
2. Pacher, P., Beckman, J. S., and Liaudet, L. Nitric Oxide and Peroxynitrite in Health and Disease. *Physiol. Rev.*, **87**, 315–424 (2007).
3. Wink, D. A., Miranda, K. M., and Espey, M. G. Cytotoxicity related to oxidative and nitrosative stress by nitric oxide. *Exp Biol Med.*, Maywood **226**, 621–623 (2001).
4. Sanina, N. A. and Aldoshin, S. M. Functional models of [Fe≈S] nitrosyl proteins. *Russ. Chem. Bull. Int. Ed.*, **53**(11), 2428–2448 (2004).
5. Rudneva, T. N., Sanina, N. A., Lysenko, K. A., Aldoshin, S. M., Antipin, M. Y., and Ovanesyan, N. S. Synthesis and structure of a water-soluble nitrosyl iron complex with cysteamine ligand. *Mend. Comm.*, **19**, 253–255 (2009).
6. Young, J. D., Leong, L. G., Di Nome, M. A., and Gohn, Z. A. A semiautomated hemolysis microassay for membrane lytic proteins. *Anal. Biochem.*, **154**, 649–654 (1986).
7. Ilani, A. and Granoth, R. The pH dependence of the hemolytic potency of bile salts. *Biomembranes*, **1027**, 199 (1990).

8. Zijlstra, W. G., Buursma, A., and Meeuwsen-van der Roest, W. P. Absorption Spectra of Human Fetal and Adult Oxyhemoglobin, De-Oxyhemoglobin, Carboxyhemoglobin, and Methemoglobin. *Clin. Chem*., **37**, 1633–1638 (1991).
9. Wallis, J. P. Nitric oxide and blood a review. *Transfusion Medicine*, **15**, 1–11 (2005).
10. Beckman, J. S., Beckman, T. W., Chen, J., Marshall, P. A., and Freeman, B. A. Apparent hydroxyl radical production by peroxynitrite Implications for endothelial injury from nitric oxide and superoxide. *PNAS*, **87**, 1620–1624. (1990).
11. Kondo, H., Takahashi, M., and Niki, E. Peroxynitrite-induced hemolysis of human erytrocytes and its inhibition by antioxidants. *FEBS Letters*, **413**, 236–238 (1997).
12. Ferrer-Sueta, G. and Radi, R. Chemical Biology of Peroxynitrite Kinetics, Diffusion, and Radicals. *ACS Chem. Biol*., **4**, 161–177 (2009).

12 Biodecomposed Polymeric Compositions on the Basis of Agriculture's Waste

I. A. Kish, D. A. Pomogova, and D. A. Sogrina

CONTENTS

12.1 INTORDUCTION

This chapter is dedicated to research the questions of the filled biodecomposed polymeric compositions creation on the basis of agriculture's waste (agroindustrial complex) and packings by way of getting raw materials and products made from it. The problem of physico-mechanical properties of the filled biodecomposed polymeric compositions is traversed. Processes of the polymeric compositions biodegradation filled with waste of agroindustrial complex have been studied.

There is a problem of polymeric waste's recycling. One of the most promising directions in the field of packing waste's recycling is creation of biodecomposed polymeric materials. Nowadays synthesis of biopolymers is expensive, and these materials have the limited usage. Therefore, the greatest interest is gathered by the filled biodecomposed compositions. Such compositions are referred to partially decomposed or punched materials. Getting to the environment, these materials are exposed to impacts of external factors and bacteria, the filler completely assimilates and polymer is destroyed. That leads to reduction of dumps by means of decomposition polymer's time reduction.

Much waste of agriculture is being accumulated now. Approximately 70% of waste is processed into forages and fertilizers, and 30% are utilized by way of dumping and thermal methods which have negative effect on our environment. That is why it is expedient to use agroindustrial complex waste as a filler for creation of biodecomposed polymeric materials, and as a polymeric matrix we should use waste of packing branch.

12.2 EXPERIMENTAL PART

Waste of polyethylene film and agricultural waste have been chosen for getting biodegradated compositions: cacao bean husk, beet bin pulp, buckwheat, rice, millet, sunflower shuck, potato, and corn mar. The maximum size of the filler's particles made less than 150 microns.

There were complexities while processing on standard laboratorial extrusion type equipment for all compositions, for example there was bad distribution of filler: its agglomeration led to extruding press's productivity decrease and to formation of defects in material.

The basic criterion is durability for creation of secondary qualitative raw polymeric materials and products made from it. It is characterized by breakdown voltage's magnitude at monoaxial stretching (σp). By data researches that were carried out this magnitude should be not less than 4 MPa.

As a result of researches in physico-mechanical characteristics compositions that have fillers like cacao bean husk, beet bin pulp, and rice schuck have been selected.

In the process of compositions research it was noticed that agroindustrial complex waste which is "incompatible" with polymeric waste, reduce durability that is connected with specialty of fillers distribution from agroindustrial complex waste in the supramolecular structure of polymer which leads to formation of non-uniform material's structure.

For clearing this lack while processing mixed compositions it is expedient to use special additives which can lead not only to improvement of compositions workability, but also to updating of secondary raw materials properties.

In this case additive's selection has been determined not only by its impact on physico-mechanical properties of polymeric compositions, but also by its ability to biodegradation. It is known that colloidal clay with an advanced surface is natural filler, a good adsorbent, and a dispergator from treatise [1]. Colloidal clay belongs to alumosilicic hydrates of layered structure [2].

12.3 DISCUSSION AND RESULTS

Investigations were conducted on compositions on basis difference concentration colloidal clay. As a result of the researches that were carried out optimum concentration colloidal clay has been established. It has made 2%. It can be explained that colloidal clay is created by uniform distribution of a filler in concentration of 2% in the conditions of processing highly filled (30–40% of agroindustrial complex waste) polymeric compositions, that the conducted researches have proved by a method of optical microscopy (Figure 1(a) and (b)). In the Figure 1(a) we can see more uniform distribution of filler in a polyethylene composition with colloidal clay in comparison with the structure of Figure 1(b). In case of concentration's increase in polymeric compositions the effect of forming of colloidal clay's own structures in polymer filled compositions is observed. Herewith, there are complexities of filler's distribution in polymer.

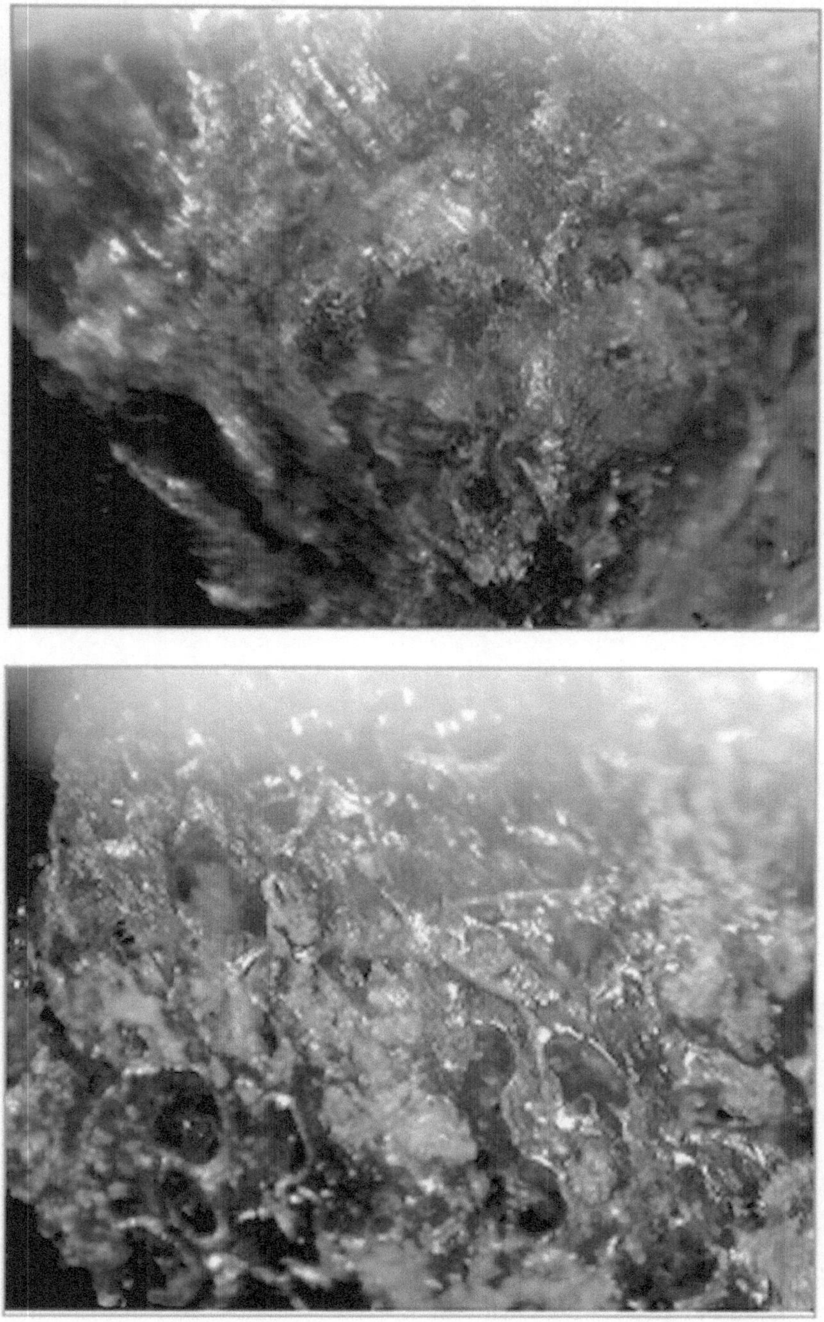

FIGURE 1 Shows optical micro photos of the filled polymeric compositions samples filled with agroindustrial complex waste. There is zooming in 250 times.

There is a composition on the basis of polyethylene waste containing 30% of agro-industrial complex waste (rice shuck) in the Figure 1(a).

There is a composition on the basis of polyethylene waste containing 30% of agro-industrial complex waste (rice shuck) and 2% of colloidal clay in the Figure 1 (b).

Consequently, infusion of colloidal clay has led not only to uniform distribution of agroindustrial complex waste, but also has allowed to increase breakdown voltage and percent elongation with rupture in 2–3 times (Table 1).

TABLE 1 Physico-mechanical properties of polymer compositions on the basis of agroindustrial complex waste and polyethylene.

Filler's name*	Colloidal clay in polymer compositions	Breakdown voltage σ_p, MPa	Percent elongation with rupture, ε_p, %
Beet bin pulp	–	1,30±0,08	9,80±0,10
	+	4,75±0,08	11,25±0,09
Rice shuck	–	2,00±0,07	8,70±0,09
	+	6,00±0,15	16,50±0,11
Cacao bean husk	–	1,60±0,08	11,00±0,12
	+	6,00±0,09	16,00±0,09

* - The amount of filler in polymer compositions is 30%.

To estimate dynamics of the filled polymeric compositions biodegradation composting was used. Samples were placed in special mallets with biohumus at temperature 23 ± 2°C and humidity 70 ± 10%. Degree of polymeric compositions biodegradation was estimated by change of physico-mechanical properties such as breakdown voltage and percent elongation with rupture. Calculation of biodegradation degree of composition was made according to the following formula:

$$\Delta = \frac{a_1 - a_0}{a_0} . 100 \text{, } (\%)$$

where, a_1 is a parameter's meaning before composting:
a_0 is parameter's meaning after composting.

In the Table 2, results of polymer compositions biodegradation research on the basis of polyethylene's containing waste, agroindustrial complex waste, and colloidal clay as filler for twelve months are presented.

TABLE 2 Changes in physico-mechanical properties after composting.

Filler's name*	Changes in physico-mechanical properties	
	Change of breakdown voltage, $\Delta\sigma_p$, %	Change of percent elongation with rupture, $\Delta\varepsilon_p$, %
Beet bin pulp	60 ± 5%	68 ± 0,1%
Rice shuck	65 ± 5%	72 ± 0,1%
Cacao bean husk	65 ± 5%	73 ± 0,1%

* - The amount of filler in polymer compositions is 30%.

It is obvious after composting breakdown voltage for all compositions has decreased on the average in 2 or 5 times and percent elongation with rupture has lowered in three times. That testifies about processing of polymer compositions biodegradation with agroindustrial complex waste. The visual estimation has allowed establishing filler's destruction on all surfaces of samples which had friable structure after composting. They were fragile and some samples can be broken into small fractions in case of withdrawal from biohumus.

Thereby, received polymeric compositions on the basis of agroindustrial complex waste are partially biodecomposed polymeric compositions. Agroindustrial complex waste, namely cacao bean husk, rice shuck, and beet bin pulp assimilates in the environment and the polymeric matrix is destroyed. Polymer's decomposition is reduced by that.

Besides, imposition of colloidal clay in polymeric compositions has multiplied physico-mechanical properties of material. Using mathematical modeling underneath breakdown voltage and term of biodecomposition on amount of filler, it has been established that filler's concentration (agroindustrial complex waste) in polymeric composition can be increased up to 40%.

It was necessary to eliminate presence of an unpleasant smell for making products from biodecomposed polymeric compositions on the basis of agroindustrial complex and polyethylene waste.

Charcoal was used as a sorbent. Samples of following structure have been received on extrusion type equipment: 58% of polyethylene, 40% of filler (cacao bean husk, rice shuck, or beet bin pulp), and 2% of sorbent and 59% of polyethylene, 40% of filler, and 1% of sorbent. As introduction of 1% charcoal reduced a smell, but did not eliminate it completely, compositions with containing 2% of coal which did not have any smell at all.

Then tests of received compositions for durability were conducted at monoaxial stretching (Table 3).

TABLE 3 Changes in physico-mechanical properties after composting.

Filler's name*	Breakdown voltage σ_p, MPa	Percent elongation with rupture, ε_p, %
Beet bin pulp	6,25±0,41	4,50±0,35
Rice shuck	5,00±0,36	6,80±0,78
Cacao bean husk	6,25±0,43	6,80±0,59

* - The amount of filler in polymer compositions is 40 %.

It has been established that breakdown voltage (σp) for all samples made more than 5 MPa that meets the requirements while manufacturing products of technical purpose, for example, trays of small extract for storage of hardware or for thin slab.

Further researches of received polymeric compositions biodegradation for six months were conducted. Biodegradation of compositions was held by a method of composting (Table 4).

TABLE 4 Changes in physico-mechanical properties after composting.

Filler's name*	Changes in physico-mechanical properties	
	Change of breakdown voltage, $\Delta\sigma_p$, %	Change of percent elongation with rupture, $\Delta\varepsilon_p$, %
Beet bin pulp	66 ± 5%	78 ± 0,1 %
Rice shuck	72 ± 5 %	74 ± 0,1 %
Cacao bean husk	67 ± 5 %	83 ± 0,1 %

* - The amount of filler in polymer compositions is 40%.

In all cases the breakdown voltage magnitude has decreased on the average in three times, the magnitude of percent elongation has decreased on the average in 4–5 times.

Also the method of optical microscopy was conducted the research of polymer's structure. Microphotos of samples before and after composting (Figure 2 (a) and (b)) were received.

Structure of biodecomposed polymeric composition contains 56% of polyethylene, 40% of filler (rice shuck), 2% of colloidal clay, and 2% of charcoal.

From the data we can see that initial structure of polymeric compositions samples containing agroindustrial complex waste, is more homogeneous, and emptiness formed after composting (dark areas in photos) tell about biodecomposition of filler in the conditions of constant temperature and humidity.

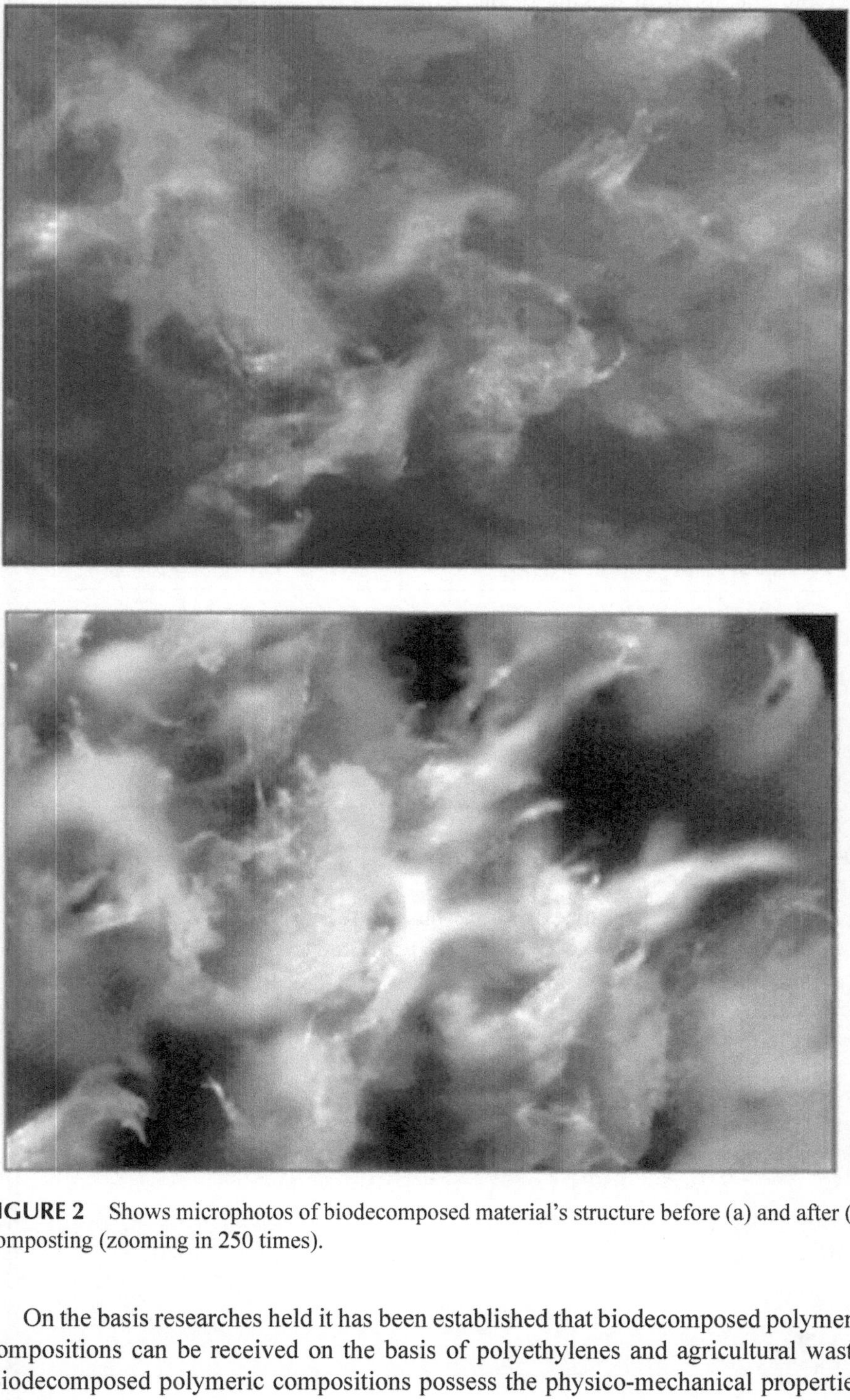

FIGURE 2 Shows microphotos of biodecomposed material's structure before (a) and after (b) composting (zooming in 250 times).

On the basis researches held it has been established that biodecomposed polymeric compositions can be received on the basis of polyethylenes and agricultural waste. Biodecomposed polymeric compositions possess the physico-mechanical properties, satisfying conditions of making secondary qualitative raw polymeric materials, and products from it.

12.4 CONCLUSION

Biodecomposed polymeric compositions on the basis of polyethylene and agricultural waste were received. Waste of sugar manufacture: beet bin pulp, confectionery manufacture: cacao bean husk, grain manufacture: rice, buckwheat, millet, sunflower shuck, starch manufactures: corn, and potato mare were used as fillers.

Dependence of structure of agroindustrial complex waste on technological properties of the filled polymeric compositions has been studied. On the basis of mathematical modeling made and researches of physico-mechanical properties the substantiation choice of agroindustrial complex waste has been made: beet bin pulp, cacao bean husk, and rice shuck were chosen as the most effective fillers for getting biodecomposed polymeric compositions.

On the basis of sorbent and colloidal clay the complex modifier has been developed for getting biodecomposed polymeric materials on the basis of polyethylenes waste filled 30–40% of agroindustrial complex waste.

KEYWORDS

- **Agriculture waste**
- **Biodecomposed polymeric materials**
- **Physico-mechanical properties**
- **Polymer compositions**
- **Charcoal**

REFERENCES

1. Anan'ev V. and Kirsh, I. *Recycle polymer materials* in Rus.. Moscow State University of Food Industry Publishing House, Moscow p. 250 (2006).
2. Andrianova, G. *Technology of polymer composition* in Rus. Kolos Publishing House, Moscow p. 270 (2008).

13 Properties Research of Secondary Polymeric Materials on the Basis of Polypropylene and Polyethylene Terephthalate Got Under the Influence of Ultrasonic Oscillations on Polymeric Melt

I. A. Kish, D. A. Pomogova, and D. A. Sogrina

CONTENTS

13.1 INTRODUCTION

The influence of ultrasonic oscillations on secondary raw materials property on the basis of polypropylene (PP) and polyethylene terephthalate (PET) during simultaneous processing is studied in this chapter.

A lot of attention is paid to recycling problem or to secondary processing of polymer materials nowadays. On the most expensive recycling's procedures is sorting of waste. Identification of polymer's waste is really difficult in some cases. If we are speaking about multi layers materials. Their identification is impossible. Waste is formed on the stage of materials processing and after packing's usage. It is utilized by storing in the dumps and polygons or it can be burned. This has a negative effect on the environment. That is why there are a lot of researches which study simultaneous processing of thermodynamically incompatible polymers and getting of new composed polymeric materials.

There is a row of difficulties during processing of such polymers as PP and PET and other polymeric compositions. They are connected with their thermodynamic incompatibility. It is known that ultrasonic oscillations can lead to reduction of molecular mass that can bring together polymers solubility parameters during their processing. However, such researches of polymeric melts had been held small [1]. That is why the purpose of our investigation is to study influence of ultrasonic oscillations on polymers properties for designing of joint polymeric waste's processing of different chemical nature.

13.2 EXPERIMENTAL PART

The PP "Kaplen" and PET TU6-05-1984-85 were chosen as objects of research. They were exposed to multiple processing with ultrasonic *vibro* attachment and without it. The compositions of PP and PET were made after processing's cycle of each polymer. The received samples were analyzed on physico-mechanical and rheological properties. It has been investigated molecular structure of the samples.

13.3 DISCUSSION AND RESULTS

The investigations were conducted on a joint recycling PP and PET. Results of tests on physical and mechanical properties are shown in Table 1.

TABLE 1 Physical and mechanical properties of PP and PET.

The number of cycles of treatment	The damaging stress, M Pa		The elongation at break,%	
	Without ultrasonic vibrations	With ultrasonic vibrations	Without ultrasonic vibrations	With ultrasonic vibrations
Polypropylene (PP)				
1	37 ± 3	33 ± 2	13 ± 1	11,44 ± 0,7
2	36 ± 10	47 ± 10	14 ± 2	11,11 ± 1,2
3	33 ± 5	63 ± 7	12 ± 3	11,52 ± 0,5
4	36 ± 7	60 ± 5	12 ± 3	10,85 ± 1,2
Polyethylene terephthalate (PET)				
1	49 ± 3	48 ± 3	302 ± 5	370 ± 2
2	73 ± 7	84 ± 2	263 ± 4	230 ± 4
3	54 ± 4	66 ± 4	28 ± 5	237 ± 2
4	28 ± 3	53 ± 2	19 ± 2	145 ± 3

The breakdown voltage of PP processed with ultrasound is gradually increasing in two times from the first to the second cycles and it does no change on the fourth cycle. Breakdown voltage of PP processed without ultrasonic *vibro* attachment does not change practically. Percent elongation with rupture of PP received with ultrasound and without it does not change practically form the first to the fourth processing's cycles.

We can see from the results that breakdown voltage of PET received with ultrasonic *vibro* attachment and without it increases in the second cycle. After first decreases on the third and on the fourth processing's cycles. We should note that the meanings of PET's physico-mechanical properties processed with ultrasonic influence are higher than for PET processed without it. Percent elongation with rupture of PET processed with ultrasonic *vibro* attachment is gradually decreasing from the second processing's cycle. Percent elongation with rupture of PET received without ultrasonic oscillations does not change practically on the first and on the second cycles, then it decreases abruptly.

The rating of rheological properties was melt's flaw index. In the Figure 1 and 2 there are a dependence of received samples melt's flaw index on the amount of processing's cycles.

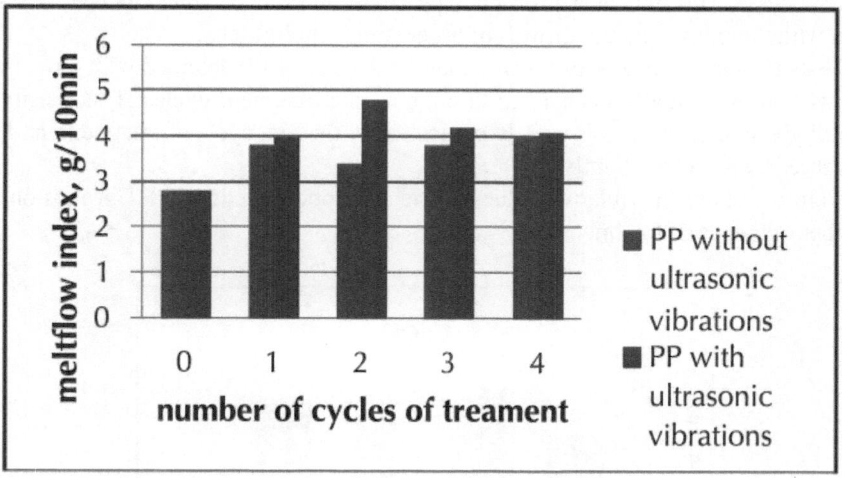

FIGURE 1 Dependence of PPs melt's flaw index on the amount of processing cycles.

Conducted researches showed that melt's flaw index of PP (Figure 1) received with ultrasonic oscillations does not change practically from cycle to processing's cycle. Whereas melt's flaw index at samples received without ultrasonic oscillations increases in two times on the second processing's cycle.

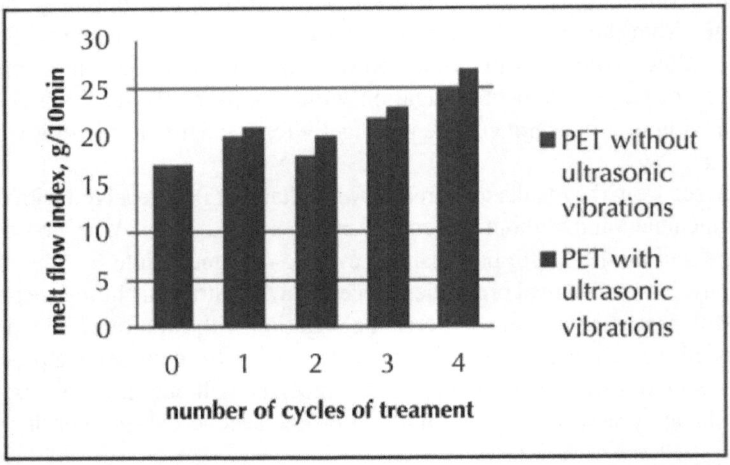

FIGURE 2 Dependence of polyethylene terephthlate's melt's flow index on the amount of processing cycles.

Conducted researches of melt's flow index of PET processed with ultrasonic *vibro* attachment and without it (Figure 2) showed that melt's flow index increases from cycle to cycle. This indicates destructive processes in PET, polymer changed its color from white to putty with the growth of processing's multiplicity.

Tests to determine density showed that the density of PP obtained with ultrasound and without it, virtually unchanged throughout the treatment cycles. The density of polyethylene in both cases is greatly reduced after the first cycle of treatment, and the next three cycles vary slightly.

Figure 3 shows the relative values of the absorption bands of PET groups on the number of cycles of treatment.

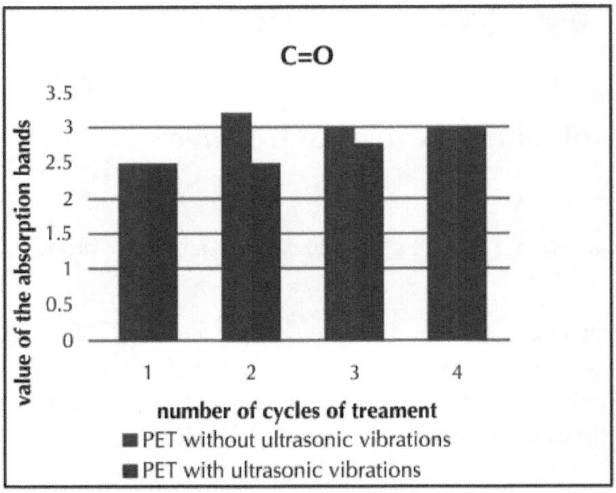

FIGURE 3 *(Continued)*

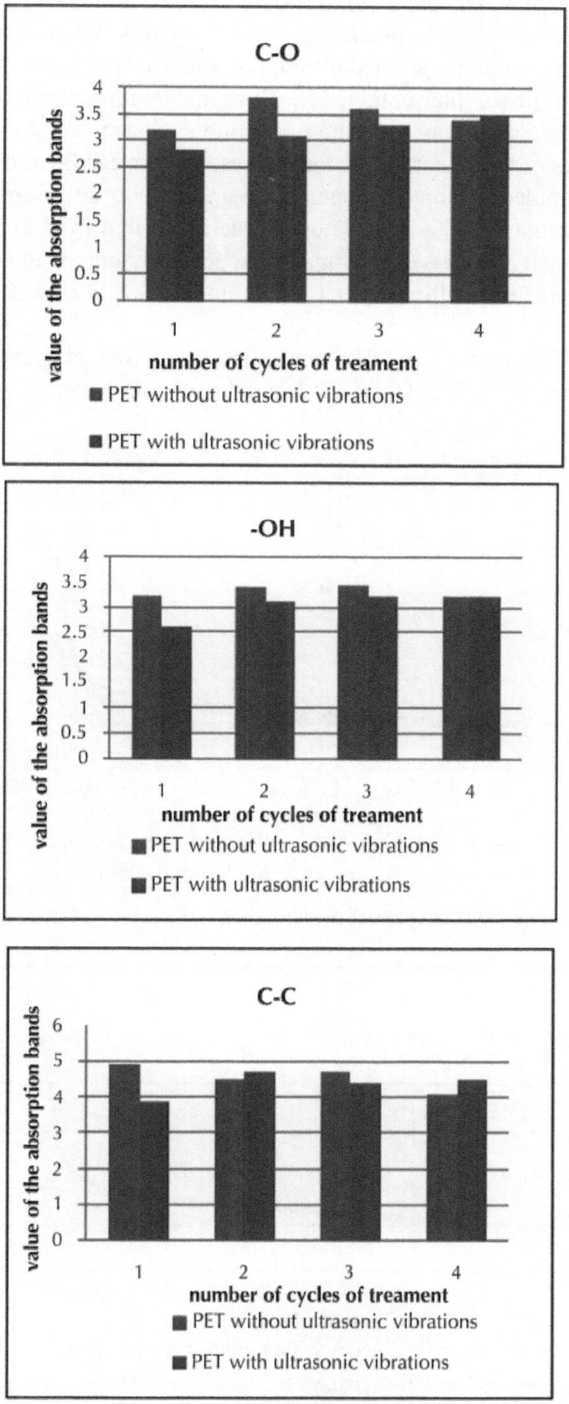

FIGURE 3 Depending on the relative values of the absorption bands of PET groups on the number of cycles of treatment.

After analyzing the data obtained, it should be noted that from cycle to cycle processing the content of oxygen containing groups increases. When comparing the values of the oxygen containing groups of samples obtained with and without ultrasound may be noted that the samples obtained by ultrasonic treatment contain fewer oxygen containing groups. Comparing the values of group C-C samples of the first cycle and subsequent cycles of treatment, reduction of these groups can be traced in the case of processing of samples without ultrasound and an increase in group C-C in samples treated with ultrasound in the melt. Thus, it was found that from cycle to cycle processing is an intensify process of degradation in polymer, and the ultrasonic treatment leads to partial recovery of the polymer macromolecules and slow degradation in recycling.

The changes in the structure of PP during multiple processing studied by thermomechanical curves (TC) .The TC of PP 1 and 4 cycles of treatment are presented in Figures 4 and 5.

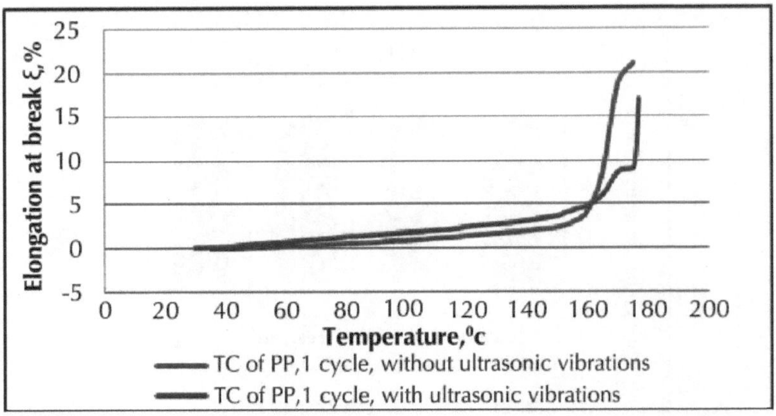

FIGURE 4 The TC of the first cycle of recycling PP.

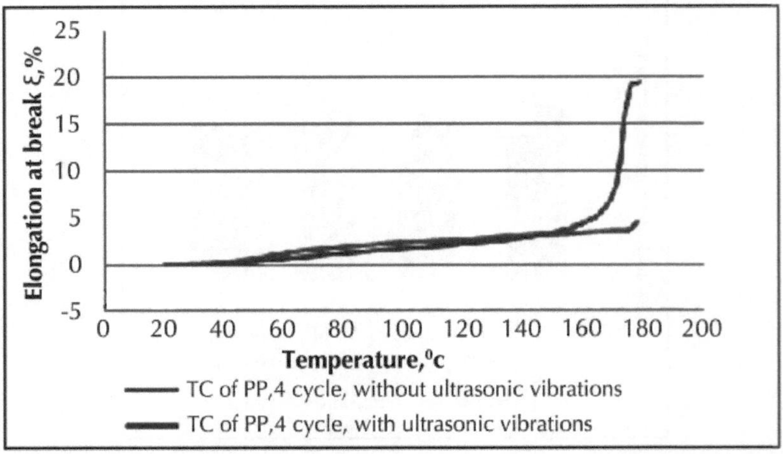

FIGURE 5 The TC of the fourth cycle of recycling PP.

From the analysis of these curves shows that with increasing multiplicity of processing an ultrasound helps reduce the flow temperature of PP. Furthermore, it should be noted that the ultrasonic treatment leads to recovery of macromolecules of the polymer, as without ultrasound PP at 180°C (fourth cycle) is irreversibly destroyed due to an increase in the number of defects in the sample when it is repeated processing, and PP treated with ultrasound is not destroyed by this temperature is due to the restoration of macromolecules under the influence of ultrasonic vibrations.

Since a mixture of PP and PET were obtained at the treatment temperature of last, then further studies were carried out flow curves of PP treated with or without ultrasound at 265°C. Figures 6 and 7 show the flow curves of PP at 230 and 265°C.

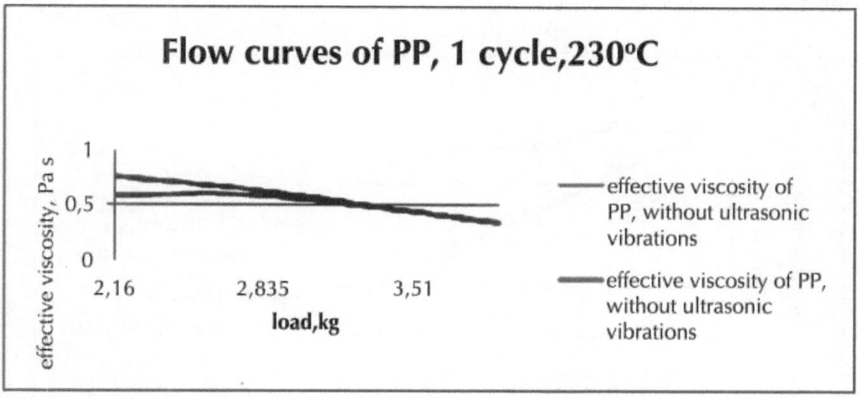

FIGURE 6 The curves of flow of the first cycle of processing of PP at 230°C.

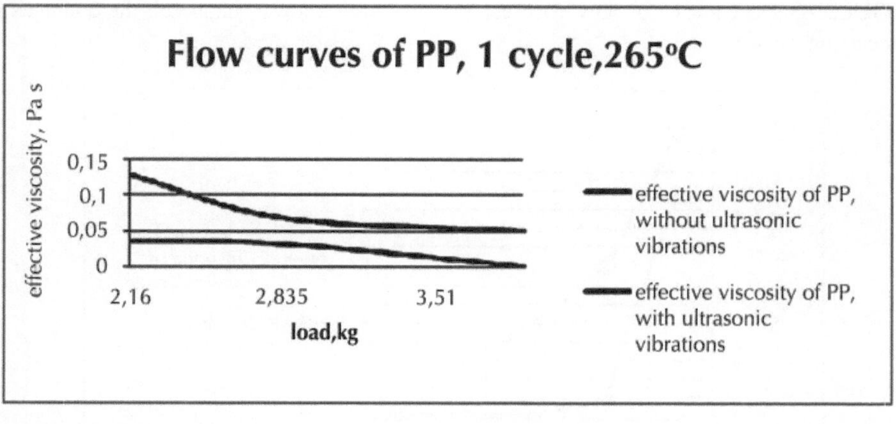

FIGURE 7 The curves of flow of the first cycle of processing of PP at 265°C.

At the temperature 230°C, the flow curves of PP treated with ultrasound and without it are in the area of the confidence interval. At a temperature of 265°C, values of the effective viscosity of the samples differ significantly. Since PP treated without ultrasound, has very low values of this characteristics and ultrasonic treatment increases the effective viscosity of the PP. When comparing the melt flow index (MFI) of PP obtained with the use of ultrasonic vibrations generator and PET found that MFI of these samples are comparable, that might improve their compatibility with the joint processing.

On the next stage of research it was necessary to receive and investigate the properties of composed mixes that were made by blending of original components and increasing of PP's content in PET. The received samples were researched on physico-mechanical properties like original components. In the Figures 8 and 9 there are dependences of breakdown voltage (Figure 8) and percent elongation with rupture (Figure 9) on PP's content in PET.

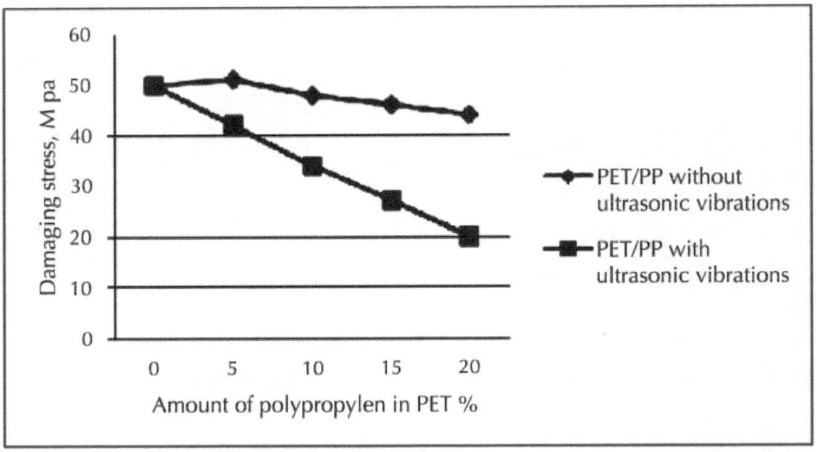

FIGURE 8 Dependence of polymeric composition's breakdown voltage on the amount of processing's cycles.

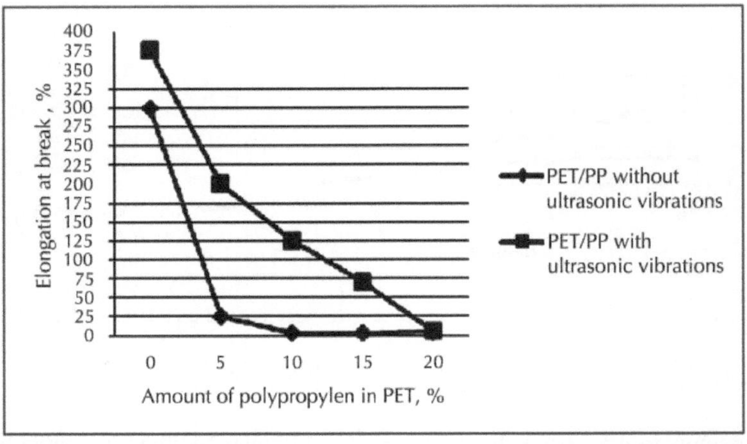

FIGURE 9 Dependence of polymeric composition's percent elongation with rupture on the amount of processing's cycles.

We can see that even a small content of PP in PET can lead to reduction of polymeric compositions deformatively strengthening indexes. It is allowed to note that percent elongation with rupture decreases to 8% with containing of PP in PET makes 10%. It is insufficient characteristic for getting secondary raw polymeric materials. During the influence of ultrasonic oscillations composition's percent elongation with rupture is considerably higher than without its impact. So, it is allowed to make a conclusion about broadening of technological compatibilities interval of such polymers like PP and PET.

13.4 CONCLUSION

Thus on the basis of these results, we can conclude about the principle possibility of changing structure of polymer materials using ultrasonic treatment. The ultrasonic treatment can improve the compatibility of mixtures of thermodynamically incompatible polymers: therefore, it can be used to develop the technology for co-processing polymer waste of different chemical nature.

KEYWORDS

- **Polyethylene terephthalate**
- **Polypropylene**
- **Secondary processing**
- **Thermomechanical curves**
- **Ultrasonic influence**

REFERENCES

1. Anan'ev, V., Gubanova, M., and Kirsh, I. Influence ultrasonic on polyetelene's properties. *Journal of Polymer Plastics* (in Rus.), **6**, 126–128 (2008).

14 An Investigation on Drying Process: An Experimental Approach

A. K. Haghi and G. E. Zaikov

CONTENTS

14.1 INTRODUCTION

The wood consists not only of water but also of polymers as cellulose, hemicellulose, and lignin.drying of wood is a complex operation involving transient transfer of heat and mass along with several rate processes, such as physical or chemical transformations, which in turn, may cause changes in product quality. The drying of wood is an important step in the processing of competitive timber. Little detailed information is available on the alternative of using different heating modes during drying of wood. This chapter presents experimental works related to the use of electric radiating technologies in wood drying operation. It includes conventional air-circulation drying, microwave drying and infrared drying of Guilan spruce woods. The experiments shows that in microwave heating, the drying time is significantly reduced while the strength were higher than the strength obtained in conventional and infrared drying.

Drying is the removal of water from wood. The effects of drying methods on mechanical properties have been studied for many materials, but wood products have seldom been the subject. However, unlike many wet materials that must be dried, wood must be dried at specified rates to avoid degrade (value loss). If degrade were not concern, lumbers could be dried in minutes. As wood dries, then, from the outside inward, it begins shrinkage.

Change in moisture content (MC) results in strain and strain induced stresses, the magnitudes of which are sufficiently large to produce configurational strain known as warp and fracture. We dry wood for several reasons. Among the most important are to minimize changes in dimension and to improve strength properties. Loss of water results provide changes in many of the properties of wood, such as strength, both thermal, and electrical conductivity. Perhaps greater importance is the fact that moisture loss from the cell walls results for shrinkage.

When wood is dried during manufacture, all the liquid water in the cell lumen is removed. The cell lumen always contains some water vapor, however. The amount of water remaining in the cell walls of a finished product depends upon the extent of drying during manufacture and the environment into which the product is later placed. After once being removed by drying, water will recur in the lumen only if the product is exposed to liquid water.

The porous materials such as wood have microscopic capillaries and pores which cause a mixture of transfer mechanisms to occur simultaneously when subjected to heating [1]. Transfer of vapor and liquids occurs in porous bodies in the form of diffusion [2]. In essence, transfer of liquids can occur by means of diffusion arising from hydrostatic pressure gradient.

Heat and mass transfer in porous media is a complicated phenomenon and a typical case is the drying of moist porous materials. Scheidegger [3] claimed 47 years ago that the structure of porous media is too complex to be described either in macroscale or microscale, not to mention the combination of water with matrix. To date, there is no credible work proving that Scheidegger was wrong.

The convective drying is usually encountered in wood industry. The study of this type of drying has attracted the attention of several authors. Among the works relating to this question it cite the works of Plumb et al [4] and Basilico and Martin [5]. Convective drying of timber is one of the oldest and time-consuming methods to prepare the wood for painting and chemical treatments. The drying method can obviously have significant effect on the mechanical properties of wood. Major disadvantages of hot air drying are low energy efficiency and lengthy drying time during the falling rate period. The desired to achieve fast thermal processing has resulted in the increasing use of radiation heating [6-8]. In this case, not only the removal of moisture is accelerated but also a smaller floor space is required, as compared to conventional heating and drying equipment. In the drying of many species, especially, medium density, heavy hardwoods, shrinkage, and accompanying distortion may increase as the temperature is raised. So with species which are prone to distort it is normal to use comparatively low kiln temperatures.

It has also been recognized that dielectric heating could perform a useful function in drying of porous materials in the leveling out moisture profiles across wet sample [9-11]. This is not surprising because water is more reactive than any other material to dielectric heating so that water removal is accelerated [12-18]. This leads to giving a temperature gradient inside the wood sample with opposite directions to that in conventional drying processes.

The objective of any drying process is to produce a dried product of desired quality at minimum cost and maximum throughput possible. High temperature and long drying times required removing water from timber in conventional hot air drying. Microwave and infrared drying could be rapid, more uniform, and energy efficient compared to conventional hot air drying. The main purpose of this study is to investigate the impacts of convective, infrared, and microwave drying on the strength of dried wood.

14.2 BACKGROUND AND BASIC CONCEPTS

Water often makes up over half of the total weight of the wood in a tree. The water or MC of wood is expressed, in percent, as the weight of water present in the wood divided by the weight of dry wood substance. It should be noted that the most of this water must be removed in order to obtain satisfactory performance from wood that is to be processed into consumer and other types of useful products. Because wood is made up of various kinds of cells, some water remains within the structure of the cell walls even after it has been manufactured into lumber or other wood based products. The physical and mechanical properties, resistance to biological deterioration, and dimensional stability of any wood based product are all affected by the amount of water present.

Although many methods of drying timber have been tried over the years only a few of these enable drying to be carried out at a reasonable cost and with minimal damage to the timber. The most common method of drying is to extract moisture in the form of water vapor [20]. To do this, heat must be supplied to the wood to provide the latent heat of vaporization. There are several ways of conveying heat to the wood and removing the evaporated moisture. Nearly all the world's timber is, in fact, dried in air. This can be carried out at ordinary atmospheric temperatures (air drying), or in a kiln at controlled temperatures raised artificially above atmospheric temperature but not usually above 100°C, the boiling point of water. Air drying and kiln drying are fundamentally the same process because, with both, air is the medium which conveys heat to the wood and carries away the evaporated moisture. When air holds the maximum possible amount of vapor, the vapor exerts which is called the saturation vapor pressure [21]. If the water vapor present is less than this maximum then the air can take up more moisture [22-24]. When a piece of wet wood is exposed to air which is not already saturated (i.e. its relative humidity is less than 100%), evaporation takes place from its surface. At a given temperature the rate of evaporation is dependent on the vapor pressure difference between the air close to the wood and that of the more mobile air above this zone. The temperature of a piece of wood and of the air surrounding it will also affect the rate of water evaporation from the wood surface. With kiln drying, warm or hot air is passed over the timber and at the start of the drying process the temperature differential between the air and the wet wood will usually be large. As a result, heat energy will be transferred from the air to the wood surface where it will raise the temperature of both the wood and the water it contains. Water, in the form of vapor, will then be lost from the wood surfaces, provided the surrounding air is not already saturated with moisture. This results in the development of a MC gradient from inside to outside of the wood. As the temperature is raised this increases not only the steepness of this moisture gradient, but also the rate of moisture movement along the gradient and the rate of loss of water vapor from the surface of the wood. When water evaporates from the surface of a piece of wet wood the MC in the outer zone is lowered and moisture begins to move outwards from the wetter interior. In practical terms this movement of moisture can be accepted as being a combination of capillary flow and moisture diffusion, a process which is resisted by the structure of the wood, particularly in dense hardwood species. If the rate of water loss by evaporation exceeds the rate at which moisture from the wet interior can pass to the surface, the moisture gradient within the wood becomes progressively steeper.

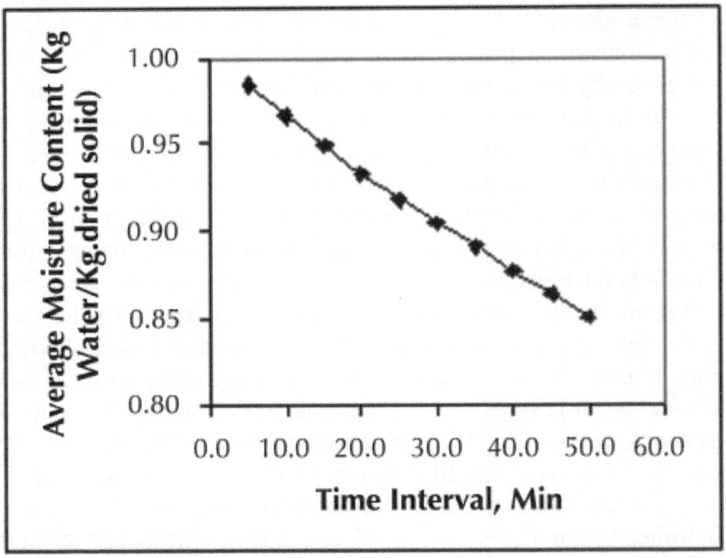

FIGURE 1 Average moisture content *versus* time (Conventional hot air-dried wood at T = 40°C).

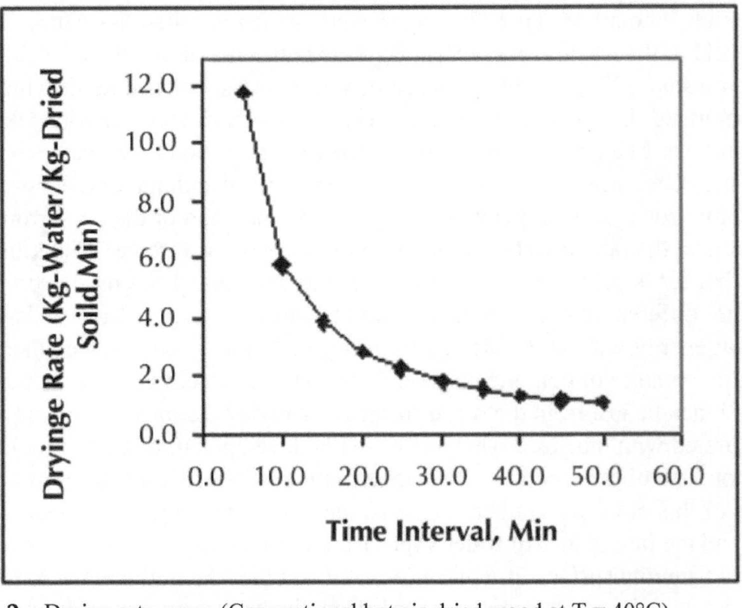

FIGURE 2 Drying rate curve (Conventional hot air-dried wood at T = 40°C).

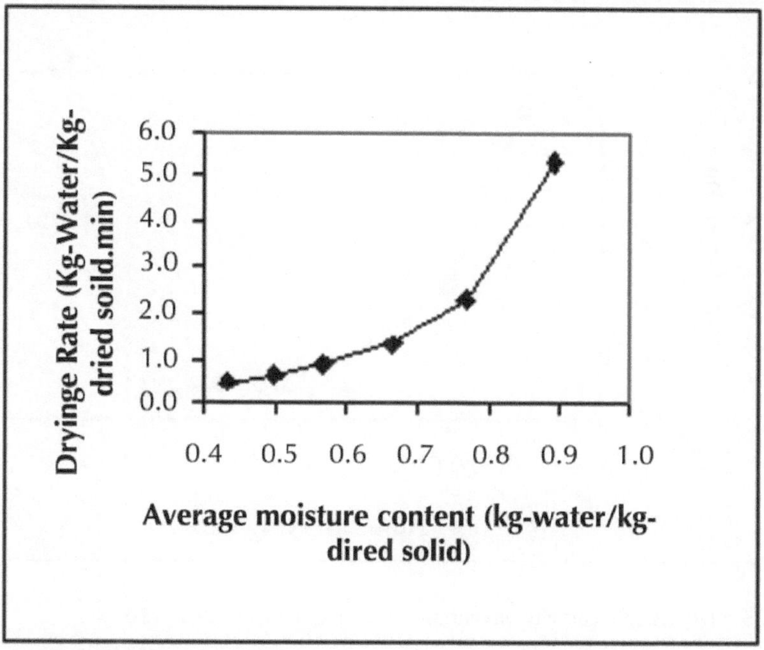

FIGURE 3 Average moisture content versus time (Conventional hot air-dried wood at T = 100°C).

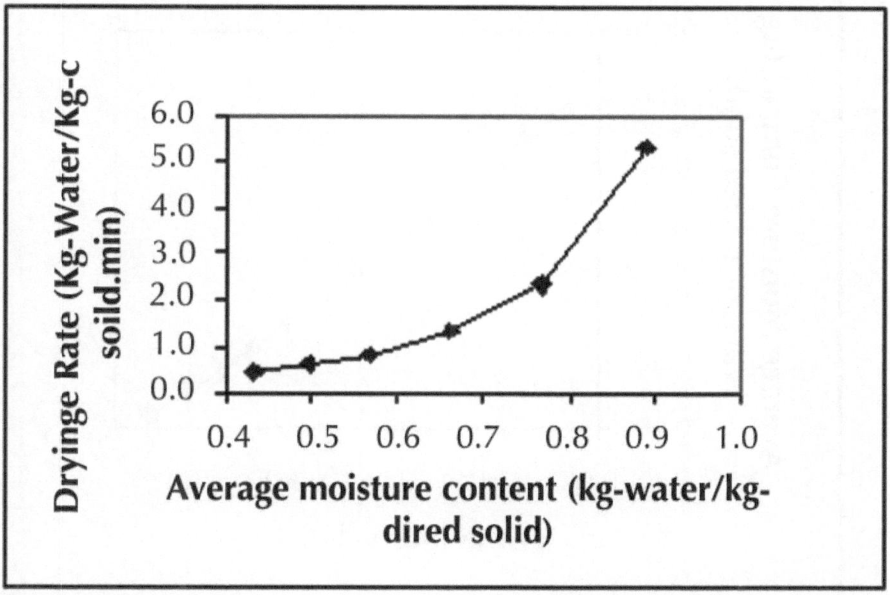

FIGURE 4 Drying rate curve (Conventional hot air-dried wood at T = 100°C).

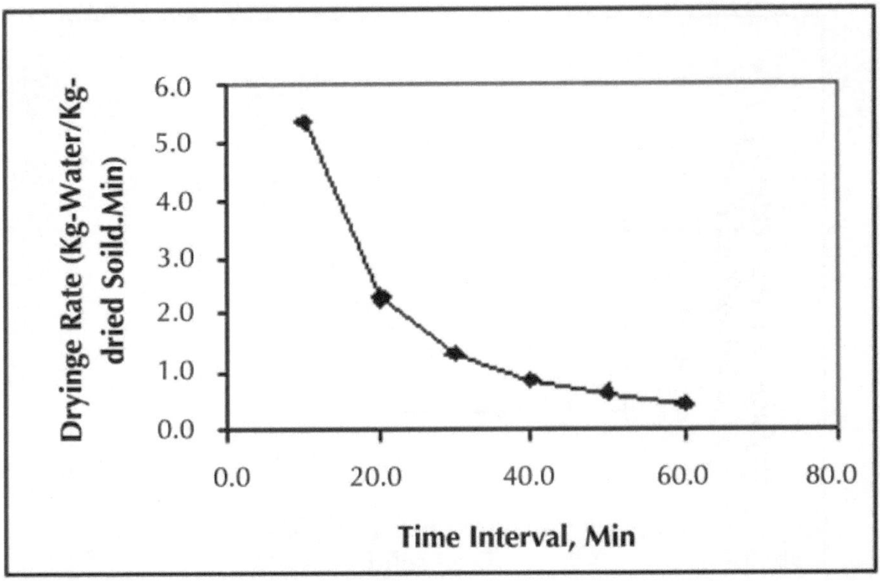

FIGURE 5 Drying rate curve (Conventional hot air-dried wood at T = 100°C).

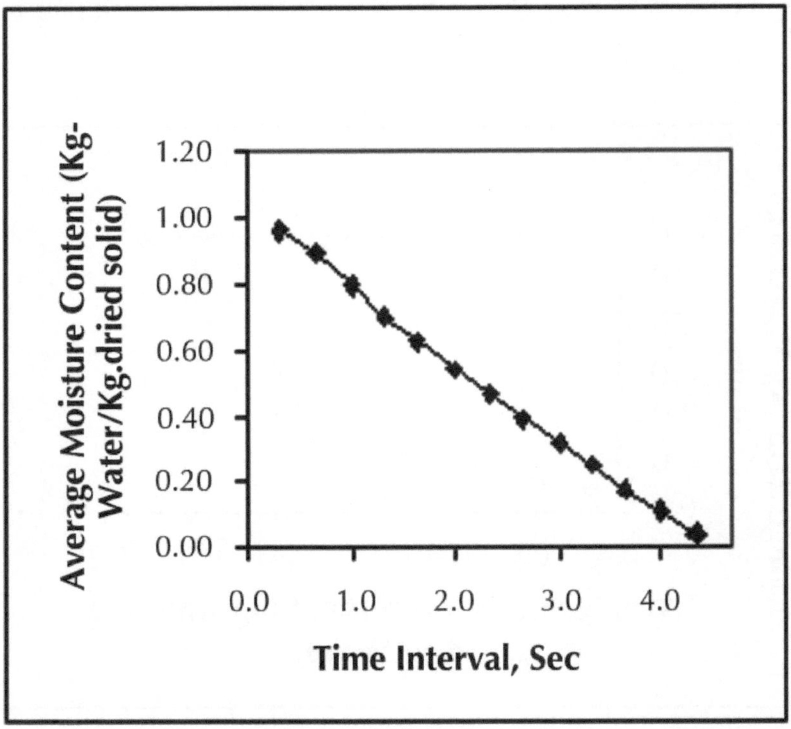

FIGURE 6 Drying curve for microwave-dried wood at 80 powers.

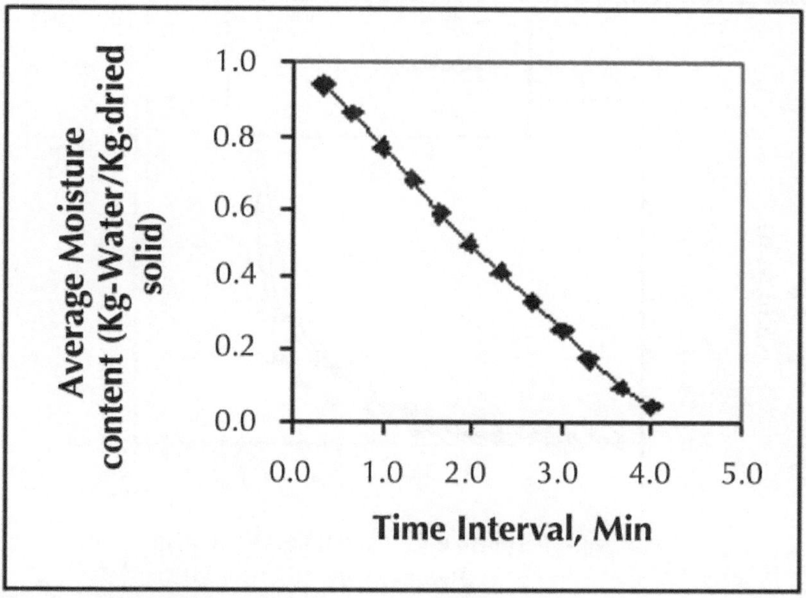

FIGURE 7 Drying curve for microwave-dried wood at 100% powers.

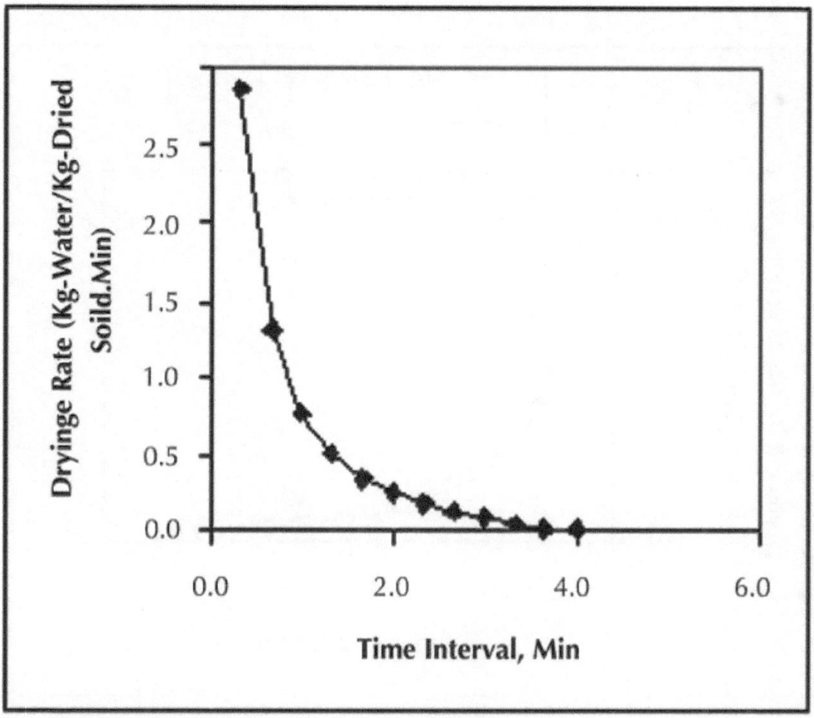

FIGURE 8 Drying rate curve for microwave-dried wood at 100% power.

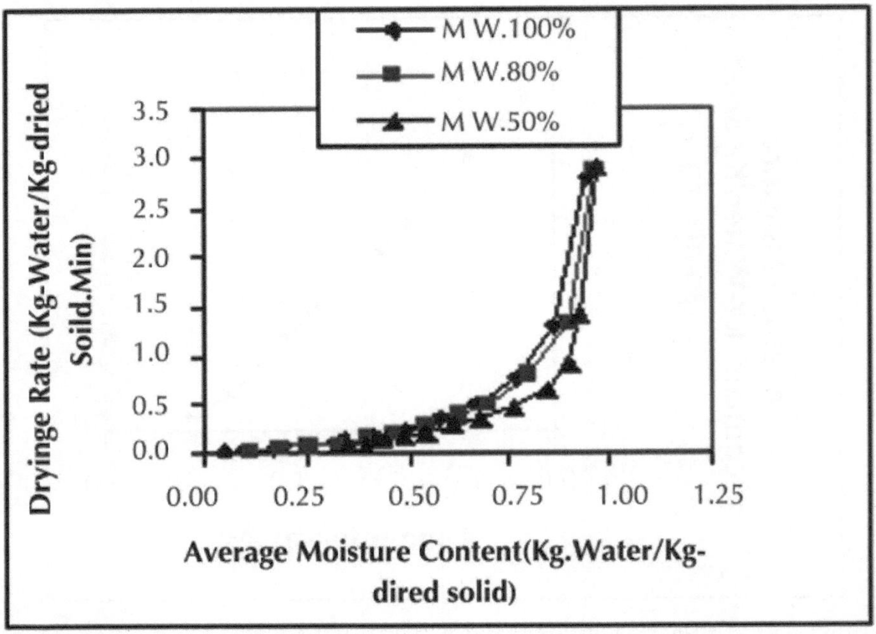

FIGURE 9 Drying rate curves for microwave-dried wood at three different powers.

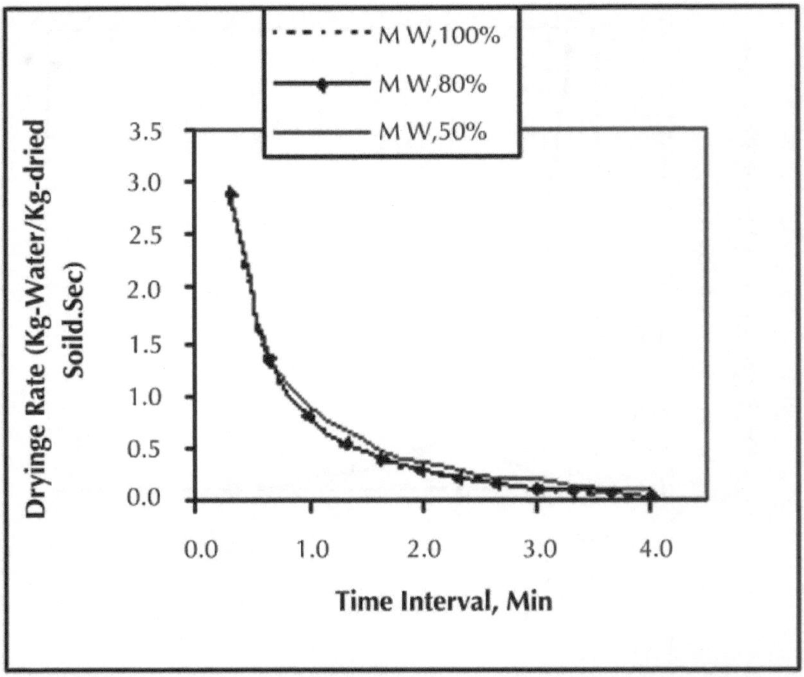

FIGURE 10 Drying rate curves for microwave-dried wood at three different powers.

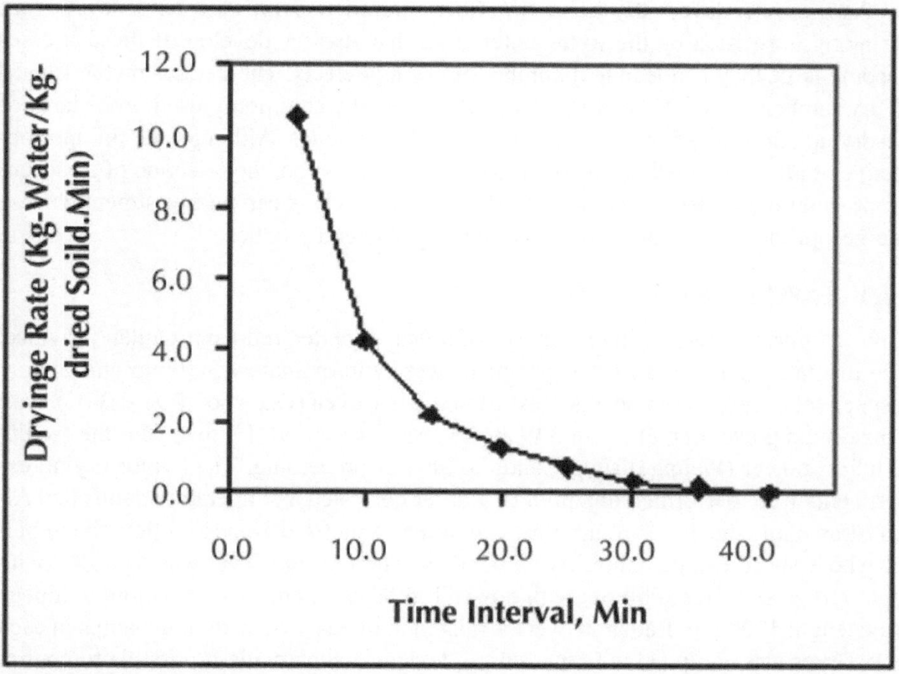

FIGURE 11 Drying curve for infrared-dried wood at 100% powers.

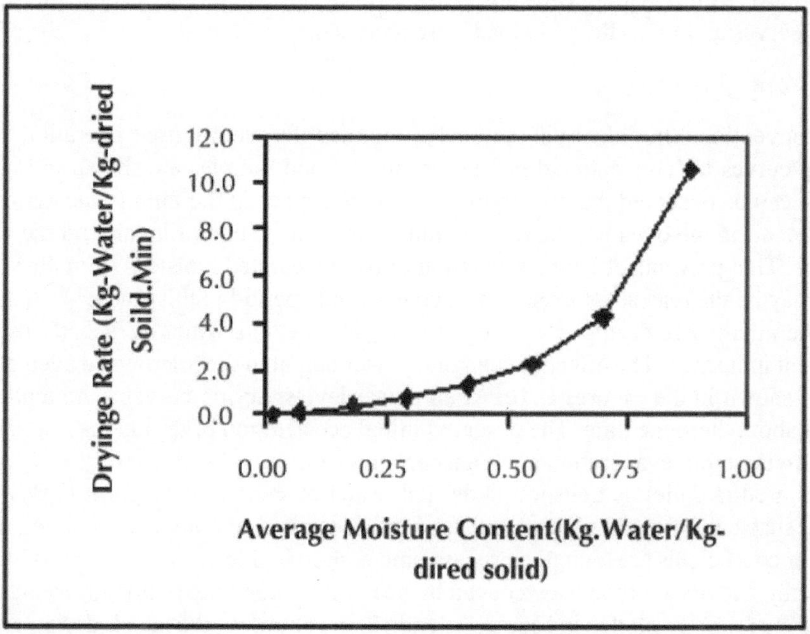

FIGURE 12 Drying curve for infrared-dried wood at 100% powers.

As the outer layers dry below the fiber saturation point their tendency to start shrinking is resisted by the wetter interior so that stresses develop. If these stresses become large they can lead to a number of drying defects. The various methods used to dry lumber can be divided into two categories: the commonly used procedures of air-drying and kiln-drying, and the specialized techniques. Although the primary objective of all drying methods is to remove water from wood, the selection of a particular procedure will depend on several other factors such as capital investment, energy sources, production capacity, drying efficiency, and end product.

14.3 EXPERIMENTAL

Fifty cylindrical green wood samples of Spruce were derived from Guilan province. The diameter and height of the specimens were approximately 300 mm and 21 mm respectively. A programmable domestic microwave oven (Daewoo, KOC-1B4K), with a maximum power output of 1000 W at 2450 MHz was used. The oven has the facility to adjust power (Wattage) supply and the time of processing. The hot air drying experiments were performed in a pilot tray dryer consisted a temperature controller. Air was drawn into the duct through a mesh guard by a motor driven axial flow fan impeller whose speed can be controlled in the duct. The infrared dryer was equipped with eight red glass lamps (Philips) with power 175 W, each emitting radiation with peak wavelength 1200 nm. Radiators were arranged in three rows, with three lamps in each row. Dryer was equipped with measuring devices, which made it possible to control air parameters. The amount of water in a piece of wood is known as its MC. All the fifty dried samples were tested on a universal tension test machine model (Hounsfield HS100KS), with a loading capacity of 100 kN. During the tensile testing, the stress-strain curves as well as the peak load were recorded.

14.4 RESULTS

The conventional hot air drying is one of the most frequently used operations. The drying curves for conventional hot air drying of wood samples are shown in Figures 1–5. It can be observed that the drying usually take place in the falling rate period. In essence, air in the oven is saturated, by time, and forms a thick film around the wood sample. That prevents effective separation of the evaporated moisture from the wood. This may be the reason for existence of constant rate period in this study.

The microwave drying is an alternative drying method, which is recently used in different industries. The effect of changing power output in the microwave oven on the MC is shown in the Figures 6–10. At all power levels, drying curves were tended to end at about the same time. The observed initial acceleration of drying may be caused by allowing rapid evaporation and transport of water.

Infrared radiation is transmitted through water at short wavelength it is absorbed on the surface. Infrared radiation has some advantages over convective heating. Heat transfer coefficients are high, the process time is short and the cost of energy is low. In this study, the drying time was reduced by nearly 34% compare to hot air drying. The drying curves were plotted in Figures 11 and 12. In contrast to the hot air drying curves which had a short constant rate period followed by a falling rate period, Figures 11 and 12 indicates that the infrared had only a falling rate period.

All dried samples were tested on a Hounsfield universal tension test machine with a loading capacity of 100 kN (Figure 13). The results of tensile loading of dried samples are presented in Figures 14–17. It is clear that the microwave dried spruce specimen with failure strength of 49.6 MPa has made a significant property improvement (Figure 14). The normal stiffness of infrared dried sample is reported as 35.0 MPa (Figure 15) whereas the oven dried sample showed strength of about 44.5 MPa (Figure 16). From Figure 17 it is revealed that the natural convection dried specimens are the strongest of about 50 MPa. In practice the drying time for this can take up months and years. In Figure 18 the strength of dried samples are compared for a better judgment.

14.5 APPLICATION OF RESULTS

In wood drying process we should note that the wood can hold moisture in the cell lumen (cavity) as liquid or "free" water, or as adsorbed or "bound" water attached to the cellulose molecules in the cell wall. Meanwhile, the occurrence of the free water does not affect the properties of wood other than its weight. Bound water, however, does affect many properties of wood, and is more difficult to remove in the drying process. Microscopically, the dimensional change with MC is anisotropic (referring to the fact that wood has very different properties parallel to the fact grain *versus* the transverse direction). As the MC decreases, wood shrinks conversely, as the MC increases, wood swells or grows larger. The process of drying focuses on producing wood with an MC about the same as the equilibrium value for the intended service environment.

For the design of dryers it is necessary to carry out drying experiments at various drying conditions. Experimentally determined drying times, transition points, and constant rate regime temperature can be used as a base case for the analytical results. Based on the information from the experimental trials, runs with lower amounts of moisture to evaporate, and higher dryer temperatures, should be expected to dry faster and reach transition point more rapidly. After an initial increase or decrease of the rate of drying, the drying process enters the constant rate period. This initial change of the rate of drying is caused by a variation of the surface temperature which in turn results into a change of vapor density.

It can be noted that time interval of drying process is solely determined by external conditions. Once the drying process has entered the falling rate period, the external conditions become relatively unimportant compared to the internal parameters.

By comparing runs with the same initial moisture, we see that as oven temperature increases, the transition points are reached more quickly and total drying times are shorter. Sample temperatures are higher because they are exposed to higher heat transfer rates, giving rise to higher mass transfer rates during the constant rate regime.

The experimental study suggests that the humidity of the free stream should be as low as possible. Partial recirculation, 100% fresh air intake, or dehumidifications are some of the possible ways to accomplish this task, but a cost analysis is imperative before deciding on any option.

Reduction of the drying time in microwave heater seems to be a motivating cost saving factor for industries. In this case a moderate mechanical property is obtained (Table 1). To minimize directional variations in use, wood needs to be dry enough to match the service environment.

14.6 CONCLUSION

Although many methods of drying timber have been tried over the years only a few of these enable drying to be carried out at a reasonable cost and with minimal damage to the timber. The most common method of drying is to extract moisture in the form of water vapor. To do this, heat must be supplied to the wood to provide the latent heat of vaporization. The temperature of a piece of wood and of the air surrounding it will also affect the rate of water evaporation from the wood surface. With kiln drying, warm or hot air is passed over the timber and at the start of the drying process the temperature differential between the air and the wet wood will usually be large. As a result, heat energy will be transferred from the air to the wood surface where it will raise the temperature of both the wood and the water it contains. Water, in the form of vapor, will then be lost from the wood surfaces, provided the surrounding air is not already saturated with moisture. This results in the development of a MC gradient from the inside to the outside of the wood. As the temperature is raised this increases not only the steepness of this moisture gradient, but also the rate of moisture movement along the gradient and the rate of loss of water vapor from the surface of the wood. At a given temperature the rate of evaporation is dependent on the vapor pressure difference between the air close to the wood and that of the more mobile air above this zone.

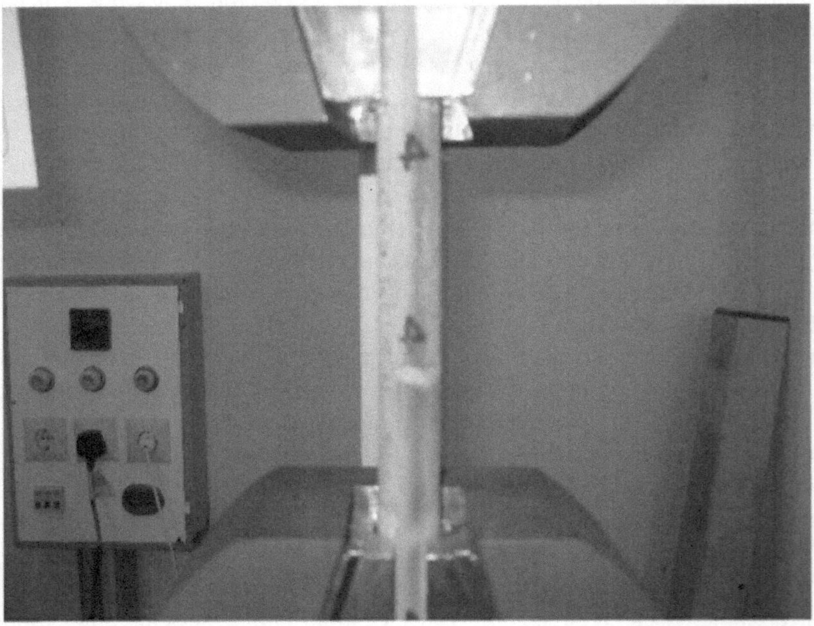

FIGURE 13 Tension test.

Unfortunately, the considerable benefits obtainable by raising the drying temperature cannot always be fully exploited because there are limits to the drying rates which various wood species will tolerate without degrade. In contrast to hot air drying a modern radiation drying provides temperature control and a steady, adequate flow of air over the timber surface. The air flow rate and direction is controlled by fans and the temperature and relative humidity of the air can be adjusted to suit the species and sizes of timber being dried. It is thus possible to make full use of the increase in drying rate which can be achieved by raising the temperature to the maximum value which a particular timber species can tolerate without excessive degrade. Contrary to the results presented [9] microwave heating improved the strength in comparison to the strength obtained in conventional hot air drying. It is also noted that the infrared drying can reduce the strength of the spruce woods significantly. It should be noted that by applying natural convection the highest strength can be obtained with the highest drying duration. This can take weeks, months, or even years [9]. The discussion suggests further investigation for future work on different specimens.

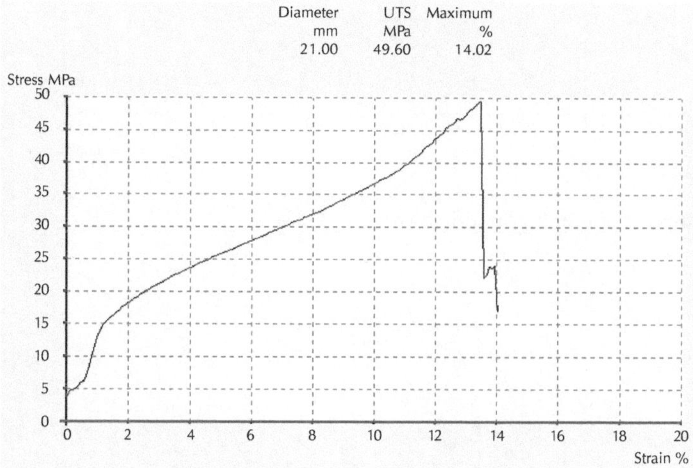

FIGURE 14 Stress-strain for microwave dried wood.

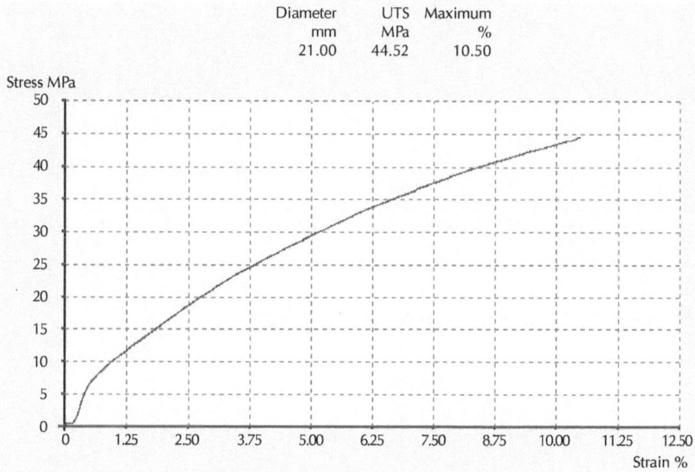

FIGURE 15 Stress-strain for infrared dried wood.

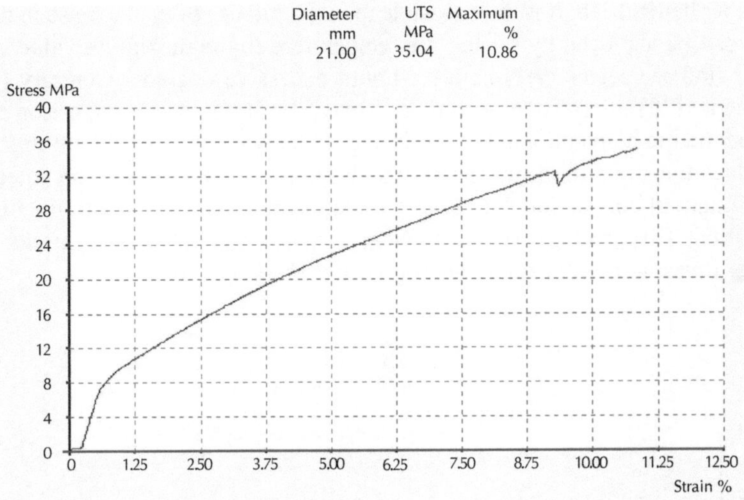

FIGURE 16 Stress-strain for hot-air dried wood.

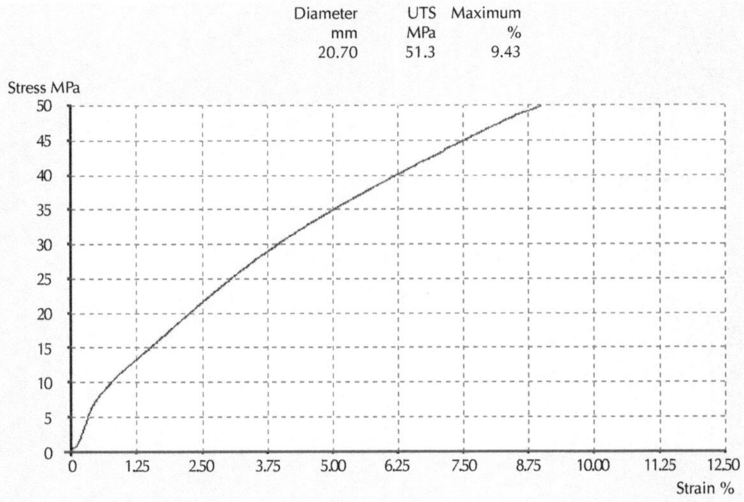

FIGURE 17 Stress-strain for natural convection dried wood.

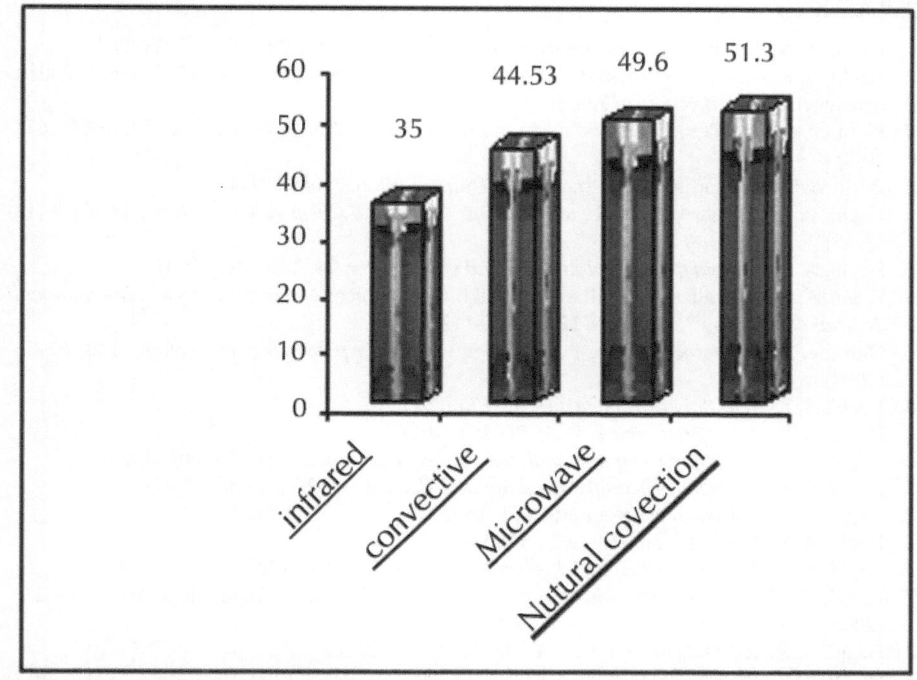

FIGURE 18 Strength of dried wood samples in three different Drying modes.

TABLE 1 Average strength properties of samples (σ values in brackets refer to standard deviations).

Drying Method	Failure Strength(Mpa)	Failure Strain %	Yield Strength(Mpa)	Modulus of elasticity(Gpa)
Natural convection Dried wood	51. 3 (σ2.44)	9. 43	12. 8 (σ 0. 615)	0. 544 (σ 0. 058)
Hot air dried wood	44. 53 (σ 1.72)	10. 5	13. 3 (σ 0. 51)	0. 424 (σ 0. 05)
Infrared dried wood	35. 04 (σ 1.16)	10. 86	10. 5 σ (0. 35)	0. 322 (σ 0. 035)
Microwave dried wood	49. 6(σ 4. 51)	14. 02	17. 0 (σ 1. 28)	0. 354 (σ0. 054)

KEYWORDS

- **Drying curves**
- **Drying process**
- **Infrared radiation**
- **Moisture content**
- **Universal tension test machine model**

REFERENCES

1. Haghi, A. K. *Theoretical Foundations of Chemical Engineering*, **39**(2), 200–203 (2005).
2. Haghi, A. K. and Ghanadzadeh, H. *Indian Journal of Chemical Technology*, **12**, 654–663 (2005).
3. Scheigger, A. E. *University of Toronto Press*, (1958).
4. Plumb, O. A., Spolek, G. A., and Olmstead, B. A. *Int. J. Heat Mass Transfer*, **28**, 1669–1678 (1985).
5. Basilico, C. and Martin, M. *Int. Heat Mass Transfer*, **27**, 657–688 (1984).
6. Haghi, A. K., Ghanadzadeh, H., and Rondot, D. *Iran. J. Chem & Chem. Eng.*, **24** (2), 1–10 (2005).
7. Haghi, A. K. *Journal of Thermal Analysis and Calorimetry*, **74**, 827–840 (2003).
8. Haghi, A. K. and Valizadeh, M. Experimental Investigation on Microwave Drying. *International Journal of Heat and Technology*, **22**(2), 167–172 (2004).
9. Oloyede, A. and Groombridge, P. *Journal of Materials processing Technology*, **100**, 67–73 (2000).
10. Haghi, A. K. *Acta Polytechnica*, **41**(1), 55–57 (2001).
11. Haghi, A. K. *Acta Polytechnica*, **41**(3), 20–25 (2001).
12. Haghi, A. K. *Journal of Computational and Applied Mechanics*, **2**(2), 195–204 (2001).
13. Haghi, A. K. *Journal of Theoretical and Applied Mechanics*, **32**(2), 47–62 (2002).
14. Haghi, A. K. *Journal Technology*, **35**(F), 1–16 (2002).
15. Haghi, A. K. *H. J. I. C.*, **30**, 261–269 (2002).
16. Haghi, A. K. *J. of Theoretical and Applied Mechanics*, **33**, 83–94 (2003).
17. Haghi, A. K. *International Journal of Applied Mechanics and Engineering*, **8**(2), 233–243 (2003).
18. Haghi, A. K. and Mohammadi, K. *JUCTM*, **38,** 85–96 (2002).
19. Haghi, A. K. and Rondot, D. *Iran. J. Chem. & Chem.*, **23**, 25–34 (2004).
20. Haghi, A. K. *14th Int. Symp. On Transport Phenomena*. pp. 209–214(2003).
21. Haghi, A. K. *Journal of Computational and Applied Mechanics,* **15**(2), 263–274 (2004).
22. Haghi, A. K. *Theoretical Foundations of Chemical Engineering*, **39**(2), 200–203 (2005).
23. Haghi, A. K. *Int. J. Applied Mech. and Eng.,* **2**, 217–226 (2005).
24. Haghi, A. K. *Asian J. of Chem*., **17**(20), 639–654 (2005).
25. Taskini, J. and Haghi, A. K. Proc. Int. Conf. on recent advances in Mech. & Mat. Eng., pp 81–86.

15 On the Composites Based on Epoxy Resin

J. Aneli, O. Mukbaniani, E. Markarashvili, G. Zaikov, and E. Klodzinska

CONTENTS

15.1 INTRODUCTION

In recent time the mineral fillers attract attention as active filling agents in polymer composites [1, 2]. Thanks to these fillers many properties of the composites are improved increases the durability and rigidity, decrease the shrinkage during hardening process and water absorption, improves thermal stability, fire proof and dielectric properties, and finally the price of composites becomes cheaper [3-5]. At the same time it must be noted that the mineral fillers at high content lead to some impair of different physical properties of composites. Therefore, the attentions of the scientists are attracted to substances, which would be remove mentioned leaks. It is known that silicon organic substances (both low and high molecular) reveal hydrophobic properties, high elasticity and durability in wide range of filling and temperatures [6, 7].

The purpose of presented work is the investigation effect of modify by tetraethyl orthosilicate (TEOS) of the mineral diatomite as main filler and same mineral with andesite (binary filler) on some physical properties of composites based on epoxy resin.

15.2 BASIC PART

The mineral diatomite as a filler was used. The organic solvents were purified by drying and distillation. The purity of starting compounds was controlled by an LKhM-8-MD gas liquid chromatography, phase SKTF-100 (10%, the NAW chromosorb, carrier gas He, 2m column). Fourier transform infrared (FTIR) spectra were recorded on a Jasco FTIR-4200 device.

The silanization reaction of diatomite surface with TEOS was carried out by means of three-necked flask supplied with mechanical mixer, thermometer, and dropping funnel. For obtaining of modified by three mass % diatomite to a solution of 50 g grind finely diatomite in 80 ml anhydrous toluene the toluene solution of 1.5 g (0.0072 mole) TEOS in 5 ml toluene was added. The reaction mixture was heated at the boiling temperature of used solvent toluene. Than the solid reaction product was filtrated, the solvents (toluene and ethyl alcohol) were eliminated, and the reaction product was dried up to constant mass in vacuum. Other product modified by 5% tetraethoxysilane was produced *via* the same method.

Following parameters were defined for obtained composites: ultimate strength (on the stretching apparatus of type "Instron"), softening temperature (Vica method), density, and water absorption (at saving of the corresponding standards).

15.3 DISCUSSION AND RESULTS

The high temperature condensation reaction between diatomite and TEOS from the one side, and between andesite and same modifier from the other one was carried out in toluene solution (~38%). The masses of TEOS were 3 and 5% from the mass of filler. The reaction systems were heated at the solvent boiling temperature (~110°C) during 5–6 hr by stirring. The reaction proceeds according to the following scheme:

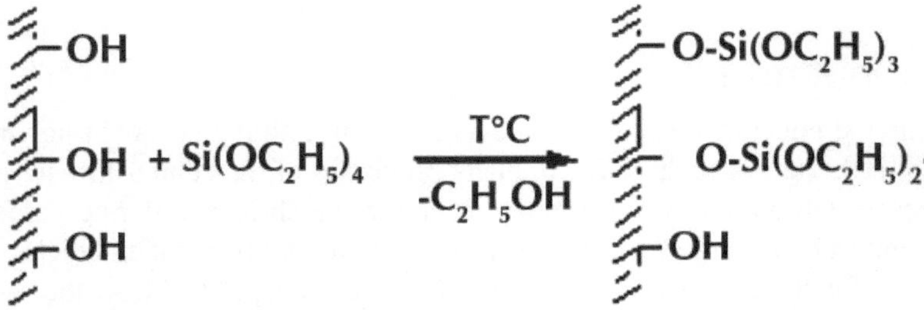

The direction of reaction defined by FTIR spectra analysis shown that after reaction between mineral surface hydroxyl, $OSi(OEt)_3$ and the $OSi(OEt)_2O$- groups are formed on the mineral particles surface.

In the FTIR spectra of modified diatomite one can observe absorption bands characteristic for asymmetric valence oscillation for linear ≡ Si-O-Si ≡ bonds at 1030 cm^{-1}. In the spectra one can see absorption bands characteristic for valence oscillation of ≡Si-O-C≡ bonds at 1150 cm^{-1} and for ° C-H bonds at 2,950-3,000 cm^{-1}. One can see also broadened absorption bands characteristic for unassociated hydroxyl groups.

On the basis of modified diatomite and epoxy resin (of type ED-20) the polymer composites with different content of filler were obtained after careful wet mixing of components in mixer. After the blends with hardening agent, (polyethylene-polyamine) were placed to the cylindrical forms (in accordance with standards ISO) for hardening at room temperature during 24 hr. The samples hardened later were exposed to temperature treatment at 120°C during 4 hr. The concentration of powder diatomite (average diameter up to 50 micron) was changed in the range 10–60 mass %.

The curves on the Figure 1 show that at increasing of filler (diatomite) concentration in the composites the density of materials essentially depends on both of diatomite contain and on the degree of concentration of modify agent TEOS. Naturally the decreasing of density of composites at increasing of filler concentration is due to increasing of micro empties because of one's localized in the filler particles (Figure 1, curve 1). The composites with modified by TEOS diatomite contain less amount of empties as they are filled with modify agent (Figure 1, curves 2 and 3).

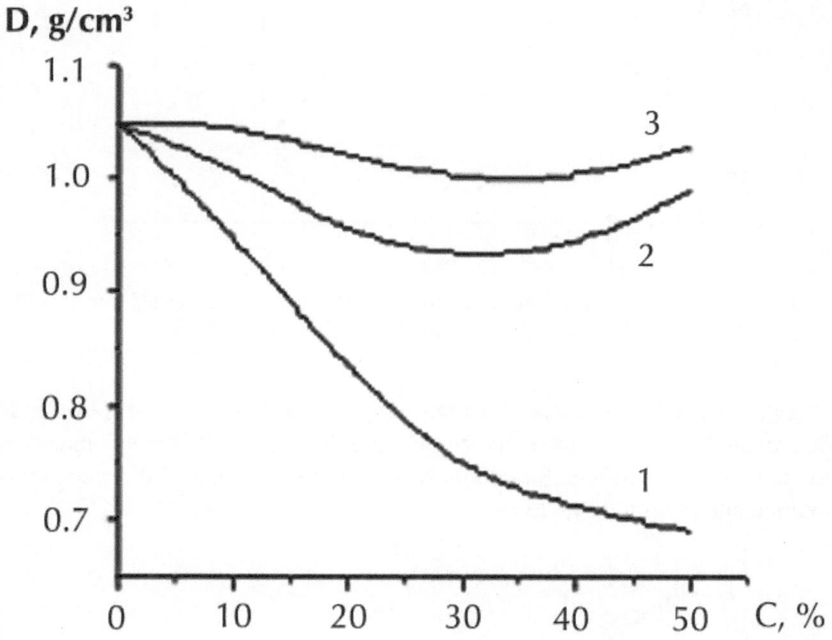

FIGURE 1 Dependence of the density of the composites based on epoxy resin on the concentration of unmodified (1), modified by 3% (2), and 5 mass % (3) tetraethoxysilane diatomite.

The dependence of ultimate strength on the content of diatomite (modified and unmodified) presented on the Figure 2 shows that it has an extreme character. However, the positions of corresponding curves maximums essentially depend on amount of modified agent TEOS. The general view of these dependences is in full conformity with well known dependence of σ C [8]. The sharing of the maximum of curve for composites containing 5% of modified diatomite from the maximum for the analogous composites containing 3% modifier to some extent is due to increasing of the amount of the bonds between filler particles and macromolecules at increasing of the concentration of the filler.

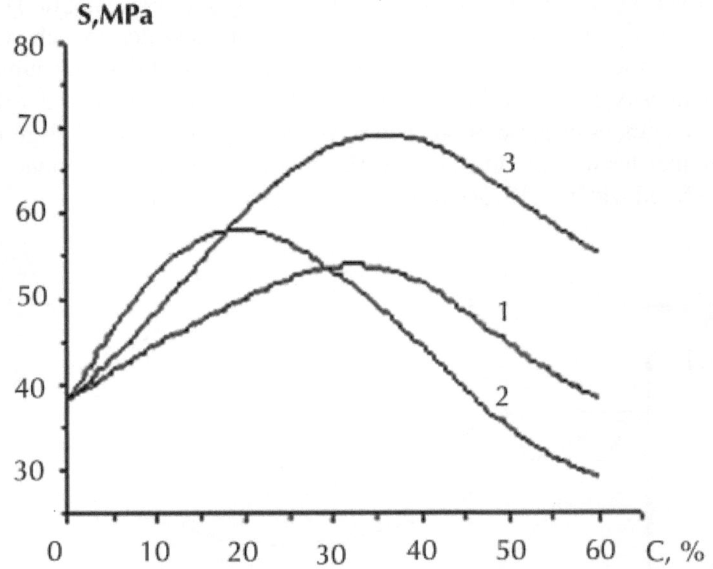

FIGURE 2 Dependence of ultimate strength of the composites based on ED-20 with unmodified (1) and modified by 3 (2), and 5 mass % (3) TEOS diatomite.

Investigation of composites softening temperature was carried out by apparatus of Vica method. Figure 3 shows the temperature dependence of the indentor deepening to the mass of the sample for composites with fixed (20 mass %) concentration of unmodified and modified by TEOS.

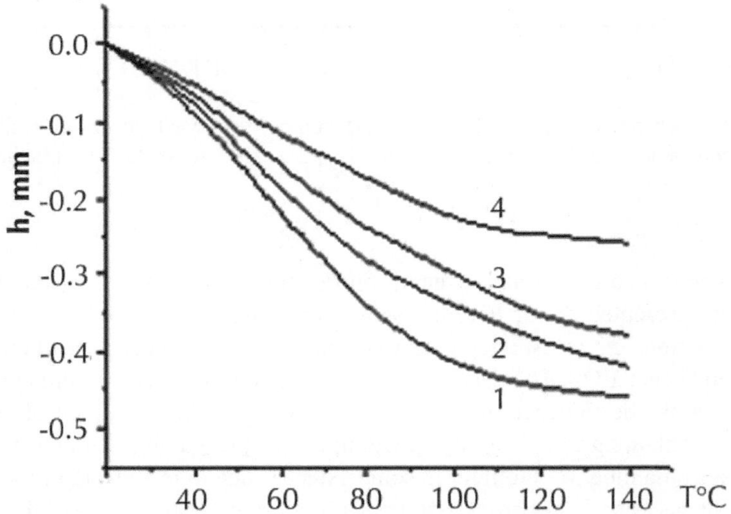

FIGURE 3 Temperature dependence of the indentor deepening in the sample for composites containing 0 (1), 20 mass % (2), 20 mass % modified by 3% TEOS (3), 20 mass % modified by 5% TEOS (4) diatomite.

Based on character of curves on the Figure 3 it may be proposed that the composites containing diatomite modified by TEOS possesses thermostability higher than in case of analogous composites with unmodified filler. Probably the presence of increased interactions between macromolecules and filler particles due to modify agent leads to increasing of thermostability of composites with modified diatomite.

Effect of silane modifier on the investigated polymer composites reveals also in the water absorption. In accordance with Figure 4 this parameter is increased at increasing of filler contain. However if the composites contain the diatomite modified by TEOS this dependence becomes weak.

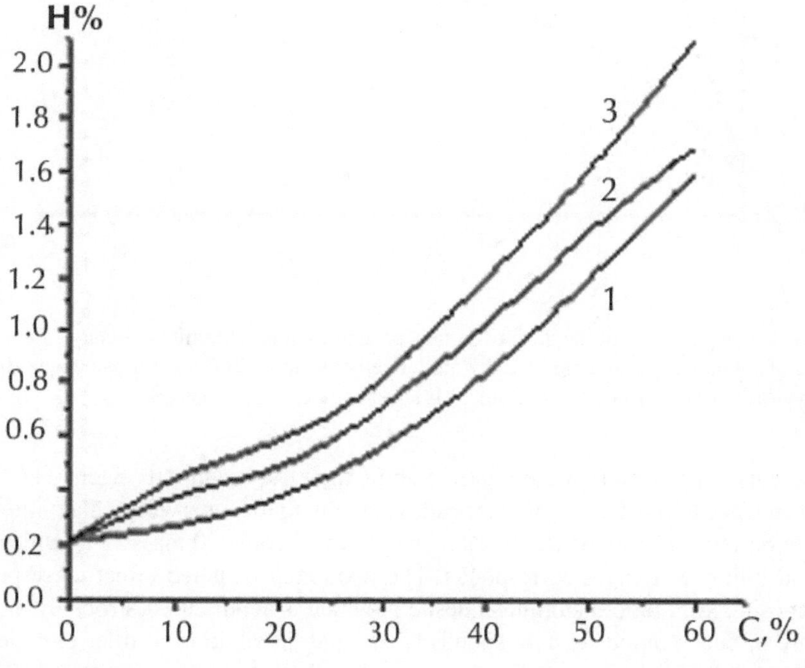

FIGURE 4 Dependence of the water absorption on the concentration of filler in the composites based on epoxy resin containing diatomite modified by 5% (1) and 3% (2) tetraethoxysilane and unmodified (3) one.

There were conducted the investigation of binary fillers on the properties of the composites with same polymer basis (ED-20). Two types of minerals diatomite and andesite with different ratios were used as fillers. It was interesting to establish effect both of ratio of the fillers and effect of modifier TEOS on the same properties of the polymer composites investigated.

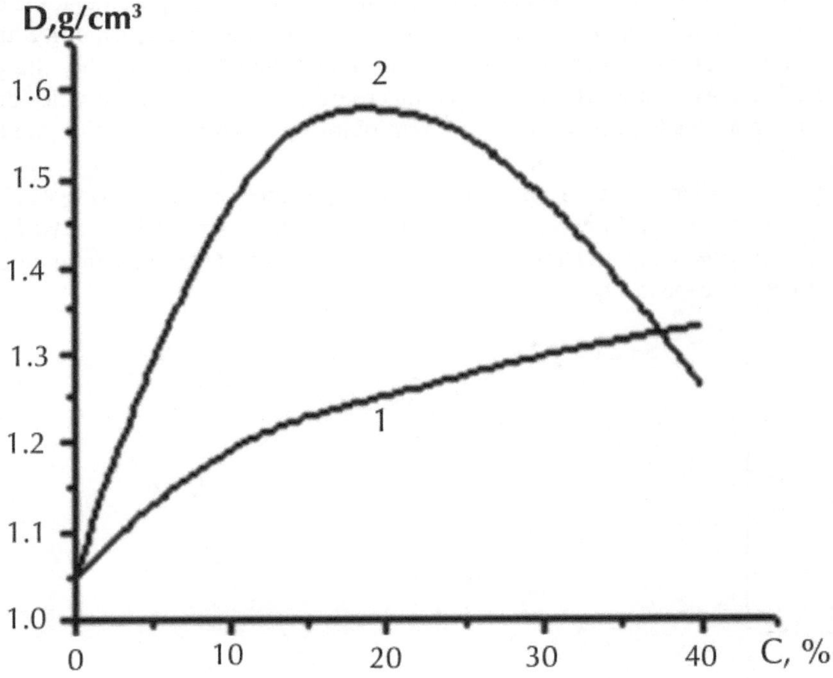

FIGURE 5 Dependence of the density on the concentration of diatomite in binary fillers with andesite (1) - unmodified and modified by 5% tetraethoxysilane (2) fillers for composites based on epoxy resin. Full concentration of binary filler in composites 50 mass %.

The curves presented on the Figure 5 show the effect of modify agent TEOS on the dependence of the density of composites containing the binary filler diatomite and andesite on ratio of lasts when the total content of fillers is 50 mass % to which the maximal ultimate strength corresponds. The maximum of noted effect corresponds to composite, filler ratio diatomite/andesite in which is about 20/30. Probably microstructure of such composite corresponds to optimal distribution of filler particles in the polymer matrix at minimal inner energy of statistical equilibration, at which the concentration of empties is minimal because of dense disposition of the composite components. It is known that such structures consists minimal amount both of micro and macro structural defects [8].

Such approach to microstructure of composites with optimal ratio of the composite ingredients allows supposing that these composites would be possessed high mechanical properties, thermostability, and low water absorption. Moreover, the composites with same concentrations of the fillers modified by TEOS possess all the noted properties better than ones for composites with unmodified by TEOS binary fillers, which may be proposed early (Figures 6–8). Indeed the curves on the Figures 6–8 show that the maximal ultimate strength, thermostability and simultaneously hydrophobicity correspond to composites with same ratio of fillers to which the maximal density corresponds.

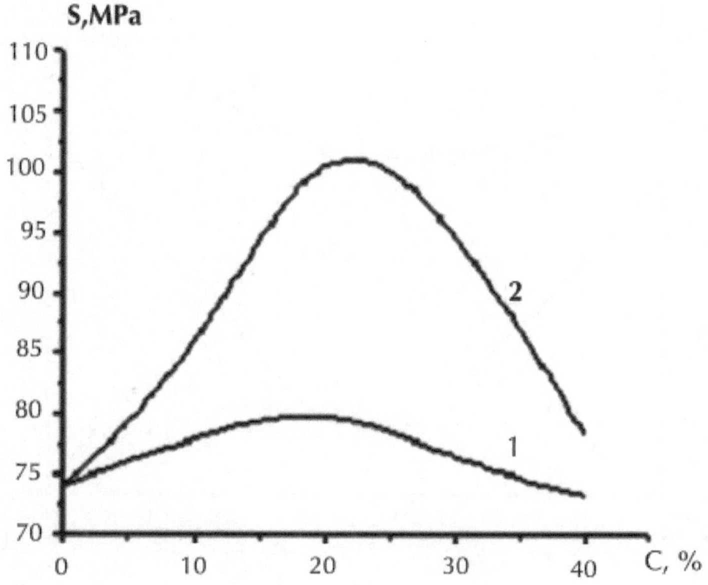

FIGURE 6 Dependence of the ultimate strength on the concentration of diatomite in binary fillers with andesite (1) - unmodified fillers and modified by 5% tetraethoxysilane (2) ones for composites based on epoxy resin. Full concentration of binary filler in composites 50 mass %.

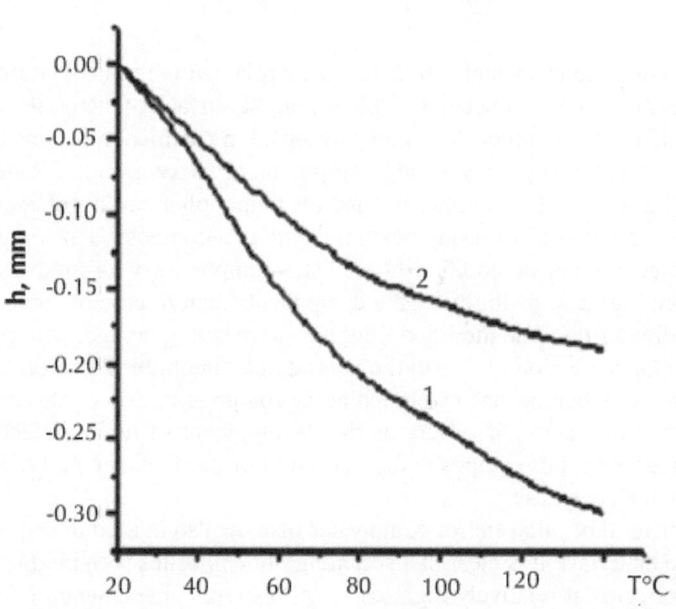

FIGURE 7 Thermostability of composites with binary fillers at ratio diatomite/andesite = 20/30.

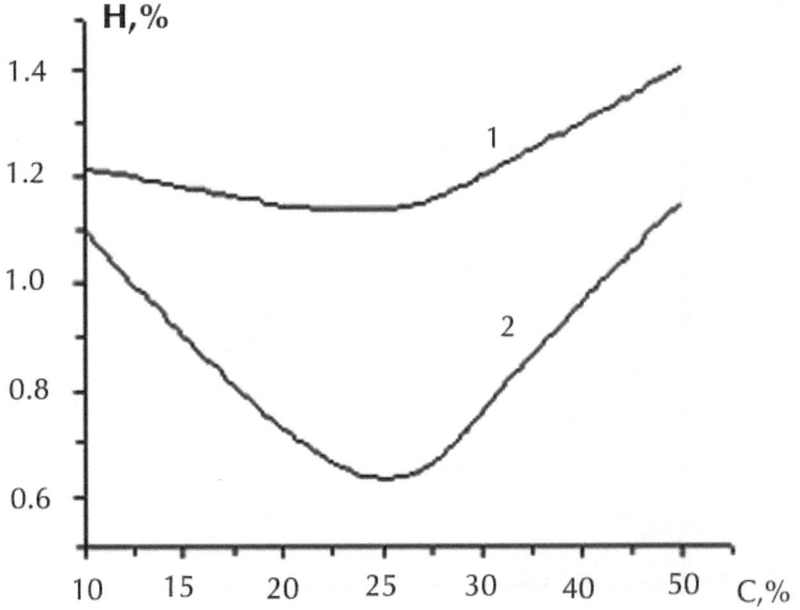

FIGURE 8 Dependence of the water-absorption of composites based on epoxy resin on the concentration of diatomite in binary fillers with andesite (1) - unmodified and modified by 5% tetraethoxysilane (2) fillers. Total concentration of binary fillers in composites 50 mass %.

The obtained experimental results may be explained in terms of composite structure peculiarities. Silane molecules displaced on the surface of diatomite and andesite particles lead to activation of them and participate in chemical reactions between active groups of TEOS (hydroxyl) and homopolymer (epoxy group). Silane molecules create the "buffer" zones between filler and the homopolymer. This phenomenon may be one of the reasons of increasing of strengthening of composites in comparison with composites containing unmodified fillers. The composites with modified diatomite display more high compatibility of the components than in case of same composites with unmodified filler. The modified filler has more strong contact with polymer matrix (thanks to silane modifier) than the unmodified diatomite. Therefore, mechanical stresses formed in composites by stretching or compressing forces absorb effectively by relatively soft silane phases that is the development of micro defects in carbon chain polymer matrix of composite districts and finishes in silane part of material the rigidity of which decreases.

The structural peculiarities of composites display also in thermomechanical properties of the materials. It is clear that softening of composites with modified by TEOS composites begins at relatively high temperatures. This phenomenon is in good correlation with corresponding composite mechanical strength. Of course, the modified filler has more strong interactions (thanks to modifier) with epoxy polymer molecules, than unmodified filler.

The amplified competition of the filler particles with macromolecules by TEOS displays well also on the characteristics of water absorption. In general loosening of microstructure because of micro empty areas is due to the increasing of filler content. Formation of such defects in the microstructure of composite promotes the water absorption processes. Water absorption of composites with modified diatomite is lower than that for one with unmodified filler to some extent. The decreasing of water absorption of composites containing silane compound is result of hydrophobic properties of ones.

Composites with binary fillers possess so called synergistic effect non-additive increasing of technical characteristics of composites at containing of fillers with definite ratio of them, which is due to creation of the dens distribution of ingredients in composites.

15.4 CONCLUSION

Comparison of the density, ultimate strength, softening temperature, and water absorption for polymer composites based on epoxy resin and unmodified and modified by tetraethoxysilane mineral fillers diatomite and andesite leads to conclusion that modify agent stipulates the formation of heterogeneous structures with higher compatibility of ingredients and consequently to enhancing of noted technical characteristics.

KEYWORDS

- **Diatomite**
- **Epoxy resin**
- **Homopolymer**
- **Silanization**
- **Tetraethyl orthosilicate**

REFERENCES

1. Katz, H. S. and Milevski, J. V. *Handbook of Fillers for Plastics*. RAPRA, (1987).
2. Mareri, P., Bastrole, S., Broda, N., and Crespi, A. *Composites Science and Technology*, **58**(5), 747–755 (1998).
3. Tolonen, H. and Sjolind, S. *Mechanics of composite materials*, **31**(4), 317–322 (1996).
4. Rothon, S. *Particulate filled polymer composites*. RAPRA, New York p. 205 (2003).
5. Lou, J. and Harinath, V. *Journal of Materials Processing Technology*, **152**(2), 185–193 (2004).
6. Khananashvili, L. M., Mukbaniani, O. V., and Zaikov, G. E. *New Concepts in Polymer Science, Elementorganic Monomers Technology, Properties, Applications*. Printed in Netherlands, Monograph, VSP, Utrecht (2006).
7. Aneli, J. N., Khananashvili, L. M., and Zaikov, G. E. *Structuring and conductivity of polymer composites*. Nova Sci. Publ., New York p. 326 (1998).
8. Zelenev, Y. V. and Bartenev, G. M. *Physics of Polymers*. M. Visshaya Shkola, p. 432 (1978) (in Russian).

16 New Trends in Polymers

A. P. Bonartsev, G. A. Bonartseva,
A. L. Iordanskii, G. E. Zaikov, and M. I. Artsis

CONTENTS

16.1 INTRODUCTION

Over the last decade an intense development of biomedical application of microbial poly((R)-3-hydroxybutyrate) (PHB) in producing of biodegradable polymer implants and controlled drug release systems need for comprehensive understanding of the PHB biodegradation process [1-3]. Examination of PHB degradation process is also necessary for development of novel friendly environment polymer packaging [4-6]. It is generally accepted that biodegradation of PHB both in living systems and in environment occurs *via* enzymatic and non-enzymatic processes that take place simultaneously under natural conditions. It is, therefore, important to understand both processes [1, 7]. Opposite to poly (glycolide) (PGA) and poly(lactide-co-glycolide) (PLGA), PHB is considered to be moderately resistant to degradation *in vitro* as well as to biodegradation in animal body. The rates of degradation are influenced by the characteristics of the polymer, such as chemical composition, crystallinity, morphology, and molecular weight [8, 9]. In spite of that PHB application *in vitro* and *in vivo* has been intensively investigated, the most of the available data are often incomplete and even contradictory. The presence of conflicting data can be partially explained by the fact that biotechnologically produced PHB with standardized properties is relatively rare and is not readily available due to a wide variety of PHB biosynthesis sources and different PHB manufacturing processes.

Contradictoriness can be explained also by excess applied trend in PHB degradation research. At most of the chapters observed in this review, PHB degradation process has been investigated in the narrow framework of development of specific medical device on the base of PHB. Depending on applied biomedical purposes biodegradation of PHB was investigated under different geometry: films and plates with various thickness [10-13], cylinders [13-16], monofilament threads [17, 18], and microspheres [19]. At these experiments PHB was used from various sources, with different molecular weight and crystallinity. Besides, different technologies of PHB devices manufacturing affect such important characteristics as polymer porosity and surface structure [11, 12]. The reports regarding the complex theoretical research of mechanisms of hydrolysis, enzymatic degradation, and biodegradation *in vivo* of PHB processes are relatively rare [10, 11, 13, 20-22] that attaches great value and importance to these investigations. Nevertheless, the effect of thickness, size, and geometry of PHB device, molecular weight, and crystallinity of PHB on the mechanism of PHB hydrolysis and biodegradation were not yet well clarified.

16.2 PHB HYDROLYSIS AND BIODEGRADATION

16.2.1 Nonenzymatic Hydrolysis of PHB *in vitro*

Examination of hydrolytic degradation of natural PHB *in vitro* is a very important step for understanding of PHB biodegradation. There are several very profound and careful examinations of PHB hydrolysis that were carried out 10–15 years [20-23]. Hydrolytic degradation of PHB was usually examined under standard experimental conditions simulating internal body fluid: in buffered solutions with pH = 7.4 at 37°C but at some cases the higher temperature (55°C, 70°C, and more) and other values of pH (from 2 to 11) were selected.

The classical experiment for examination of PHB hydrolysis in comparison with hydrolysis of other widespread biopolymer, polylactic acid (PLA), was carried out by Koyama N. and Doi Y. [20]. They selected films (10 × 10 mm size, 50 μm thickness, 5 mg initial mass) from PHB (M_n = 300 kDa, M_w = 650 kDa) and PLA (M_n = 9 kDa, M_w = 21 kDa) prepared by solvent casting and aged for 3 weeks to reach equilibrium crystallinity. It was shown that hydrolytic degradation of natural PHB is very slow process. The mass of PHB film remained unchanged at 37°C in 10 mm phosphate buffer (pH = 7.4) over a period of 150 days, while the mass of the PLA film rapidly decreased with time and reached 17% of the initial mass after 140 days. The rate of decrease in the M_n of the PHB was also much slower than the rate of decrease in the M_n of PLA. The M_n of the PHB decreased approximately to 65% of the initial PHB M_n after 150 days, while the M_n of the PLA decreased to 20% (2 kDa) of the initial PLA M_n at the end of same time point. As PLA used at this research was with low molecular weight we should compare this data with the data of investigation of hydrolysis of PLA with the same molecular weight as observed PHB. We examined rates of *in vitro* mass loss of polymer films with the same thickness (40 μm) from PLA and PHB with the same molecular weight (M_w = 450 kDa). It was shown that the mass of PLA film decreased to 87%, whereas the mass of PHB film remained unchanged at 37°C in 25 mm phosphate buffer (pH = 7.4) over a period of 84 days [24, 25].

The cleavage of polyester chains is known to be catalyzed by the carboxyl end groups, and its rate is proportional to the concentrations of water and ester bonds that may be constant during the hydrolysis, owing to the presence of a large excess of water molecules and of ester bonds of polymer chains. Thus, the kinetics of non-enzymatic hydrolysis can be expressed by the following equation [26, 27]:

$$\ln M_n = \ln M_n^{\circ} - kt \qquad (1)$$

where M_n and M_n° are the number average molecular weights of a polymer component at time t and zero, respectively.

The average number of bond cleavage per original polymer molecule, N, is given by Equation 2:

$$N = (M_n^{\circ}/M_n) - 1 = k_d P_n^{\circ} t, \qquad (2)$$

where k_d is the rate constant of hydrolytic degradation, and P_n° is the number average degree of polymerization at time zero. Thus, if the chain scission is completely random, the value of N is linearly dependent on time.

The molecular weight decrease with time is the distinguishing feature of mechanism in non-enzymatic hydrolysis condition in contrast to enzymatic hydrolysis condition of PHB when M_n values remained almost unchanged. It was supposed also that water-soluble oligomers of PHB may accelerate chain scissions of PHB homopolymer [20]. In contrast, Freier T. et al. [11] showed that PHB hydrolysis were not accelerated by the addition of predegraded PHB: the rate of mass and M_w loss of blends (70/30) from high molecular PHB (M_w = 641 kDa) and low molecular PHB (M_w = 3 kDa) were the same with degradation rate of pure high molecular PHB. Meanwhile, the addition of amorphous atactic PHB (at PHB) (Mw = 10 kDa) to blend with high molecular PHB caused significant acceleration of PHB hydrolysis: the mass loss of PHB/at PHB blends was 7% in comparison with 0% mass loss of pure PHB, the decrease of M_w was 88% in comparison with 48% M_w decrease of pure PHB [11, 28]. We have shown that the rate of hydrolysis of PHB films depends on M_w of PHB. The films from PHB of high molecular weight (450 and 1000 kDa) degraded slowly as it was described whereas films from PHB of low molecular weight (150 and 300 kDa) lost weight relatively gradually and more rapidly [24, 25].

To enhance the hydrolysis of PHB a higher temperature was selected for degradation experiments: 55°C, 70°C, and more [20]. It was showed by the same research team that the weight of films (12 mm diameter, 65 μm thick) from PHB (M_n = 768 and 22 kDa, M_w = 1460 and 75 kDa) were unchanged at 55°C in 10 mm phosphate buffer (pH = 7.4) over a period of 58 days. The M_n value decreased from 768 to 245 kDa for 48 days. The film thickness increased from 65 to 75 μm for 48 days, suggesting that water permeated the polymer matrix during the hydrolytic degradation. Examination of the surface and cross section of PHB films before and after hydrolysis showed that surface after 48 days of hydrolysis was apparently unchanged, while the cross section of the film exhibited a more porous structure (pore size<0.5 μm). It was shown also that the rate of hydrolytic degradation is not dependent upon the crystallinity of PHB film. The observed data indicates that the non-enzymatic hydrolysis of PHB in the aqueous media proceeds *via* a random bulk hydrolysis of ester bonds in the polymer chain films and occurs throughout the whole film, since water permeates the polymer matrix [20, 21]. Moreover, as the molecular weight distribution was unimodal over the whole degradation time which, together with the observed first order kinetics, indicates a random chain scission both in the crystalline and the amorphous regions of PHB [11, 29]. For synthetic amorphous atactic PHB it was shown that the hydrolysis of PHB is the two step process. First, the random chain scission proceeds. The scission accompanies by a molecular weight decrease. Then, at a molecular weight of about 10,000, mass loss begins [23].

The analysis of literature data shows a great spread in values of rate of PHB hydrolytic degradation *in vitro*. It can be explained by different thickness of PHB films or geometry of PHB devices used for experiment as well as by different sources, purity degree and molecular weight of PHB (Table 1). At 37°C and pH = 7.4 the weight loss of PHB (unknown M_w) films (500 μm thick) was 3% after 40 days incubation [32], 0% after 52 weeks (364 days) and after 2 years (730 days) incubation (640 kDa PHB, 100 μm films) [11, 12], 0% after 150 days incubation (650 kDa PHB, 50 μm film) [20], 7.5% after 50 days incubation (279 kDa PHB, unknown thickness of films) [31],, 0% after 3 months (84 days) incubation (450 kDa PHB, 40 μm films), 12% after 3 months (84 days) incubation (150 kDa PHB, 40 μm films) [24, 25], 0% after 180 days incubation of monofilament threads (30 μm in diameter) from PHB (470 kDa) [17, 18]. The molecular weight of PHB dropped to 36% of the initial values after two years (730 days) of storage in buffer solution [12], to 87% of the initial values after 98 days [32], and 58% of the initial values after 84 days [24, 25] (Table 1).

At acidic or alkaline aqueous media PHB degrades more rapidly: 0% after 20 weeks (140 days) incubation in 0.01 NaOH (pH = 11) (200 kDa PHB, 100 μm films) with surface changing [33], 0% after 180 days incubation of PHB threads in phosphate buffer (pH = 5.2 and 5.9) [18], complete PHB films biodegradation after 19 days (pH = 13) and 28 days (pH = 10) [31]. It was demonstrated that after 20 weeks of exposure to NaOH solution, the surfaces of PHB samples became rougher, along with an increased density of whole formation on their surfaces. From these results, it can be surmised that the non-enzymatic degradation of PHAs progresses on their surfaces before noticeable weight loss occurs (Figure 1) [33].

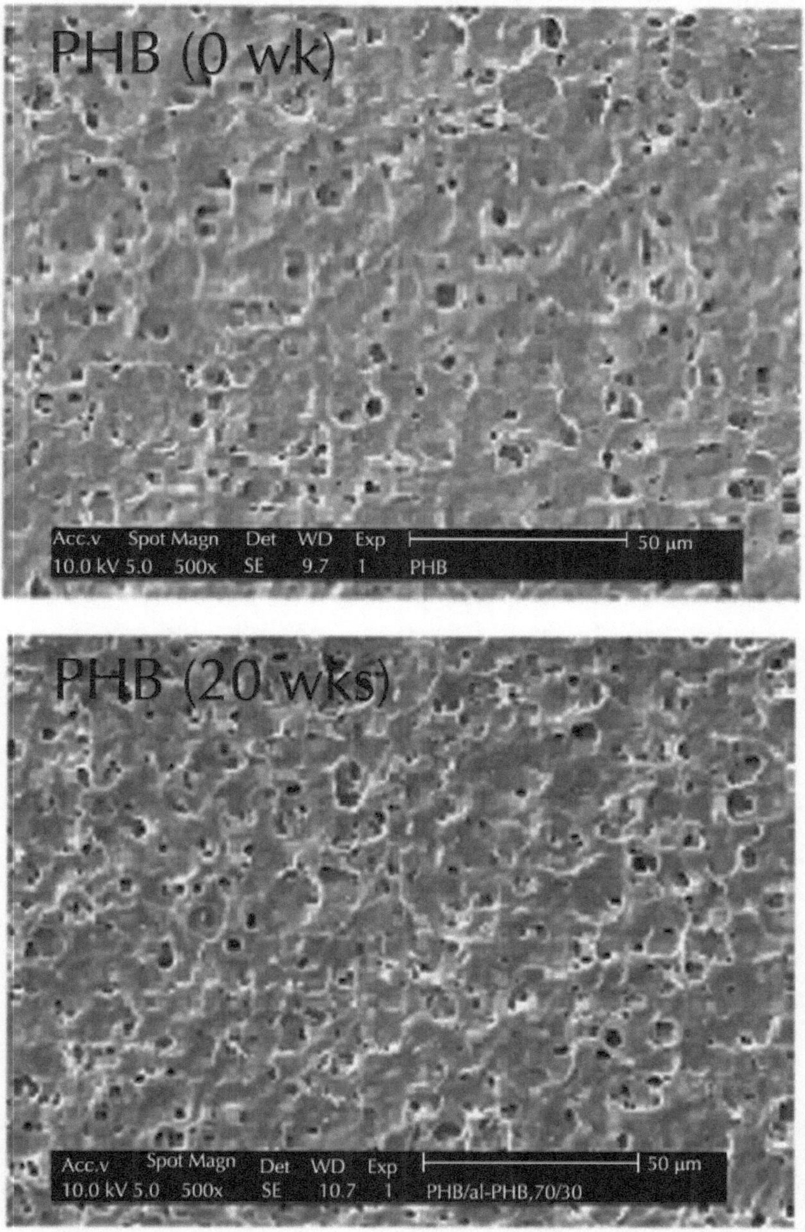

FIGURE 1 Scanning electron microscopy (SEM) photographs of PHB films both before (initial sample, panel on the left) and after 20 weeks (panel on the right) of non-enzymatic hydrolysis in 0.01 N NaOH solution (scale bars, 10 μm) [33].

It was shown also that the treatment of PHB film with 1M NaOH caused to reduce in pore size on film surface from 1–5 µm to around 1 µm that indicates a partially surface degradation of PHB in alkaline media [34, 35]. At higher temperature no weight loss of PHB films and threads was observed after 98 and 182 days incubation in phosphate buffer (pH = 7.2) at 55°C and 70°C, respectively [17], 12 and 39% of PHB (450 and 150 kDa, respectively) films after 84 days incubation at 70°C [35,40], 50 and 25% after 150 days incubation of microspheres (250–850 µm diameter) from PHB (50 kDa and 600 kDa, respectively) [36].

TABLE 1 Nonenzymatic hydrolysis of PHB *in vitro.*

Type of device	Initial M_w of PHB, kDa	Size/ Thickness, µm	Conditions	Relative mass loss of PHB, %	Relative decrease of PHB M_w, %	Time, days	Links
film	650	50	37°C, pH=7.4	0	35	150	20
film	640	100	37°C, pH=7.4	0	64	730	12
film	640	100	37°C, pH=7.4	0	45	364	11
film	450	40	37°C, pH=7.4	0	42	84	24–25
film	150	40	37°C, pH=7.4	12	63	84	24–25
film	279	–	37°C, pH=7.4	7.5	–	50	31
plate	–	500	37°C, pH=7.4	3	–	40	30
plate	380	1000	37°C, pH=7.4	0	–	28	37
plate	380	2000	37°C, pH=7.4	0	8	98	32
thread	470	30	37°C, pH=7.0	0	–	180	18
thread	–	–	37°C, pH=7.2	0	–	182	17
micro-spheres	50	250–850	37°C, pH=7.4	0	0	150	36
thread	470	30	37°C, pH=5.2	0	–	180	17

TABLE 1 *(Continued)*

Type of device	Initial M_w of PHB, kDa	Size/ Thickness, µm	Conditions	Relative mass loss of PHB, %	Relative decrease of PHB M_w, %	Time, days	Links
film	279	–	37°C, pH=10	100	–	28	31
film	279	–	37°C, pH=13	100	–	19	31
film	650	50	55°C, pH=7.4	0	68	150	20
plate	380	2000	55°C, pH=7.4	0	61	98	32
film	640	100	70°C, pH=7.4	–	55	28	11
film	150	40	70°C, pH=7.4	39	96	84	24–25
film	450	40	70°C, pH=7.4	12	92	84	24–25
micro-spheres	50	250–850	85°C, pH = 7.4	50	68	150	36
micro-spheres	600	250–850	85°C, pH=7.4	25	–	150	36

During degradation of PHB monofilament threads, films and plates *in vitro,* the change of mechanical properties was observed under different conditions [17, 37]. It was shown that a number of mechanical indices of threads became worse: load at break lost 36%, strain at break lost 33%, Young's modulus did not change, tensile strength lost 42% after 182 days incubation in phosphate buffer (pH = 7.2) at 70°C. But at 37°C the changes were more complicated: at first load at break increased from 440 g to 510 g (16%) at 90th day and then decreased to the initial value at 182nd day, strain at break increased rapidly from 60 to 70% (in 17%) at 20th day and then gradually increased to 75% (in 25%) at 182nd day, Young's modulus did not change [17]. For PHB films demonstrated a gradual 32% decrease in Young's modulus and 77% fall in tensile strength during 120 days incubation in phosphate buffer (pH = 7.4) at 37°C [37]. For PHB plates more complicated changes were observed: at first tensile strength dropped in 13% for 1st day and then increased to the initial value at 28th day, Young's modulus dropped in 32% for 1st day and then remain unchanged up to 28th day, stiffness decreased sharply also in 40% for 1st day and then remain unchanged up to 28th day [38].

16.2.2 Enzymatic Degradation of PHB *in vitro*

The examination of enzymatic degradation of PHB *in vitro* is having the following important step for understanding of PHB operation in animal tissues and in environment. The most studies observed degradation of PHB by depolymerases of its own bacterial producers. The degradation of PHB *in vitro* by depolymerase was thoroughly examined and mechanism of enzymatic PHB degradation was perfectly clarified by Doi Y. [20, 21]. At these early works it was shown that 68–85% and 58% mass loss of PHB (Mw = 650–768 and 22 kDa, respectively) films (50–65 μm thick) occurred for 20 hr under incubation at 37°C in phosphate solution (pH = 7.4) with depolymerase (1.5–3 μg/ml) isolated from *A. faecalis*. The rate (k_e) of enzymatic degradation of films from PHB (M_n = 768 and 22 kDa) was 0.17 and 0.15 mg/h, respectively. The thickness of polymer films dropped from 65 to 22 μm (32% of initial thickness) during incubation. The SEM examination showed that the surface of the PHB film after enzymatic degradation was apparently blemished by the action of PHB depolymerase, while no change was observed inside the film. Moreover, the molecular weight of PHB remains almost unchanged after enzymatic hydrolysis: the M_n of PHB decreased from 768 to 669 kDa or unchanged (22 kDa) [20, 21].

The extensive literature data on enzymatic degradation of PHB by specific PHB depolymerases were collected in detail in review of Sudesh K., Abe H., and Doi Y. [39]. We would like to summarize some of the most important data. But at first it is necessary to note that PHB depolymerase is very specific enzyme and the hydrolysis of polymer by depolymerase is a unique process. But in animal tissues and even in environment the enzymatic degradation of PHB is occurred mainly by nonspecific esterases [19, 40]. Thus, in the frameworks of this review, it is necessary to observe the fundamental mechanisms of PHB enzymatic degradation.

The rate of enzymatic erosion of PHB by depolymerase is strongly dependent on the concentration of the enzyme. The enzymatic degradation of solid PHB polymer is heterogeneous reaction involving two steps, namely, adsorption and hydrolysis. The first step is adsorption of the enzyme onto the surface of the PHB material by the binding domain of the PHB depolymerase, and the second step is hydrolysis of polyester chains by the active site of the enzyme. The rate of enzymatic erosion for chemosynthetic PHB samples containing both monomeric units of (R)- and (S)-3-hydrohybutyrate is strongly dependent on both the stereo composition and on the tacticity of the sample as well as on substrate specificity of PHB depolymerase. The water soluble product of random hydrolysis of PHB by enzyme shows a mixture of monomers and oligomers of (R)-3-hydrohybutirate. The rate of enzymatic hydrolysis for melt-crystallized PHB films by PHB depolymerase decreased with an increase in the crystallinity of the PHB film, while the rate of enzymatic degradation for PHB chains in an amorphous state was approximately twenty times higher than the rate for PHB chains in a crystalline state. It was suggested that the PHB depolymerase predominantly hydrolyzes polymer chains in the amorphous phase and then, subsequently, erodes the crystalline phase. The surface of the PHB film after enzymatic degradation was apparently blemished by the action of PHB depolymerase, while no change was observed inside the film. Thus, depolymerase hydrolyses of the polyester chains in the surface layer of the film and polymer erosion proceeds in surface layers, while dissolution, the enzymatic degradation of PHB are affected by many factors as monomer composition, molecular weight and degree of crystallinity [39].

At the next step it is necessary to observe enzymatic degradation of PHB under the conditions that modeled the animal tissues and body fluids containing nonspecific esterases. *In vitro* degradation of PHB films in the presence of various lipases as non-specific esterases was carried out in buffered solutions containing lipases [41, 42], in digestive juices (for example, pancreatin) [11], biological media (serum, blood etc.) [18] and crude tissue extracts containing a mixture of enzymes [19] to examine the mechanism of nonspecific enzymatic degradation process. It was noted that a Ser. His. Asp triad constitutes the active center of the catalytic domain of both PHB depolymer-ase [43] and lipases [44]. The serine is part of the pentapeptide Gly X1-Ser-X2-Gly, which has been located in all known PHB depolymerases as well as in lipases, ester-ases and serine proteases [43].

On the one hand, it was shown that PHB was not degraded for 100 days with a quantity of lipases isolated from different bacteria and fungi [41, 42]. On the other hand, the progressive PHB degradation by lipases was shown [24, 25, 34, 35]. The PHB enzymatic biodegradation was studied also in biological media: it was shown that with pancreatin addition no additional mass loss of PHB was observed in compar-ison with simple hydrolysis [11], the PHB degradation process in serum and blood was demonstrated to be similar to hydrolysis process in buffered solution [24, 25], whereas progressive mass loss of PHB sutures was observed in serum and blood: 16 and 25%, respectively, after 180 days incubation [18], crude extracts from liver, muscle, kidney, heart, and brain showed the activity to degrade the PHB: from 2 to 18% mass loss of PHB microspheres after 17 hr incubation at pH 7.5 and 9.5 [19]. The degradation rate in solution with pancreatin addition, obtained from the decrease in M_w of pure PHB, was accelerated about threefold: 34% decrease in M_w after incubation for 84 days in pancreatin (10 mg/ml in Sorensen buffer) *versus*. 11% decrease in Mw after incubation in phosphate buffer [11].The same data was obtained for PHB biodegradation in buff-ered solutions with porcine lipase addition: 72% decrease in M_w of PHB (450 kDa) after incubation for 84 days with lipase (20 U/mg, 10 mg/ml in Tris buffer) *versus*. 39% decrease in Mw after incubation in phosphate buffer [24, 25]. This observation is in contrast with enzymatic degradation by PHB depolymerases which was reported to proceed on the surface of the polymer film with an almost unchanged molecular weight [20, 21]. It has been proposed that for depolymerases the relative size of the en-zyme compared with the void space in solvent cast films is the limiting factor for diffu-sion into the polymer matrix [Jesudason J. J. et al., 1993] whereas lipases can penetrate into the polymer matrix through pores in PHB film [34, 35]. It was shown that lipase (0.1 g/l in buffer) treatment for 24 hr caused significant morphological change in PHB film surface: transferring from native PHB film with many pores ranging from 1 to 5 μm in size into a pore free surface without producing a quantity of hydroxyl groups on the film surface. It was supposed that the pores had a fairly large surface exposed to lipase, thus it was degraded more easily (Figure 2) [34, 35]. It indicates also that lipase can partially penetrate into pores of PHB film but the enzymatic degradation proceeds mainly on the surface of the coarse polymer film achievable for lipase. Two additional effects reported for depolymerases could be of importance. It was concluded that segmental mobility in amorphous phase and polymer hydrophobicity play an important role in enzymatic PHB degradation by nonspecific esterases [11]. Significant impairment of the tensile strength and other mechanical properties were observed during enzymatic biodegradation of PHB threads in serum and blood. It was shown that load at break lost 29%, Young's modulus lost 20%, and tensile strength did not change after 180 days of threads incubation, the mechanical properties changed gradually [18].

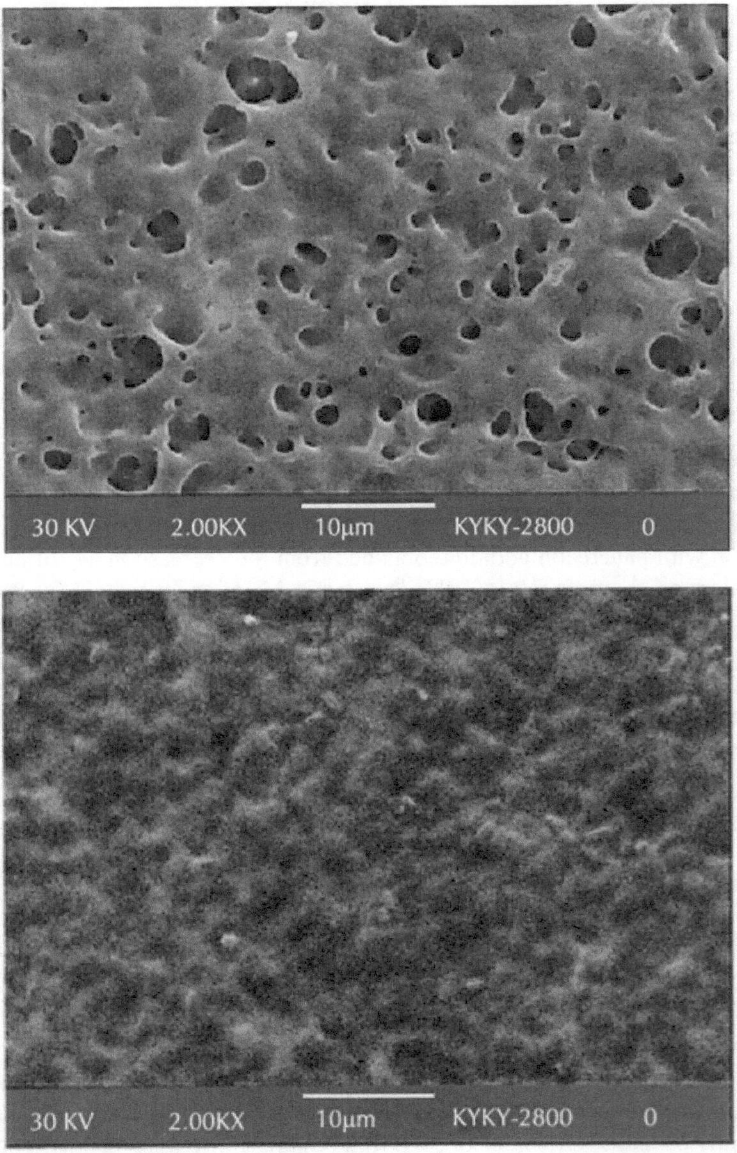

FIGURE 2 The SEM photographs documented the surface structure of PHB polymer films: (a) PHB film; (b) PHB film treated with lipase (0.1 g/l at 30°C and pH = 7.0 for 24 hr) [35].

16.2.3 Biodegradation of PHB by Soil Microorganisms

Polymers exposed to the environment are degraded through their hydrolysis, mechanical, thermal, oxidative, and photochemical destruction, and biodegradation [4, 32, 45, 46]. One of the valuable properties of PHB is its biodegradability, which can be evaluated using various field and laboratory tests. Requirements for the biodegradability of PHB may vary in accordance with its applications. The most attractive property of PHB with respect to ecology is that it can be completely degraded by microorganisms finally to CO_2 and H_2O. This property of PHB allows to manufacture biodegradable polymer objects for various applications (Figure 3) [2].

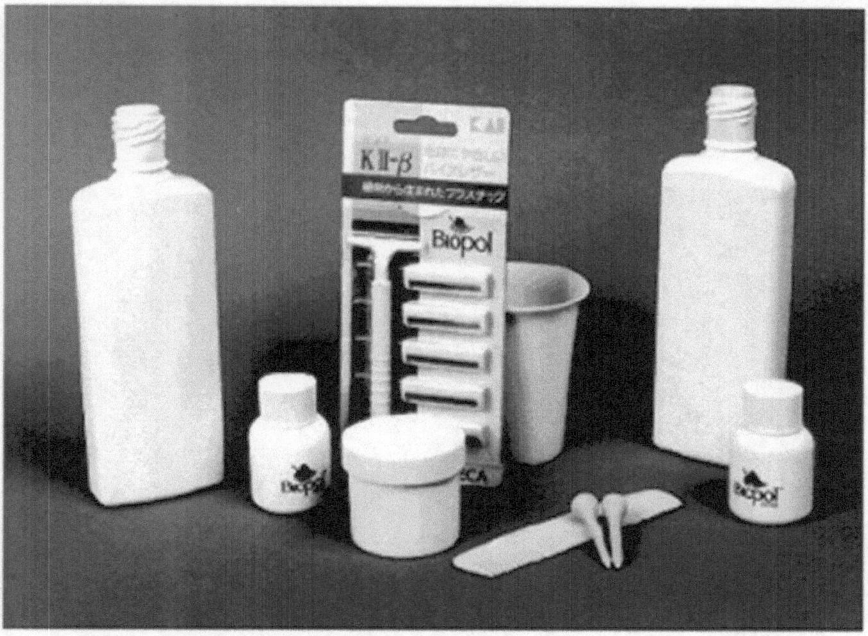

FIGURE 3 Moulded PHB objects for various applications. In soil burial or composting experiments such objects biodegrade in about three months [2].

The degradation of PHB and its composites in natural ecosystems, such as soil, compost, and bodies of water, was described [2, 32, 45, 46]. Maergaert et al. isolated from soil more than 300 microbial strains capable of degrading PHB *in vitro* [45]. The bacteria detected on the degraded PHB films were dominated by the genera *Pseudomonas*, *Bacillus, Azospirillum, Mycobacterium,* and *Streptomyces* and so on. The samples of PHB have been tested for fungicidity and resistance to fungi by estimating the growth rate of test fungi from the genera *Aspergillus, Aureobasidium, Chaetomium, Paecilomyces, Penicillum,* and *Trichoderma* under optimal growth conditions. The PHB film did not exhibit neither fungicide properties, nor the resistance to fungal damage, and served as a good substrate for fungal growth [47].

It was studied that biodegradability of PHB films under aerobic, microaerobic, and anaerobic condition in the presence and absence of nitrate by microbial populations of soil, sludge from anaerobic and nitrifying/denitrifying reactors, and sediment of a sludge deposit site, as well as to obtain active denitrifying enrichment culture degrading PHB (Figure 4) [48]. Changes in molecular mass, crystallinity, and mechanical properties of PHB have been studied. A correlation between the PHB degradation degree and the molecular weight of degraded PHB was demonstrated. The most degraded PHB exhibited the highest values of the crystallinity index. As it has been shown by Spyros et al., The PHAs contain amorphous and crystalline regions, of which the former are much more susceptible to microbial attack [49]. If so, the microbial degradation of PHB must be associated with a decrease in its molecular weight and an increase in its crystallinity, which was really observed in the experiments. Moreover, microbial degradation of the amorphous regions of PHB films made them more rigid. However, further degradation of the amorphous regions made the structure of the polymer much looser [48].

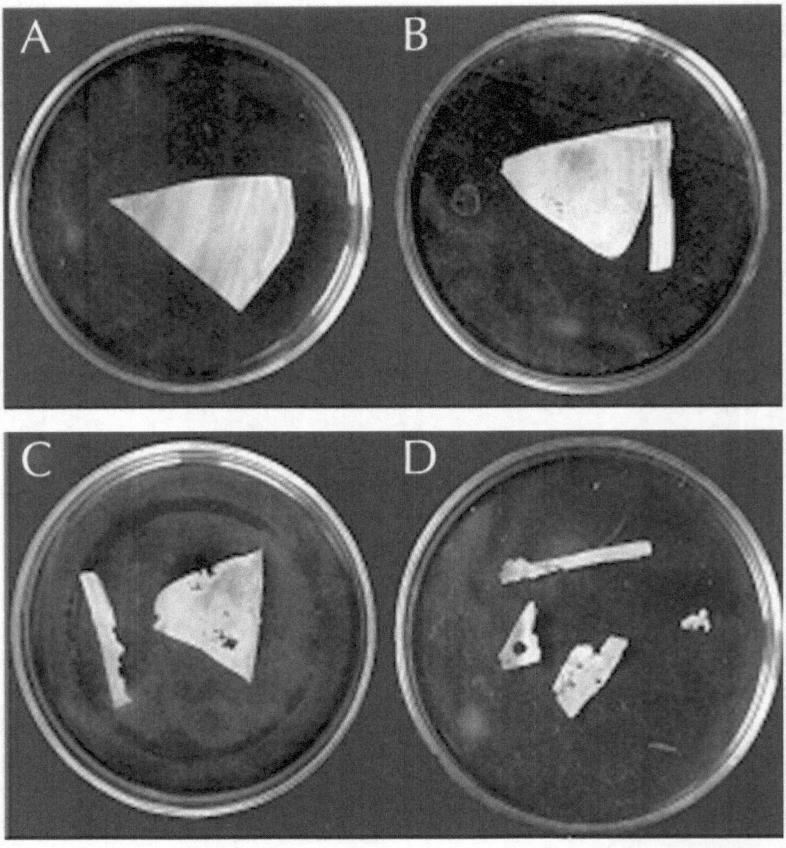

FIGURE 4 Undegraded PHB film (A) and PHB films with different degrees of degradation after 2 months incubation in soil suspension: anaerobic conditions without nitrate (B), microaerobic conditions without nitrate (C), and microaerobic conditions with nitrate (D) [24, 48].

The PHB biodegradation in the enriched culture obtained from soil on the medium used to cultivate denitrifying bacteria (Gil'tai medium) has been also studied. The dominant bacterial species, *Pseudomonas fluorescens* and *Pseudomonas stutzeri*, have been identified in this enrichment culture. Under denitrifying conditions, PHB films were completely degraded for seven days. Both the film weight and M_w of PHB decreased with time. In contrast to the data of Doi et al. [21] who found that M_w of PHB remained unchanged upon enzymatic biodegradation in an aquatic solution of PHB depolymerase from *Alcaligenes faecalis,* in our experiments, the average viscosity molecular weight of the higher and lower molecular polymers decreased gradually from 1540 to 580 kDa and from 890 to 612 kDa, respectively. The "exo"-type cleavage of the polymer chain, that is a successive removal of the terminal groups, is known to occur at a higher rate than the "endo"-type cleavage, that is, a random breakage of the polymer chain at the enzyme binding sites. Thus, the former type of polymer degradation is primarily responsible for changes in its average molecular weight. However the "endo"-type attack plays the important role at the initiation of biodegradation, because at the beginning, a few polymer chains are oriented so that their ends are accessible to the effect of the enzyme [50]. Biodegradation of the lower molecular polymer, which contains a higher number of terminal groups, is more active, probably, because the "exo"-type degradation is more active in lower than in higher molecular polymer [48, 51].

16.2.4 Biodegradation of PHB *in vivo* in Animal Tissues

The first scientific works on biodegradation of PHB *in vivo* in animal tissues were carried out 15–20 years ago by Miller N. D. et al. and Saito T. et al. [17, 19]. They are high qualitative researches that disclosed many important characteristics of this process. As it was noted that the both enzymatic and non-enzymatic processes of biodegradation of PHB *in vivo* can occur simultaneously under normal conditions. But it does not mean that polymer biodegradation *in vivo* is a simple combination of non-enzymatic hydrolysis and enzymatic degradation. Moreover, *in vivo* the biodegradation (decrease of molecular weight and mass loss) of PHB is a controversial subject in the literature. As it was noted for *in vitro* PHB hydrolysis, the main reason for the controversy is the use of samples made by various processing technologies and the incomparability of different implantation and animal models. The most of researches on PHB biodegradation was carried out with use of prototypes of various medical devices on the base of PHB: solid films and plates [10, 13, 24, 52], porous patches [11, 12], nonwoven patches consisted of fibers [53-57], screws [24], cylinders as nerve guidance channels and conduits [13, 15, 16], monofilament sutures [17, 18], microspheres [19, 58]. *In vivo* biodegradation researches were carried out on various laboratory animals: rats [11, 16-19, 24], mice [13, 58], rabbits [10, 52, 59], minipigs [12], cats [15], calves [53], sheep [54-56], and patients [57]. It is obvious that these animals differ in level of metabolism are very much: for example, only weight of these animals differs from 10–20 g (mice) to 50 kg (calves). The implantation of devices from PHB was carried out through different ways: subcutaneously [10, 13, 17, 18, 24, 59], intraperitoneally on a bowel [11], subperiostally on the osseus skull [12, 52], nerve wraparound [14-16], intramuscularly [58, 59], into the pericardium [54-57], into the atrium [53], and intravenously [19]. The terms of implantation were also different: 2.5 hr, 24 hr, 13 days, and 2 months [19], 7, 14, and 30 days [16], 2, 7, 14, 21, 28, 55, 90, and 182 days [17], 1, 3, and 6 months [10, 13, 14], 3, 6, and 12 months [53], 6 and 12 months [15], 6 and 24 months [57], and 3, 6, 9, 12, 18, and 24 months [56] (Table 2).

In the entire study of PHB *in vivo* biodegradation was fulfilled by Gogolewski S. et al. and Qu X. H. et al. [10, 13]. It was shown that PHB lost about 1.6% (injection molded film, 1.2 mm thick, M_w of PHB = 130 kDa) [13] and 6% (solvent casting film, 40 μm thick, M_w = 534 kDa) [10] of initial weight after 6 months of implantation. But the observed small weight loss was partially due to the leaching out of low molecular weight fractions and impurities present initially in the implants. The M_w of PHB decreased from 130 kDa to 74 kDa (57% of initial M_w) [13] and from 534 kDa to 216 kDa (40% of initial Mw) [10] after 6 months of implantation. The polydispersity of PHB polymers narrowed following implantation. The PHB showed a constant increase in crystallinity (from 60.7 to 64.8%) up to 6 months [13] or an increase (from 65.0 to 67.9%) after 1 month and then fall again (to 64.5%) after 6 month of implantation [10] which suggests the degradation process had not affected the crystalline regions. This data is in accordance with data of PHB hydrolysis [20] and enzymatic PHB degradation by lipases *in vitro* [11] where M_w decrease was observed. The initial biodegradation of amorphous regions of PHB *in vivo* is similar to PHB degradation by depolymerase [39].

Thus, the observed biodegradation of PHB showed coexistence of two different degradation mechanisms in hydrolysis in the polymer: enzymatically or non-enzymatically catalyzed degradation. Although non-enzymatical catalysis occurred randomly in homopolymer, indicated by M_w loss rate in PHB, at some point in a time, a critical molecular weight is reached whereupon enzyme-catalyzed hydrolysis accelerated degradation at the surface because easier enzyme/polymer interaction becomes possible. However considering the low weight loss of PHB, the critical molecular weight appropriate for enzymes predominantly does not reach, yet resulting low molecular weight and crystallinity in PHB could provide some sites for the hydrolysis of enzymes to accelerate the degradation of PHB [10, 13]. Additional data revealing the mechanism of PHB biodegradation in animal tissues were obtained by Kramp B. et al. in long term implantation experiments. A very slow, clinically not recordable degradation of films and plates was observed during 20 month (much more than in experiments mentioned). A drop in the PHB weight loss evidently took place between the 20th and 25th month. Only initial signs of degradation were to be found on the surface of the implant until 20 months after implantation but no more test body could be detected after 25 months [52]. The complete biodegradation *in vivo* in the wide range from 3–30 months of PHB was shown by other researches [53, 55-57, 60], whereas almost no weight loss and surface changes of PHB during 6 months of biodegradation *in vivo* was shown [13, 17]. Residual fragments of PHB implants were found after 30 months of the patches implantation [54, 56]. A reduction of PHB patch size in 27% was shown in patients after 24 months after surgical procedure on pericardial closure with the patch [57]. Significantly more rapid biodegradation *in vivo* was shown by other researches [10, 15, 18, 37, 53]. It was shown that 30% mass loss of PHB sutures occurred gradually during 180 days of *in vivo* biodegradation with minor changes in the microstructure on the surface and in volume of sutures [18]. It was shown that PHB nonwoven patches (made to close atrial septal defect in calves) was slowly degraded by polynucleated macrophages, and 12 months postoperatively no PHB device was identifiable but only small particles of polymer were still seen. The absorption time of PHB patches was long enough to permit regeneration of a normal tissue [53]. The PHB sheets progressive biodegradation was demonstrated qualitatively at 2, 6. and 12 months after implantation as weakening of the implant surface, tearing/cracking of the implant, fragmentation and a decrease in the volume of polymer material [15, 37, 59]. The complete biodegradation of PHB (M_w = 150–1000 kDa) thin films (10–50 μm) for 3–6 months was shown and degradation process was described. The process of PHB biodegradation consists of several phases. At initial phase PHB films was covered by fibrous capsule. At second phase capsulated PHB films very slowly lost their weight with simultaneous increase of crystallinity and decrease of M_w and mechanical properties of PHB. At third phase PHB films were rapidly disintegrated and then completely degraded. At forth phase empty fibrous capsule resolved (Figure 5) [24, 25]. Interesting data were obtained for biodegradation *in vivo* of PHB microspheres (0.5–0.8 μm in diameter). It was demonstrated indirectly that PHB loss about 8% of weight of microspheres accumulated in liver after 2 month of intravenous injection. It was demonstrated also a presence of several types PHB degrading enzymes in the animal tissues extracts [19].

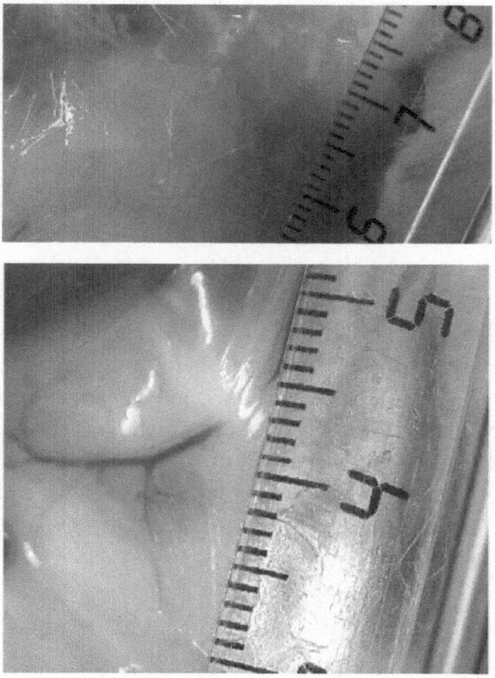

FIGURE 5 Biodegradation of PHB films *in vivo*. Connective tissue capsule with PHB thin films (outlined with broken line) 2 weeks (98% residual weight of the film) (left photograph) and 3 months (0% residual weight of the film) (left photograph) after subcutaneous implantation [24, 25].

Some of the researchers studied a biodegradation of PHB threads with a tendency of analysis of its mechanical properties *in vivo* [17, 18]. It was shown that at first load at break index decreased rapidly from 440 to 390 g (12%) at 15th day and then gradually increased to the initial value at 90th day and remain almost unchanged up to 182nd day [17] or gradually decreased in 27% during 180 days [18], strain at break decreased rapidly from 60 to 50% (in 17% of initial value) at 10th day and then gradually increased to 70% (in 17% of initial value) at 182nd day [17] or did not change significantly during 180 days [18].

It was demonstrated that the primary reason of PHB biodegradation *in vivo* was a lysosomal and phagocytic activity of polynucleated macrophages and giant cells of foreign body reaction. The activity of tissue macrophages and nonspecific enzymes of a body liquids made a main contribution to significantly more rapid rate of PHB biodegradation *in vivo* in comparison with rate of PHB hydrolysis *in vitro.* The PHB material was encapsulated by degrading macrophages. Presence of PHB stimulated uniform macrophage infiltration, which is important for not only the degradation process but also the restoration of functional tissue. The long absorption time produced a foreign body reaction, which was restricted to macrophages forming a peripolymer layer [18, 53, 56, 59]. Very important data that clarifies the tissue response that contributes to biodegradation of PHB was obtained by Lobler M. It was demonstrated a significant increase of expression of two specific lipases after 7 and 14 days of PHB contact with animal tissues. Moreover, liver specific genes were induced with similar results. It is striking that pancreatic enzymes are induced in the gastric wall after contact with biomaterials [40]. Saito T. et al. suggested the presence of at least two types of degradative enzymes in rat tissues: liver serine esterases with the maximum of activity in alkaline media (pH = 9.5) and kidney esterases with the maximum of activity in neutral media [19]. The mechanism of PHB biodegradation by macrophages was demonstrated at cultured macrophages incubated with particles of low-molecular weight PHB [61]. It was shown that macrophages and, to a lesser level, fibroblasts have the ability to take up (phagocytize) PHB particles (1–10 μm). At high concentrations of PHB particles (>10 μg/ml) the phagocytosis is accompanied by toxic effects and alteration of the functional status of the macrophages but not the fibroblasts. This process is accompanied by cell damage and cell death. The elevated production of nitric oxide (NO) and tumor necrosis factor alfa (TNF-α) by activated macrophages were observed. It was suggested that the cell damage and cell death may be occur due to phagocytosis of large amounts of PHB particles: after phagocytosis, polymer particles may fill up the cells, it cause cell damage, and cell death. It was demonstrated also that phagocytized PHB particles disappeared in time due to an active PHB biodegradation process (Figure 6) [61].

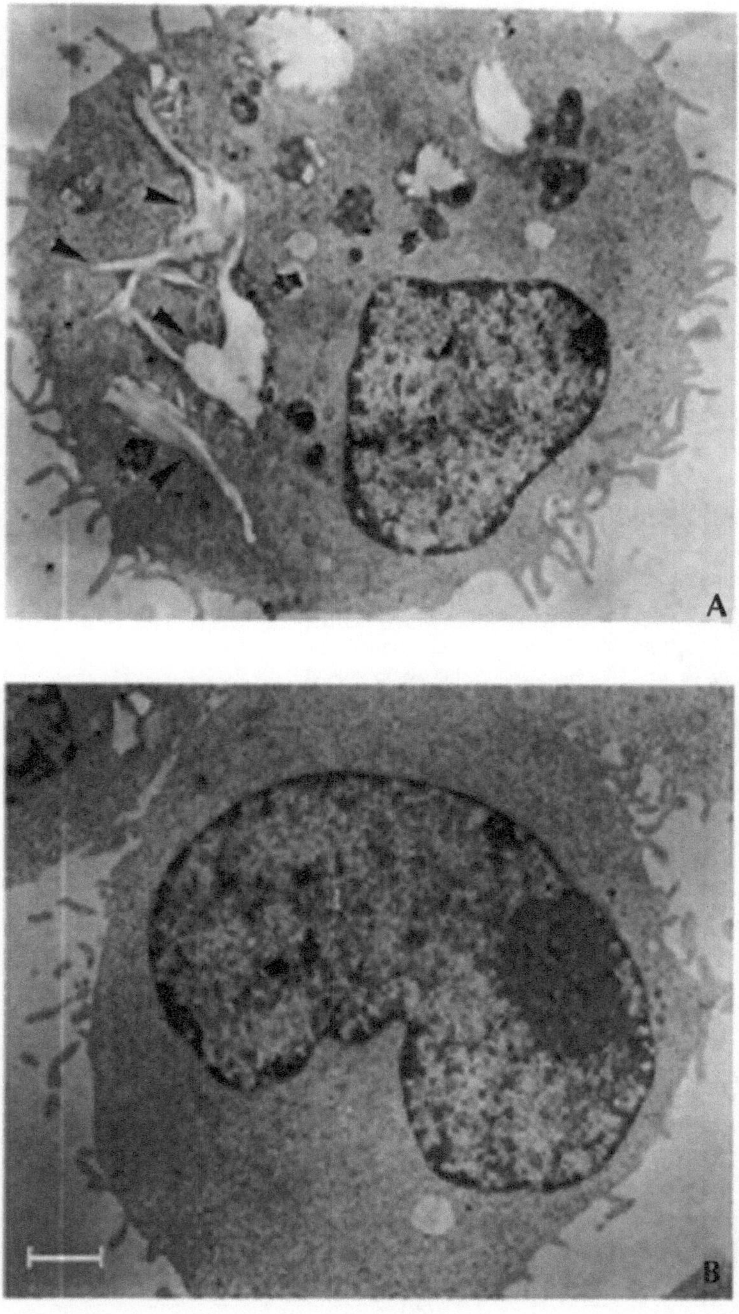

FIGURE 6 Phagocytosis of microparticles of PHB in macrophages. The TEM analysis of cultured macrophages in the presence (A) or absence (B) of 2 μg PHB microparticles/ml for 24 hr. Bar in B represents 1 μm, for A and B.

TABLE 2 Biodegradation of PHB *in vivo* (data for comparison).

Type of device	Thickness/ diameter, μm	Animal	Site of implantation/ surgical procedure	Relative mass loss of PHB, %	Relative loss of PHB molecular weight, %	Time, months	Links
film (injection molded)	1200	mouse	subcutaneously	1.6	43	6	13
film (solvent casting)	150–200	rabbit	subcutaneously (dorsal side)	6	60	6	10
film (solvent casting)	50	rat	subcutaneously (ventral side)	100	100	3	24,25
porous PHB/ atactic PHB patch	100	rat	intraperitoneally to repair a bowel defect	>90*	62	6.5	11
porous PHB/ atactic PHB patch	250	minipigs	with contact to bone and dura mater to cover rhinobasal skull defects	>50*	65	6.5	12
Films and plates	100–1000	rabbit	subperiostally on the osseus skull or respectively on cut trough zygomatic arches	100	–	25	52
Films and plates	100 and 500	rabbit	subperiostally on the osseus skull or respectively on cut trough zygomatic arches	<10*	–	20	52
Plates and screws	500 and 1500	rabbit	subperiostally on the osseus skull or respectively trough osseus skull	0	–	12	25
cylinder (nerve conduits)	150 (of wall)	rat	nerve wraparound to bridge an irreducible nerve gap	0	–	1	16
cylinder (nerve conduits)	150 (of wall)	cat	nerve wraparound to bridge an irreducible nerve gap	>25*	–	12	15
monofila-ment suture	–	rat	subcutaneously (dorsal side)	0	–	6	17
monofila-ment suture	30	rat	subcutaneously (fold of neck)	30	–	6	18

TABLE 2 *(Continued)*

Type of device	Thickness/ diameter, μm	Animal	Site of implantation/ surgical procedure	Relative mass loss of PHB, %	Relative loss of PHB molecular weight, %	Time, months	Links
thin films and ground particles	–	rabbit	subcutaneously and intramuscularly in the legs	>30*	–	2	59
nonwoven patch (consisted of fibers)	200–600 (of patch) 2–20 (of fibers)	sheep	on the wall of pericardium to close artificial defect and prevent pericardial adhesions	>90*	–	24	55
nonwoven patch (consisted of fibers)	200–600 (of patch) 2–20 (of fibers)	sheep	as transannular patches on the wall of right ventricular outflow tract and pulmonary artery	>99*	–	12	56
nonwoven patch (consisted of fibers)	200–600 (of patch) 2–20 (of fibers)	calve	on the septal of right atrium to close artificial septal defect	>99*	–	12	53
nonwoven patch (consisted of fibers)	200–600 (of patch) 2–20 (of fibers)	patient	on the wall of pericardium to close artificial defect and prevent pericardial adhesions	27	–	24	57
microspheres	0.5–0.8	rat	intravenously	8*	–	2	19
microspheres	100–300	mice	intramuscularly in the legs	0*	–	2	58
Rivet shaped plate	2300	rabbit	intraosseously, into the lateral condyle of femur	<10*	–	6	37

* Indirect data

16.3 APPLICATION OF PHB

16.3.1 Medical Devices on the Base of PHB and PHB *in vivo* Biocompatibility

The perspective area of PHB application is development of implanted medical devices for dental, craniomaxillofacial, orthopedic, cardiovascular, hernioplastic, and skin surgery. A number of potential medical devices on the bases of PHB: bioresorbable surgical sutures [17, 18, 62, 63], biodegradable screws and plates for cartilage and bone fixation [24, 52], biodegradable membranes for periodontal treatment, surgical meshes with PHB coating for hernioplastic surgery [24], wound coverings [64], patches for repair of a bowel, pericardial, and osseous defects [11, 12, 53-57], nerve guidance channels, conduits [15, 16] and so on was developed (Figure 7).

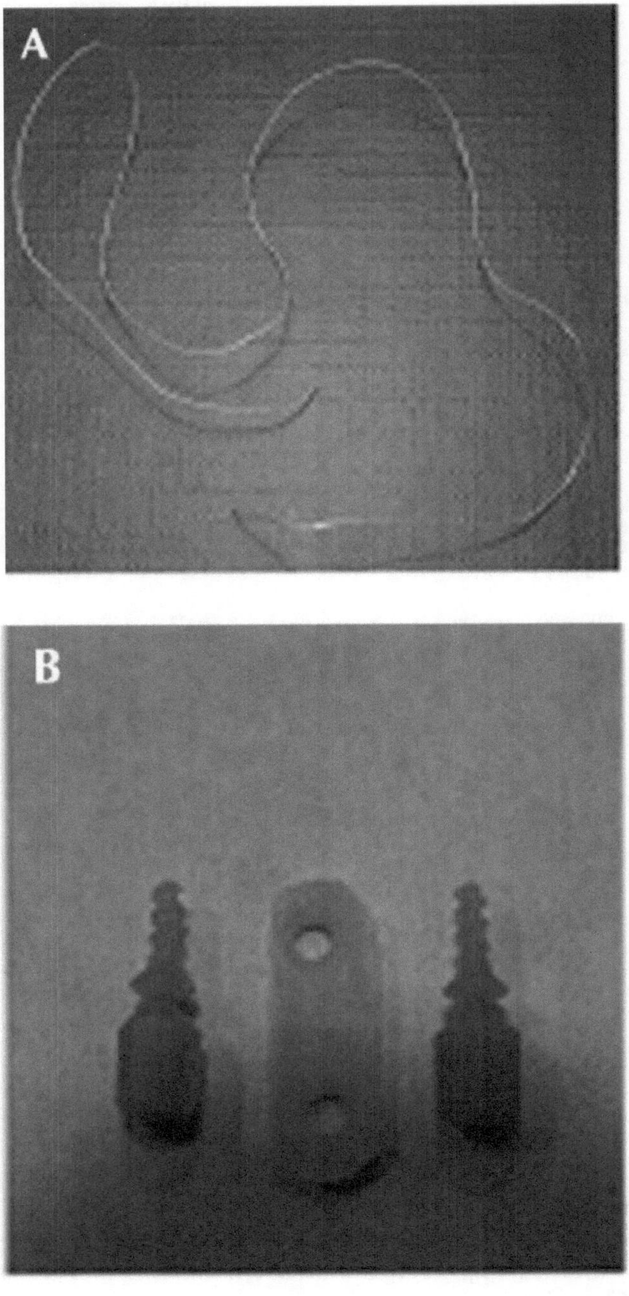

FIGURE 7 *(Continued)*

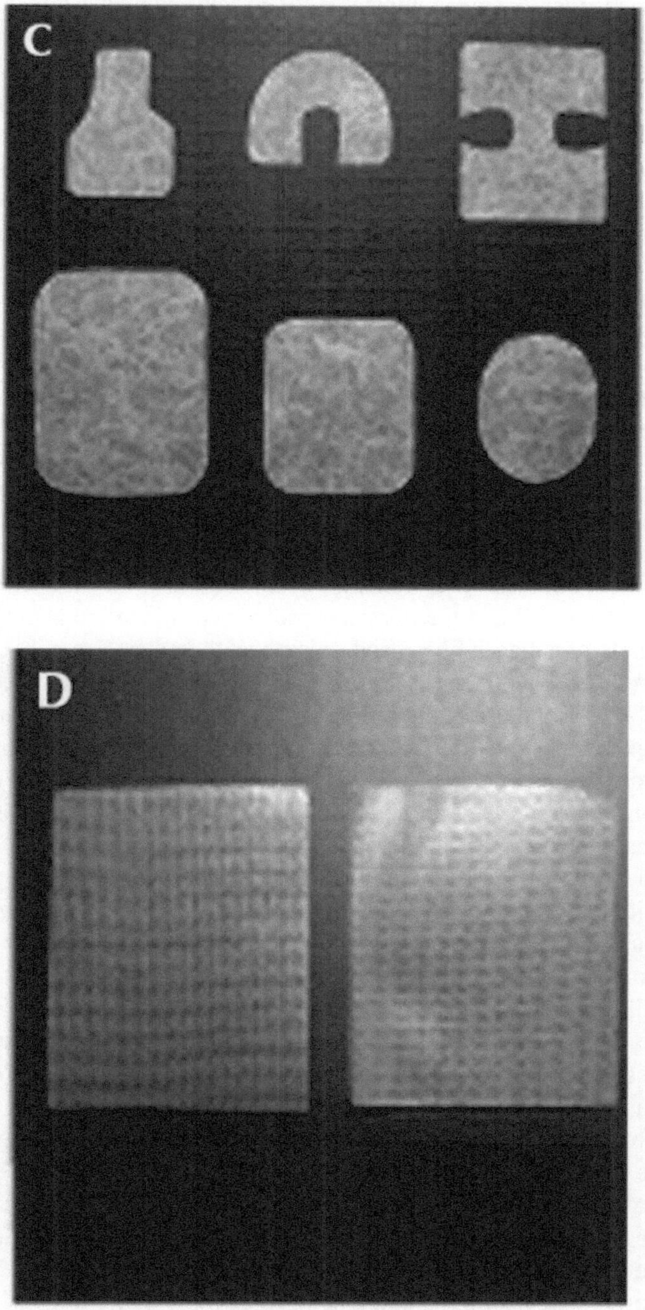

FIGURE 7 Medical devices on the base of PHB. (A) bioresorbable surgical suture, (B) biodegradable screws and plate for cartilage and bone fixation, (C) biodegradable membranes for periodontal treatment, and (D) surgical meshes with PHB coating for hernioplastic surgery, pure (left) and loaded with antiplatelet drug, dipyridamole (right) [24].

The tissue reaction *in vivo* to implanted PHB films and medical devices was studied. In most cases a good biocompatibility of PHB was demonstrated. In general, no acute inflammation, abscess formation, or tissue necrosis was observed in tissue surrounding of the implanted PHB materials. In addition, no tissue reactivity or cellular mobilization occurred in areas remote from the implantation site [10, 13, 24, 58]. On the one hand, it was shown that PHB elicited similar mild tissue response as PLA did [13], but on the other hand the use of implants consisting of polylactic acid, polyglicolic acid and their copolymers is not response without a number of sequelae related with the chronic inflammatory reactions in tissue [65-69].

Subcutaneous implantation of PHB films for 1 month has shown that the samples were surrounded by a well-developed, homogeneous fibrous capsule of 80–100 μm in thickness. The vascularized capsule consists primarily of connective tissue cells (mainly, round, immature fibroblasts) aligned parallel to the implant surface. A mild inflammatory reaction was manifested by the presence of mononuclear macrophages, foreign body cells, and lymphocytes. Three months after implantation, the fibrous capsule has thickened to 180–200 μm due to the increase in the amount of connective tissue cells and a few collagen fiber deposits. A substantial decrease in inflammatory cells was observed after 3 months, tissues at the interface of the polymer were densely organized to form bundles. After 6 months of implantation, the number of inflammatory cells had decreased and the fibrous capsule, now thinned to about 80–100 μm, consisted mainly of collagen fibers, and a significantly reduced amount of connective tissue cells. A little inflammatory cells effusion was observed in the tissue adherent to the implants after 3 and 6 months of implantation [10, 13]. The biocompatibility of PHB has been demonstrated *in vivo* under subcutaneous implantation of PHB films. Tissue reaction to films from PHB of different molecular weight (300, 450, 1000 kDa) implanted subcutaneously was relatively mild and did not change from tissue reaction to control glass plate [24].

At implantation of PHB with contact to bone the overall tissue response was favorable with a high rate of early healing and new bone formation with some indication of an osteogenic characteristic for PHB compared with other thermoplastics, such as polyethylene. Initially there was a mixture of soft tissue, containing active fibroblasts, and rather loosely woven osteonal bone seen within 100 μm of the interface. There was no evidence of a giant cell response within the soft tissue in the early stages of implantation. With time this tissue became more orientated in the direction parallel to the implant interface. The dependence of the bone growth on the polymer interface is demonstrated by the new bone growing away from the interface rather than towards it after implantation of 3 months. By 6 months postimplantation the implant is closely encased in new bone of normal appearance with no interposed fibrous tissue. Thus, PHB-based materials produce superior bone healing [37].

Regeneration of a neointima and a neomedia, comparable to native arterial tissue, was observed at 3–24 months after implantation of PHB nonwoven patches as transannular patches into the right ventricular outflow tract and pulmonary artery. In the control group, a neointimal layer was present but no neomedia comparable to native arterial tissue. Three layers were identified in the regenerated tissue: neointima with an endothelium-like lining, neomedia with smooth muscle cells, collagenous, and elastic tissue, and a layer with polynucleated macrophages surrounding istets of PHB, capillaries and collagen tissue. Lymphocytes were rare one. It were concluded that PHB nonwoven patches can be used as a scaffold for tissue regeneration in low pressure systems. The regenerated vessel had structural and biochemical qualities in common with the native pulmonary artery [56]. Biodegradable PHB patches implanted in atrial septal defects promoted formation of regenerated tissue that macroscopically and microscopically resembled native atrial septal wall. The regenerated tissue was found to be composed of three layers: monolayer with endothelium-like cells, a second layer with fibroblasts and some smooth muscle cells, collagenous tissue and capillaries, and a third layer with phagocytizing cells isolating and degrading PHB. The neointima contained a complete endothelium-like layer resembling the native endothelial cells. The patch material was encapsulated by degrading macrophages. There was a strict border between the collagenous and the phagocytizing layer. Presence of PHB seems to stimulate uniform macrophage infiltration, which was found to be important for the degradation process and the restoration of functional tissue. Lymphocytic infiltration as foreign body reaction, which is common after replacement of vessel wall with commercial woven Dacron patch, was wholly absent when PHB. It was suggested that the absorption time of PHB patches was long enough to permit regeneration of a tissue with sufficient strength to prevent development of shunts in the atrial septal position [53]. The prevention of postoperative pericardial adhesions by closure of the pericardium with absorbable PHB patch was demonstrated. The regeneration of mesothelial layer after implantation of PHB pericardial patch was observed. The complete regeneration of mesothelium, with morphology and biochemical activity similar to finding in native mesothelium, may explain the reduction of postoperative pericardial adhesions after operations with insertion of absorbable PHB patches [55]. The regeneration of normal filament structure of restored tissues was observed by immunohistochemical methods after PHB devices implantation [54]. The immunohistochemical demonstration of cytokeratine, an intermediate filament, which is constituent of epithelial and mesodermal cells, agreed with observations on intact mesothelium. Heparan sulfate proteoglycan, a marker of basement membrane, was also identified [54].

The PHB patches for the gastrointestinal tract were tested using animal model. The patches are made from PHB sutured and PHB membranes were implanted to close experimental defects of stomach and bowel wall. The complete regeneration of tissues of stomach and bowel wall was observed at 6 months after patch implantation without strong inflammatory response and fibrosis [11, 70].

Recently an application of biodegradable nerve guidance channels (conduits) for nerve repair procedures and nerve regeneration after spinal cord injury was demonstrated. Polymer tubular structures from PHB can be modulated for this purpose. Successful nerve regeneration through a guidance channel was observed as early as after 1 month. Virtually all implanted conduits contained regenerated tissue cables centrally located within the channel lumen and composed of numerous myelinated axons and Schwann cells. The inflammatory reaction had not interfered with the nerve regeneration process. The progressive angiogenesis was present at the nerve ends and through the walls of the conduit. The results demonstrate good quality nerve regeneration in PHB guidance channels [16, 71].

Biocompatibility of PHB was evaluated by implanting microspheres from PHB (M_w = 450 kDa) into the femoral muscle of rats. The spheres were surrounded by one or two layers of spindle cells, and infiltration of inflammatory cells and mononuclear cells into these layers was recognized at 1 week after implantation. After 4 weeks, the number of inflammatory cells had decreased and the layers of spindle cells had thickened. No inflammatory cells were seen after 8 weeks, and the spheres were encapsulated by spindle cells. The toxicity of PHB microspheres was evaluated by weight change and survival times in L1210 tumor-bearing mice. No differences were observed in the weight change or survival time compared with those of control. These results suggest that inflammation accompanying microsphere implantation is temporary as well as toxicity to normal tissues is minimal [58].

The levels of tissue factors, inflammatory cytokines, and metabolites of arachidonic acid were evaluated. Growth factors derived from endothelium and from macrophages were found. These factors most probably stimulate both growth and regeneration occurring when different biodegradable materials were used as grafts [40, 53, 55, 70]. The positive reaction for thrombomodulin, a multifunctional protein with anticoagulant properties, was found in both mesothelial and endothelial cells after pericardial PHB patch implantation. Prostacycline production level, which was found to have cytoprotective effect on the pericardium and prevent adhesion formation, in the regenerated tissue was similar to that in native pericardium [53, 55]. The PHB patch seems to be highly biocompatible, since no signs of inflammation were observed macroscopically and also the level of inflammation associated cytokine mRNA did not change dramatically, although a transient increase of interleukin-1β and interleukin-6 mRNA through days 1–7 after PHB patch implantation was detected. In contrast, tumor necrosis factor-α mRNA was hardly detectable throughout the implantation period, which agrees well with a observed moderate fibrotic response [40, 70].

16.3.2 PHB as Tissue Engineering Material and PHB *in vitro* Biocompatibility

The biopolymer PHB is promising material in tissue engineering due to high biocompatibility *in vitro*. Cell cultures of various origins including murine and human fibroblasts [12, 34, 72-74], human mesenchymal stem cells [75], rabbit bone marrow cells (osteoblasts) [30, 73, 76], human osteogenic sarcoma cells [77], human epithelial cells [74, 77], human endothelial cells [78, 79], rabbit articular cartilage chondrocytes [80, 81] and rabbit smooth muscle cells [82] in direct contact with PHB when cultured on polymer films and scaffolds exhibited satisfactory levels of cell adhesion, viability and proliferation. Moreover, it was shown that fibroblasts, endothelium cells, and isolated hepatocytes cultured on PHB films exhibited high levels of cell adhesion and growth (Figure 8) [83].

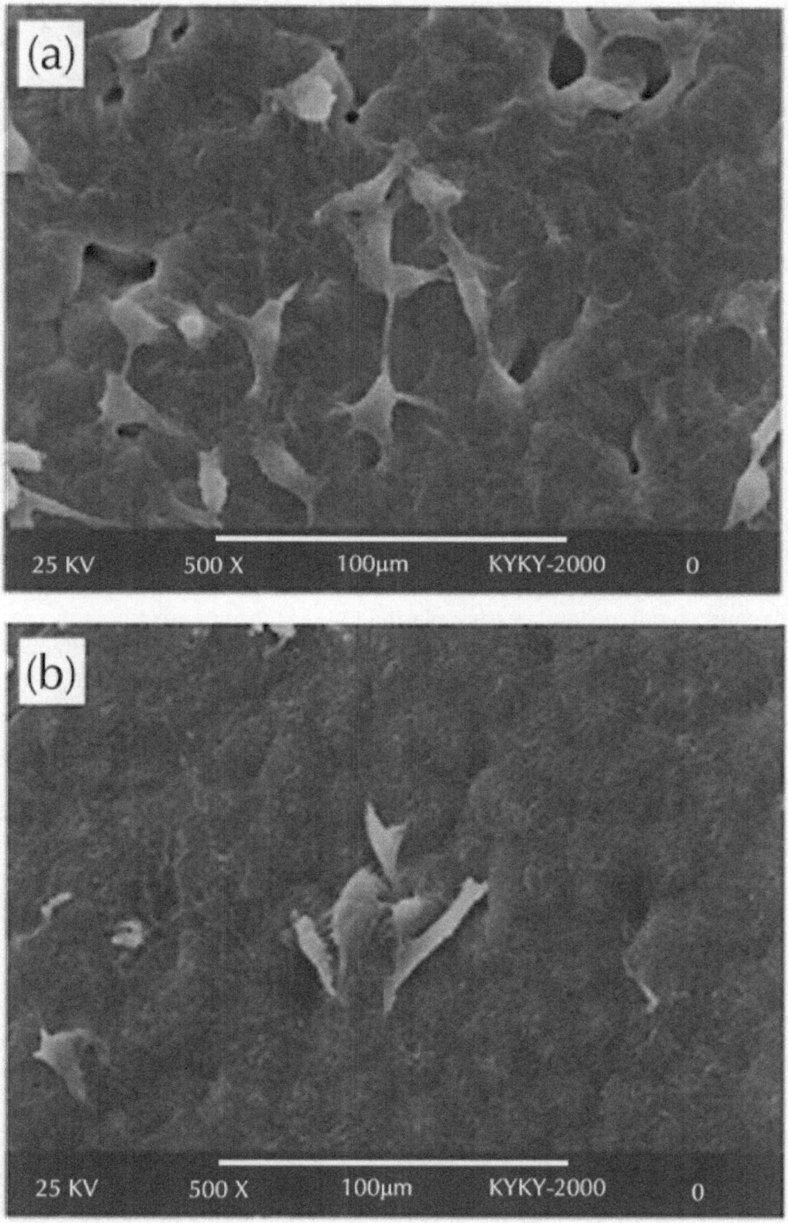

FIGURE 8 The SEM image of 2 days growth of fibroblast cells on films made of (a) PHB, (e) PLA, (500 x). Cell density of fibroblasts grown on PHB film is significantly higher *versus* cell density of fibroblasts grown on PLA film [73].

It was shown also that cultured cells produced collagen II and glycosaminoglycan, the specific structural biopolymers formed the extracellular matrix [77, 80, 81]. A good viability and proliferation level of macrophages and fibroblasts cell lines was obtained under culturing in presence of particles from short-chain low-molecular PHB [61]. However it was shown that cell growth on the PHB films was relatively poor: the viable cell number ranged from 1×10^3 to 2×10^5 [34, 73, 81]. An impaired interaction between PHB matrix and cytoskeleton of cultured cells was also demonstrated [77]. It was reported that a number of polymer properties including chemical composition, surface morphology, surface chemistry, surface energy, and hydrophobicity play important roles in regulating cell viability and growth [84]. The investigation showed that this biomaterial can be used to make scaffolds for *in vitro* proliferous cells [34, 76, 80].

The most widespread methods to manufacture the PHB scaffolds for tissue engineering by means of improvement of cell adhesion and growth on polymer surface are change of PHB surface properties and microstructure by salt leaching methods and enzymatic/chemical/physical treatment of polymer surface [34, 76, 80, 85]. Adhesion to polymer substrates is one of the key issues in tissue engineering, because adhesive interactions control cell physiology. One of the most effective techniques to improve adhesion and growth of cells on PHB films is treatment of polymer surface with enzymes, alkali, or low pressure plasma [34, 85]. Lipase treatment increases the viable cell number on the PHB film from 100 to 200 times compare to the untreated PHB film. The NaOH treatment on PHB film also indicated an increase of 25 times on the viable cell number compared with the untreated PHB film [34]. It was shown that treatment of PHB film surface with low pressure ammonia plasma improved growth of human fibroblasts and epithelial cells of respiratory mucosa due to increased hydrophylicity (but with no change of microstructure) of polymer surface [74]. It was suggested that the improved hydrophilicity of the films after PHB treatment with lipases, alkali and plasma allowed cells in its suspension to easily attach on the polymer films compared to that on the untreated ones. The influence of hydrophilicity of biomaterial surface on cell adhesion was demonstrated earlier [86].

But a microstructure of PHB film surface can be also responsible for cell adhesion and cell growth [87–89]. Therefore, modification of polymer film surface after enzymatic and chemical treatment (in particular, reduced pore size, and a surface smoothing) is expected to play an important role for enhanced cell growth on the polymer films [34]. Different cells prefer different surface. For example, osteoblasts preferred rougher surfaces with appropriate size of pores [87, 88] while fibroblast prefer smoother surface, yet epithelial cells only attached to the smoothest surface [89]. This appropriate roughness affects cell attachment as it provides the right space for osteoblast growth, or supplies solid anchors for filapodia. A scaffold with appropriate size of pores provided better surface properties for anchoring type II collagen filaments and for their penetration into internal layers of the scaffolds implanted with chondrocytes. This could be illuminated by the interaction of extracellular matrix proteins with the material surface. The right surface properties may also promote cell attachment and proliferation by providing more spaces for better gas/nutrients exchange or more serum protein adsorption. [30, 76, 80]. Additionally, Sevastianov et al. found that PHB films in contact with blood did not activate the hemostasis system at the level of cell response, but they did activate the coagulation system and the complement reaction [90].

The high biocompatibility of PHB may be due to several reasons. First of all, PHB is a natural biopolymer involved in important physiological functions both prokaryotes and eukaryotes. The PHB from bacterial origin has property of stereospecificity that is inherent to biomolecules of all living things and consists only from residues of D(-)-3-hydrohybutyric acid [91]. Low molecular weight PHB (up to 150 resides of 3-hydrohybutyric acid), complexed to other macromolecules (cPHB), was found to be a ubiquitous constituent of both prokaryotic and eukaryotic organisms of nearly all phyla [92-96]. Complexed cPHB was found in a wide variety of tissues and organs of mammals (including human): blood, kidney, vessels, nerves, vessels, eye, brain, as well as in organelles, membrane proteins, lipoproteins, and plaques. The cPHB concentration ranged from 3 to 4 μg/g wet tissue weight in nerves and brain to 12 μg/g in blood plasma [97, 98]. In humans, total plasma cPHB ranged from 0.60 to 18.2 mg/l, with a mean of 3.5 mg/l. [98]. It was shown that cPHB is a functional part of ion channels of erythrocyte plasma membrane and hepatocyte mitochondria membrane [99,100]. The singular ability of cPHB to dissolve salts and facilitate their transfer across hydrophobic barriers defines a potential physiological niche for cPHB in cell metabolism [94]. However a mechanism of PHB synthesis in eukaryotic organisms is not well clarified that requires additional studies. Nevertheless, it could be suggested that cPHB is one of the products of symbiotic interaction between animals and gut microorganisms. It was shown, for example, that *E.coli* is able to synthesize low molecular weight PHB and cPHB plays various physiological roles in bacteria cell [96, 101].

Intermediate product of PHB biodegradation, D(-)-3-hydroxybutyric acid is also a normal constituent of blood at concentrations between 0.3 and 1.3 mM and contains in all animal tissues [102, 103]. As it was noted that PHB has a rather low degradation rate in the body in comparison to, for example, poly(lactic-co-glycolic) acids, that prevent from increase of 3-hydroxybutyric acid concentration in surrounding tissues [10, 13], whereas PLA release, following local pH decrease in implantation area and acidic chronic irritation of surrounding tissues is a serious problem in application of medical devices on the base of poly(lactic-co-glycolic) acids [104, 105]. Moreover, chronic inflammatory response to polylactic and polyglycolic acids that was observed in a number of cases may be induced by immune response to water-soluble oligomers that released during degradation of synthetic polymers [105-107].

16.3.3 Novel Drug Dosage Forms on the Base of PHB

An improvement of medical devices on the basis of biopolymers by encapsulating different drug opens up the wide prospects in applications of these new devices with pharmacological activity in medicine. The design of injection systems for sustained drug delivery in the forms of microparticles (microspheres, microcapsules) prepared on the basis of biodegradable polymers is extremely challenging in the modern pharmacology. The fixation of pharmacologically active component with the biopolymer and following slow drug release from the microparticles provides an optimal level of drug concentration in local target organ during long-term period (up to several months), that provides effective pharmaceutical action. At curative dose the prolonged delivery of drugs from the systems into organism permits to eliminate the shortcomings in peroral, injectable, aerosol, and the other traditional methods of drug administration. Among those shortcomings hypertoxicity, instability, pulsative character of rate delivery, ineffective expenditure of drugs should be pointed out. Alternatively, applications of therapeutical polymer systems provide orderly, and purposefully the deliverance for an optimal dose of agent that is very important at therapy of acute or chronic diseases [108]. An ideal biodegradable microsphere formulation would consist of a free-flowing powder of uniform-sized microspheres less than 125 μm in diameter and with a high drug loading. In addition, the drug must be released in its active form with an optimized release profile. The manufacturing method should produce such microspheres in a process that is reproducible, scalable, and benign to some often delicate drugs, with high encapsulation efficiency [109, 110].

The PHB as biodegradable and biocompatible is a promising material for producing of polymer systems for controlled drug release. A number of drugs with various pharmacological activities were used for development of polymer controlled release systems on the base of PHA, mainly on the base of poly(3-hydroxybutyrate-co-3-hydroxyvalerate) and poly(3-hydroxybutyrate-co-4- hydroxybutyrate) copolymers: model drugs (2,7-dichlorofluorescein [111], dextran-FITC [112], methyl red [113, 114], 7-hydroxethyltheophylline [115, 116]), antibiotics and antibacterial drugs (rifampicin [117, 118], tetracycline [119], cefoperazone and gentamicin [120], sulperazone. and duocid [121-124], sulbactam and cefoperazone [125]), anticancer drugs (5-fluorouracil [126], 2',3'-diacyl-5-fluoro-2'-deoxyuridine [58]), anti-inflammatory drug (indomethacin [127]), analgesics (tramadol [128]), vasodilator, and antithrombotic drugs (dipyridamole [24, 127, 129], NO donor [130, 131]). The biocompatibility and pharmacological activity of some of these systems was studied [24, 58, 117, 123-125, 128, 131]. But only a few drugs were used for production of drug controlled release systems on the base of PHB homopolymer: 7-hydroxethyltheophylline, methyl red, 2',3'-diacyl-5-fluoro-2'-deoxyuridine, rifampicin, tramadol, indomethacin, and dipyridamole [58, 113-118, 127-131].

The first drug sustained delivery system on the base of PHB was developed by Korsatko W. et al., who observed a rapid release of encapsulated drug, 7-hydroxethyltheophylline, from tablets of PHB (M_w = 2000 kDa), as well as weight losses of PHB tablets containing the drug after subcutaneous implantation. It was suggested that PHB with molecular weight greater than 100 kDa was undesirable for long-term medication dosage [115].

Pouton C.W. and Akhtar S. describing the release of low molecular drugs from PHB matrices reported that the latter have the tendention of enhanced water penetration and pore formation [132]. The entrapment and release of model drug, methyl red, from melt-crystallized PHB matrices was found to be a function of polymer crystallization kinetics and morphology whereas overall degree of crystallinity was shown to cause no effect on drug release kinetics. Methyl red released from PHB films for more than 7 days with initial phase of rapid release ("burst effect") and second phase with relatively uniform release. Release profiles of PHB films crystallized at 110°C exhibited a greater burst effect when compared to those crystallized at 60°C. This was explained by better trapping of drug within polymeric spherulites with the more rapid rates of PHB crystallization at 110°C [113, 114].

Kawaguchi T. et al showed that chemical properties of drug and polymer molecular weight had a great impact on drug delivery kinetics from PHB matrix. Microspheres (100–300 μm in diameter) from PHB of different molecular weight (65, 135, and 450 kDa) were loaded with prodrugs of 5-fluoro-2'deoxyuridine (FUDR) synthesized by esterification with aliphatic acids (propionate, butyrate, and pentanoate). Prodrugs have different physico-chemical properties, in particular, solubility in water (from 70 mg/ml for FUDR to 0.1 mg/ml for butyryl-FUDR). The release rates from the spheres depended on both the lipophilicity of the prodrug and the molecular weight of the polymer. Regardless of the polymer, the relative release rates were propionyle-FUDR>butyryl- FUDR>pentanoyl-FUDR. The release of butyryl- FUDR and pentanoyl-FUDR from the spheres consisting of low-molecular-weight polymer (M_w = 65 kDa) was faster than that from the spheres of higher molecular weight (M_w = 135 or 450 kDa). The effect of drug content on the release rate was also studied. The higher the drug content in the PHB microspheres, the faster was the drug release. The release of FUDR continued for more than 5 days [58].

Kassab A. C. developed a well-managed technique for manufacture of PHB microspheres loaded with drugs. Microspheres were obtained within a size of 5–100 μm using a solvent evaporation method by changing the initial polymer/solvent ratio, emulsifier concentration, stirring rate, and initial drug concentration. Very high drug loading of up to 408 g rifampicin/g PHB were achieved. Drug release rates were rapid: the maximal duration of rifampicin delivery was 5 days. Both the size and drug content of PHB microspheres were found to be effective in controlling the drug release from polymer microspheres [118].

The sustained release of analgesic drug, tramadol, from PHB microspheres was demonstrated by Salman M. A. et al. It was shown that 58% of the tramadol (the initial drug content in PHB matrix = 18%) was released from the microspheres (7.5 μm in diameter) in the first 24 hr. Drug release decreased with time. From 2 to 7 days the drug release was with zero order rate. The entire amount of tramadol was released after 7 days [128].

The kinetics of different drug release from PHB films and microspheres was studied by our team [24, 127]. It was found that the release occurs *via* two mechanisms, diffusion and degradation, operating simultaneously. Vasodilator and antithrombotic drug, dipyridamole, and anti-inflammatory drug, indomethacin, diffusion processes determine the rate of the release at the early stages of the contact of the system with the environment (the first 6–8 days). The coefficient of the release diffusion of a drug depends on its nature, the thickness of the PHB films containing the drug, the weight ratio of dipyridamole and indomethacin in polymer, and the molecular weight of PHB. Thus, it is possible to regulate the rate of drug release by changing of molecular weight of PHB, for example. A number of other drugs have been also used for development polymeric systems of sustained drug delivery: antibiotics (rifampicin, metronidazole, ciprofloxacin, levofloxacin), anti-inflammatory drugs (flurbiprofen, dexamethasone, prednisolone), and antitumor drugs (paclitaxel) [127]. The biodegradable microspheres on the base of PHB designed for controlled release of dipyridamole were kinetically studied. The profiles of release from the microspheres with different diameters 4, 9, 63, and 92 μm present the progression of nonlinear and linear stages. Diffusion kinetic equation describing both linear (PHB hydrolysis) and nonlinear (diffusion) stages of the dipyridamole release profiles from the spherical subjects has been written down as the sum of two terms: desorption from the homogeneous sphere in accordance with diffusion mechanism and the zero order release. In contrast to the diffusivity dependence on microsphere size, the constant characteristics of linearity are scarcely affected by the diameter of PHB microparticles. The view of the kinetic profiles as well as the low rate of dipyridamole release are in satisfactory agreement with kinetics of weight loss measured *in vitro* for the PHB films and observed qualitatively for PHB microspheres. Taking into account kinetic results, it was supposed that the degradation of both films and PHB microspheres is responsible for the linear stage of dipyridamole release profiles. Thus, a good method for production of systems with sustained drug release was demonstrated. The sustained invariable drug release is an essential property of injectable therapeutic polymer systems that allows to keep constant the adjusted drug dosing. The PHB films and microspheres with sustained uniform drug release for more that 1 month were developed (Figure 9) [24,127, 129].

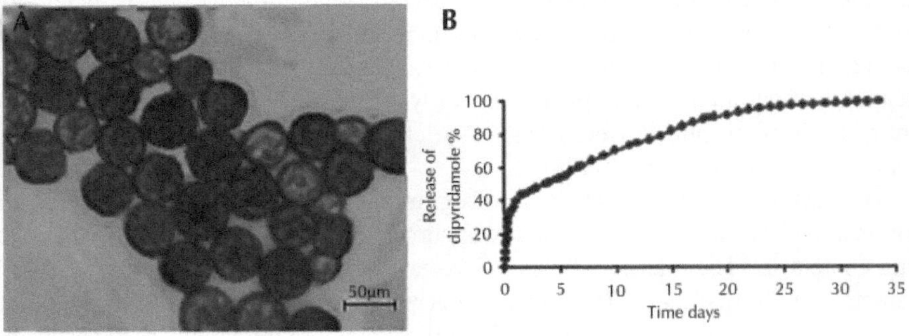

FIGURE 9 The PHB microspheres for sustained delivery of drugs (A) PHB microspheres (average diameter = 63 μm, PHB M_w = 1000 kDa) loaded with dipyridamole (10% w/w), (B) sustained delivery of dipyridamole from PHB microspheres for more than 1 month [24, 131].

The biocompatibility and pharmacological activity of advanced drug delivery systems on the base of PHB was studied [24, 58, 117, 128]. It was shown that implanted PHB films loaded with dipyridamole and indomethacin caused the mild tissue reaction. The inflammation accompanying implantation of PHB matrices is temporary and additionally toxicity relative to normal tissues is minimal [24]. No signs of toxicity were observed after administration of PHB microspheres loaded with analgesic, tramadol, [128]. A single intraperitoneal injection of PHB (M_w = 450 kDa) microspheres containing anticancer prodrugs, butyryl-FUDR and pentanoyl-FUDR, resulted in high antitumor effects against P388 leukemia in mice over a period of five days [58]. Embolization with PHB microspheres *in vivo* on dogs as test animals has been studied by Kasab et al. Renal angiograms obtained before and after embolization and also the histopathological observations showed the feasibility of using these microspheres as an alternative chemoembolization agent [117]. Epidural analgesic effects of tramadol released from PHB microspheres were observed for 21 hr, whereas an equal dose of free tramadol was effective for less than 5 hr. It was suggested that controlled release of tramadol from PHB microspheres *in vivo* is possible, and pain relief during epidural analgesia is prolonged by this drug formulation compared with free tramadol [128].

16.4 CONCLUSION

The observed data indicates that the wide prospects in applications of drug-loaded medical devices and microspheres on the base of PHB as implantable and injectable therapeutic systems in medicine for treatment of various diseases: cancer, cardiovascular diseases, tuberculosis, osteomyelitis, arthritis, and so on. Besides application of PHB for producing of medical devices and systems of sustained drug delivery, PHB can be used for production of systems for controlled release of activators or inhibitors of enzymes. The use of these systems allows the development of the physiological models for prolonged local activation or inhibition of enzymes *in vivo*. The PHB is a perspective tool in design of novel physiological models due to minimal adverse inflammatory tissue reaction to PHB matrices implantation or PHB microspheres administration. A system of sustained NO donor delivery on basis of PHB was developed. This system can be used for study of prolonged NO action on normal tissues of blood vessels *in vivo*. The development of *in vivo* model of prolonged NO local action on vascular tissues is a difficult problem, because NO donors deliver NO at most only for a few minutes. We have developed a model of prolonged local NO action on appropriate artery on basis of PHB loaded with a new effective NO donor, FPTO [133]. It has been shown that FPTO loaded PHB cylinders can release FPTO (and consequently NO) for up to 1 month with relatively constant rate. The FPTO loaded PHB cylinders with sustained FPTO delivery were implanted around left carotid artery of wistar rats, pure PHB cylinders were implanted around right carotid artery as control. At 1st, 4th, and 10th days after implantation arteries and cylinders have been isolated. The elevated levels of the main metabolic products of NO, nitrites and nitrates, in arterial tissues were observed that indicates the possibility of application of this system for production of physiological model of NO prolonged action on arterial tissues *in vivo* [24, 130, 131].

KEYWORDS

- **Biodegradation**
- **Cell culture**
- **Enzymatic degradation**
- **Polylactic acid**
- **Scanning electron microscopy**

REFERENCES

1. Chen, G. Q. and Wu, Q. The application of polyhydroxyalkanoates as tissue engineering materials. *Biomaterials*, **26**(33), 6565–6578 (2005).
2. Lenz, R. W. and Marchessault, R. H. Bacterial Polyesters Biosynthesis, Biodegradable Plastics and Biotechnology. *Biomacromolecules*, **6**(1), 1–8 (2005).
3. Anderson, A. J. and Dawes, E. A. Occurrence, metabolism, metabolic role, and industrial uses of bacterial polyhydroxyalkanoates. *Microbiological Reviews*, **54**(4), 450–472 (1990).
4. Jendrossek, D. and Handrick, R. Microbial degradation of polyhydroxyalkanoates. *Annu Rev Microbiol.*, **56**, 403–432 (2002).
5. Kim, D. Y. and Rhee, Y. H. Biodegradation of microbial and synthetic polyesters by fungi. *Appl. Microbiol. Biotechnol.*, **61**, 300–308 (2003).
6. Steinbuchel, A. and Lutke-Eversloh, T. Metabolic engineering and pathway construction for biotechnological production of relevant polyhydroxyalkanoates in microorganisms. *Biochem. Eng. J.*, **16**, 81–96 (2003).
7. Marois, Y., Zhang, Z., Vert, M., Deng, X., Lenz, R., and Guidoin, R. Hydrolytic and enzymatic incubation of polyhydroxyoctanoate (PHO) a short-term in vitro study of a degradable bacterial polyester. *J. Biomater. Sci. Polym. Ed.*, **10**, 483–499 (1999).
8. Abe, H. and Doi, Y. Side-chain effect of second monomer units on crystalline morphology, thermal properties, and enzymatic degradability for random copolyesters of (R)-3-hydroxybutyric acid with (R)-3-hydroxyalkanoic acids. *Biomacromolecules*, **3**(1), 133–138 (2002).
9. Renstad, R., Karlsson, S., and Albertsson, A. C. The influence of processing induced differences in molecular structure on the biological and non-biological degradation of poly(3-hydroxybutyrate-co-3-hydroxyvalerate), P(3-HB-co-3-HV). *Polym. Degrad. Stab.*, **63**, 201–211 (1999).
10. Qu, X. H., Wu, Q., Zhang, K. Y., and Chen, G. Q. In vivo studies of poly(3-hydroxybutyrate-co-3-hydroxyhexanoate) based polymers biodegradation and tissue reactions. *Biomaterials*, **27**(19), 3540–3548 (2006).
11. Freier, T., Kunze, C., Nischan, C., Kramer, S., Sternberg, K., Sass, M., Hopt, U. T., and Schmitz, K. P. In vitro and in vivo degradation studies for development of a biodegradable patch based on poly(3-hydroxybutyrate). *Biomaterials*, **23**(13), 2649–2657 (2002).
12. Kunze, C., Edgar Bernd, H., Androsch, R., Nischan, C., Freier, T., Kramer, S., Kramp, B., and Schmitz, K. P. In vitro and in vivo studies on blends of isotactic and atactic poly (3-hydroxybutyrate) for development of a dura substitute material. *Biomaterials*, **27**(2), 192–201 (January, 2006).
13. Gogolewski, S., Jovanovic, M., Perren, S. M., Dillon, J. G., and Hughes, M. K. Tissue response and in vivo degradation of selected polyhydroxyacids polylactides (PLA), poly(3-hydroxybutyrate) (PHB), and poly(3-hydroxybutyrate-co-3-hydroxyvalerate) (PHB/VA). *J. Biomed. Mater. Res.*, **27**(9), 1135–1148(1993).
14. Borkenhagen, M., Stoll, R. C., Neuenschwander, P., Suter, U. W., and Aebischer, P. In vivo performance of a new biodegradable polyester urethane system used as a nerve guidance channel. *Biomaterials*, **19**(23), 2155–2165 (1998).
15. Hazari, A., Johansson-Ruden, G., Junemo-Bostrom, K., Ljungberg, C., Terenghi, G., Green, C., and Wiberg, M. A new resorbable wrap-around implant as an alternative nerve repair technique. *Journal of Hand Surgery British and European*, **24**(3), 291–295 (a) (1999).
16. Hazari, A., Wiberg, M., Johansson-Rudén, G., Green, C., and Terenghi, G. A resorbable nerve conduit as an alternative to nerve autograft. *British Journal of Plastic Surgery*, **52**, 653–657 (b) (1999).
17. Miller, N. D. and Williams, D. F. On the biodegradation of poly-beta-hydroxybutyrate (PHB) homopolymer and poly-beta-hydroxybutyrate-hydroxyvalerate copolymers. *Biomaterials*, **8**(2), 129–137 (March, 1987).
18. Shishatskaya E. I., Volova T. G., Gordeev S. A., and Puzyr A. P. Degradation of P(3HB) and P(3HB-co-3HV) in biological media. *J Biomater. Sci. Polym. Ed.*, **16**(5), 643–657 (2005).

19. Saito, T., Tomita, K., Juni, K., and Ooba, K. In vivo and in vitro degradation of poly(3-hydroxybutyrate) in rat. *Biomaterials*, **12**(3), 309–312 (1991).
20. Koyama, N. and Doi, Y. Morphology and biodegradability of a binary blend of poly((R)-3-hydroxybutyric acid) and poly((R,S)-lactic acid). *Can. J. Microbiol.*, **41**(1), 316–322 (1995).
21. Doi, Y., Kanesawa, Y., Kunioka, M., and Saito, T. Biodegradation of microbial copolyesters poly(3-hydroxybutyrate-co-3-hydroxyvalerate) and poly(3-hydroxybutyrate-co-4- hydroxybutyrate). *Macromolecules*, **23**, 26–31 (a) (1990).
22. Holland, S. J., Jolly, A. M., Yasin, M., and Tighe, B. J. Polymers for biodegradable medical devices. II. Hydroxybutyrate-hydroxyvalerate copolymers: hydrolytic degradation studies. *Biomaterials*, **8**(4), 289–295 (1987).
23. Kurcok, P., Kowalczuk, M., Adamus, G., Jedlinrski, Z., and Lenz, R. W. Degradability of poly (b-hydroxybutyrate)s. Correlation with chemical microstructure. *JMS-Pure Appl. Chem.*, **32**, 875–880 (1995).
24. Bonartsev, A. P., Myshkina, V. L., Nikolaeva, D. A., Furina, E. K., Makhina, T. A., Livshits, V. A., Boskhomdzhiev, A. P., Ivanov, E. A., Iordanskii, A. L., and Bonartseva G. A. Biosynthesis, biodegradation, and application of poly(3-hydroxybutyrate) and its copolymers - natural polyesters produced by diazotrophic bacteria. *Communicating Current Research and Educational Topics and Trends in Applied Microbiology*, A. Méndez-Vilas, Formatex (Ed.). Spain, **1**, 295–307 (2007).
25. Bonartsev, A. P. private communication.
26. Cha, Y. and Pitt, C. G. The biodegradability of polyester blends. *Biomaterials*, **11**(2), 108–112 (1990).
27. Schliecker, G., Schmidt, C., Fuchs, S., Wombacher, R., and Kissel, T. Hydrolytic degradation of poly(lactide-co-glycolide) films: effect of oligomers on degradation rate and crystallinity. *Int. J. Pharm.*, **266**(1-2), 39–49 (2003).
28. Scandola, M., Focarete, M. L., Adamus, G., Sikorska, W., Baranowska, I., Swierczek, S., Gnatowski, M., Kowalczuk, M., and Jedlinr ski, Z. Polymer blends of natural poly(3-hydroxybutyrate-co-hydroxyvalerate) and a synthetic atactic poly(3-hydroxybutyrate). Characterization and biodegradation studies. *Macromolecules*, **30**, 2568–2574 (1997).
29. Doi, Y, Kanesawa, Y., Kawaguchi, Y., and Kunioka, M. Hydrolytic degradation of microbial poly(hydroxyalkanoates). *Makrom. Chem. Rapid. Commun.*, **10**, 227–230 (1989).
30. Wang, Y. W., Yang, F., Wu, Q., Cheng, Y. C., Yu, P. H., Chen, J., and Chen, G. Q. Evaluation of three-dimensional scaffolds made of blends of hydroxyapatite and poly(3-hydroxybutyrate-co-3-hydroxyhexanoate) for bone reconstruction. *Biomaterials*, **26**(8), 899–904 (a) (2005).
31. Muhamad, I. I., Joon, L. K., and Noor, M. A. M. Comparing the degradation of poly-β-(hydroxybutyrate), poly-β-(hydroxybutyrate-co-valerate)(PHBV) and PHBV/Cellulose triacetate blend. *Malaysian Polymer Journal*, **1**, 39–46 (2006).
32. Mergaert, J., Webb, A., Anderson, C., Wouters, A., and Swings, J. Microbial degradation of poly(3-hydroxybutyrate) and poly(3-hydroxybutyrate-co-3-hydroxyvalerate) in soils. *Applied and environmental microbiology*, **59**(10), 3233–3238 (1993).
33. Choi, G. G., Kim, H. W., and Rhee, Y. H. Enzymatic and non-enzymatic degradation of poly(3-hydroxybutyrate-co-3-hydroxyvalerate) copolyesters produced by Alcaligenes sp. MT-16. *The Journal of Microbiology*, **42**(4), 346–352 (December, 2004).
34. Yang, X., Zhao, K., and Chen, G. Q. Effect of surface treatment on the biocompatibility of microbial polyhydroxyalkanoates. *Biomaterials*, **23**(5), 1391–1397 (2002).
35. Zhao, K., Yang, X., Chen, G. Q., and Chen, J. C. Effect of lipase treatment on the biocompatibility of microbial polyhydroxyalkanoates. *J. material science materials in medicine*, **13**, 849–854 (2002).
36. Wang, H. T., Palmer, H., Linhardt, R. J., Flanagan, D. R., and Schmitt, E. Degradation of poly(ester) microspheres. *Biomaterials*, **11**(9), 679–685 (1990).
37. Doyle, C., Tanner, E. T., and Bonfield, W. In vitro and in vivo evaluation of polyhydroxybutyrate and of polyhydroxybutyrate reinforced with hydroxyapatite. *Biomaterials*, **12**, 841–847 (1991).

38. Coskun, S., Korkusuz, F., and Hasirci, V. Hydroxyapatite reinforced poly(3-hydroxybutyrate) and poly(3-hydroxybutyrate-co-3-hydroxyvalerate) based degradable composite bone plate. *J. Biomater. Sci. Polymer Edn.*, **16**(12), 1485–1502 (2005).
39. Sudesh, K., Abe, H., and Doi, Y. Synthesis, structure and properties of polyhydroxyalkanoates: biological polyesters. *Prog. Polym. Sci.*, **25**, 1503–1555 (2000).
40. Lobler, M., Sass, M., Kunze, C., Schmitz, K. P., and Hopt, U. T. Biomaterial patches sutured onto the rat stomach induce a set of genes encoding pancreatic enzymes. *Biomaterials*, **23**, 577–583 (2002).
41. Tokiwa, Y., Suzuki, T., and Takeda, K. Hydrolysis of polyesters by Rhizopus arrhizus lipase. *Agric. Biol. Chem*., **50**, 1323–1325 (1986).
42. Hoshino, A. and Isono, Y. Degradation of aliphatic polyester films by commercially available lipases with special reference to rapid and complete degradation of poly(L-lactide) film by lipase PL derived from Alcaligenes sp. *Biodegradation*, **13**, 141–147 (2002).
43. Jendrossek, D., Schirmer, A., and Schlegel, H. G. Biodegradation of polyhydroxyalkanoic acids. *Appl. Microbiol. Biotechnol*., **46**, 451–463 (1996).
44. Winkler, F. K., D'Arcy, A., and Hunziker, W. Structure of human pancreatic lipase. *Nature,* **343**, 771–774 (1990).
45. Mergaert, J., Anderson, C., Wouters, A., Swings, J., and Kersters, K. Biodegradation of polyhydroxyalkanoates. *FEMS Microbiol. Rev*., **9**(2-4), 317–321 (1992).
46. Tokiwa, Y. and Calabia, B. P. Degradation of microbial polyesters. *Biotechnol. Lett*., **26**(15), 1181–1189 (2004).
47. Mokeeva, V., Chekunova, L., Myshkina, V., Nikolaeva, D., Gerasin, V., and Bonartseva, G. Biodestruction of poly(3-hydroxybutyrate) by microscopic fungi: tests of polymer on resistance to fungi and fungicidal properties. *Mikologia and Fitopatologia*, **36**(5), 59–63 (2002).
48. Bonartseva, G. A., Myshkina, V. L., Nikolaeva, D. A., Kevbrina, M. V., Kallistova, A. Y., Gerasin, V. A., Iordanskii, A. L., and Nozhevnikova, A. N. Aerobic and anaerobic microbial degradation of poly-beta-hydroxybutyrate produced by Azotobacter chroococcum. *Appl. Biochem. Biotechnol.* **109**(1-3), 285–301 (2003).
49. Spyros, A., Kimmich, R., Briese, B. H., and Jendrossek, D. 1H NMR Imaging Study of Enzymatic Degradation in Poly(3-hydroxybutyrate) and Poly(3-hydroxybutyrate-co-3-hydroxyvalerate). Evidence for Preferential Degradation of the Amorphous Phase by PHB Depolymerase B from Pseudomonas lemoignei. *Macromolecules*, **30**(26), 8218–8225 (1997).
50. Hocking, P. J., Marchessault, R. H., Timmins, M. R., Lenz, R. W., and Fuller, R. C. Enzymatic Degradation of Single Crystals of Bacterial and Synthetic Poly(-hydroxybutyrate). *Macromolecules*, **29**(7), 2472–2478 (1996).
51. Bonartseva, G. A., Myshkina, V. L., Nikolaeva, D. A., Rebrov, A. V., Gerasin, V. A., and Makhina, T. K. The biodegradation of poly-beta-hydroxybutyrate (PHB) by a model soil community: the effect of cultivation conditions on the degradation rate and the physicochemical characteristics of PHB. *Mikrobiologiia*, **71**(2), 258–263 (2002) Russian.
52. Kramp, B., Bernd, H. E., Schumacher, W. A., Blynow, M., Schmidt, W., Kunze, C., Behrend, D., and Schmitz, K. P. Poly-beta-hydroxybutyric acid (PHB) films and plates in defect covering of the osseus skull in a rabbit model. *Laryngorhinootologie*, **81**(5), 351–356 (2002) [Article in German].
53. Malm, T., Bowald, S., Karacagil, S., Bylock, A., and Busch, C. A new biodegradable patch for closure of atrial septal defect An experimental study. *Scand J Thorac Cardiovasc Surg*., **26**(1), 9–14(a) (1992).
54. Malm, T., Bowald, S., Bylock, A., Saldeen, T., and Busch, C. Regeneration of pericardial tissue on absorbable polymer patches implanted into the pericardial sac. An immunohistochemical, ultrastructural and biochemical study in the sheep. *Scandinavian Journal of Thoracic and Cardiovascular Surgery*, **26**(1), 15–21(b) (1992).
55. Malm, T., Bowald, S., Bylock, A., and Busch, C. Prevention of postoperative pericardial adhesions by closure of the pericardium with absorbable polymer patches. An experimental study. *The Journal of Thoracic and Cardiovascular Surgery*, **104**, 600–607 (c) (1992).

56. Malm, T., Bowald, S., Bylock, A., Busch, C., and Saldeen, T. Enlargement of the right ventricular outflow tract and the pulmonary artery with a new biodegradable patch in transannular position. *European Surgical Research*, **26**, 298–308 (1994).
57. Duvernoy, O., Malm, T., Ramström, J., and Bowald, S. A biodegradable patch used as a pericardial substitute after cardiac surgery: 6- and 24-month evaluation with CT. *Thorac Cardiovasc. Surg.*, **43**(5), 271–274 (October, 1995).
58. Kawaguchi, T., Tsugane, A., Higashide, K., Endoh, H., Hasegawa, T., Kanno, H., Seki, T., Juni, K., Fukushima, S., and Nakano, M. Control of drug release with a combination of prodrug and polymer matrix: antitumor activity and release profiles of 2',3'-Diacyl-5-fluoro-2'-deoxyuridine from poly(3-hydroxybutyrate) microspheres. *Journal of Pharmaceutical Sciences*, **87**(6), 508–512 (1992).
59. Baptist, J. N. (Assignor to W.R. Grace Et Co., New York), US Patent No. 3 225 766, (1965).
60. Holmes, P. Biologically produced (R)-3-hydroxy-alkanoate polymers and copolymers. In D. C. Bassett (Ed.). *Developments in crystalline polymers*, Elsevier, London **I**(2), 1–65 (1988).
61. Saad, B., Ciardelli, G., Matter, S., Welti, M., Uhlschmid, G. K., Neuenschwander, P., and Suterl, U. W. Characterization of the cell response of cultured macrophages and fibroblasts td particles of short-chain poly[(R)-3-hydroxybutyric acid]. *Journal of Biomedical Materials Research*, **30**, 429–439 (1996).
62. Fedorov, M., Vikhoreva, G., Kildeeva, N., Maslikova, A., Bonartseva, G., and Galbraikh, L. Modeling of surface modification process of surgical suture. *Chimicheskie volokna*, **6**, 22–28 (2005) [Article in Russian].
63. Rebrov, A. V., Dubinskii, V. A., Nekrasov, Y. P., Bonartseva, G. A., Shtamm, M., and Antipov, E. M. Structure phenomena at elastic deformation of highly oriented polyhydroxybutyrate. *Vysokomol. Soedin.*, (Russian) **44**, 347–351 (2002) [Article in Russian].
64. Kil'deeva, N. R., Vikhoreva, G. A., Gal'braikh, L. S., Mironov, A. V., Bonartseva, G. A., Perminov, P. A., and Romashova, A. N. Preparation of biodegradable porous films for use as wound coverings. *Prikl. Biokhim. Mikrobiol.*, **42**(6), 716–720 (2006) [Article in Russian].
65. Solheim, E., Sudmann, B., Bang, G., and Sudmann, E. Biocompatibility and effect on osteogenesis of poly(ortho ester) compared to poly(DL-lactic acid). *J. Biomed. Mater. Res.*, **49**(2), 257–263 (2000).
66. Bostman, O. and Pihlajamaki, H. Clinical biocompatibility of biodegradable orthopaedic implants for internal fixation a review. *Biomaterials*, **21**(24), 2615–2621 (2000).
67. Lickorish, D., Chan, J., Song, J., and Davies, J. E. An in-vivo model to interrogate the transition from acute to chronic inflammation. *Eur. Cell. Mater.*, **8**, 12–19 (2004).
68. Khouw, I. M., van Wachem, P. B., de Leij, L. F., and van Luyn, M. J. Inhibition of the tissue reaction to a biodegradable biomaterial by monoclonal antibodies to IFN-gamma. *J. Biomed. Mater. Res.*, **41**, 202–210 (1998).
69. Su, S. H., Nguyen, K. T., Satasiya, P., Greilich, P. E., Tang, L., and Eberhart, R. C. Curcumin impregnation improves the mechanical properties and reduces the inflammatory response associated with poly(L-lactic acid) fiber. *J. Biomater. Sci. Polym. Ed.*, **16**(3), 353–370 (2005).
70. Lobler, M., Sass, M., Schmitz, K. P., and Hopt, U. T. Biomaterial implants induce the inflammation marker CRP at the site of implantation. *J. Biomed. Mater. Res.*, **61**, 165–167 (2003).
71. Novikov, L. N., Novikova, L. N., Mosahebi, A., Wiberg, M., Terenghi, G., and Kellerth, J. O. A novel biodegradable implant for neuronal rescue and regeneration after spinal cord injury. *Biomaterials*, **23**, 3369–3376 (2002).
72. Cao, W., Wang, A., Jing, D., Gong, Y., Zhao, N., and Zhang, X. Novel biodegradable films and scaffolds of chitosan blended with poly(3-hydroxybutyrate). *J. Biomater. Sci. Polymer Edn.*, **16**(11), 1379–1394 (2005).
73. Wang, Y. W., Yang, F., Wu, Q., Cheng, Y. C., Yu, P. H., Chen, J., and Chen, G. Q. Effect of composition of poly(3-hydroxybutyrate-co-3-hydroxyhexanoate) on growth of fibroblast and osteoblast. *Biomaterials*, **26**(7), 755–761 (b) (2005).

74. Ostwald, J., Dommerich, S., Nischan, C., and Kramp, B. In vitro culture of cells from respiratory mucosa on foils of collagen, poly-L-lactide (PLLA) and poly-3-hydroxy-butyrate (PHB). *Laryngorhinootologie*, **82**(10), 693–699 (2003) [Article in German].
75. Wollenweber, M., Domaschke, H., Hanke, T., Boxberger, S., Schmack, G., Gliesche, K., Scharnweber, D., and Worch, H. Mimicked bioartificial matrix containing chondroitin sulphate on a textile scaffold of poly(3-hydroxybutyrate) alters the differentiation of adult human mesenchymal stem cells. *Tissue Eng.*, **12**(2), 345–359 (February, 2006).
76. Wang, Y. W, Wu, Q., and Chen, G. Q. Attachment, proliferation and differentiation of osteoblasts on random biopolyester poly(3-hydroxybutyrate-co-3-hydroxyhexanoate) scaffolds. *Biomaterials*, **25**(4), 669–675 (2004).
77. Nebe, B., Forster, C., Pommerenke, H., Fulda, G., Behrend, D., Bernewski, U., Schmitz, K. P., and Rychly, J. Structural alterations of adhesion mediating components in cells cultured on poly-beta-hydroxy butyric acid. *Biomaterials*, **22**(17), 2425–2434 (2001).
78. Qu, X. H., Wu, Q., and Chen, G. Q. In vitro study on hemocompatibility and cytocompatibility of poly(3-hydroxybutyrate-co-3-hydroxyhexanoate). *J. Biomater. Sci. Polymer Edn.*, **17**(10), 1107–1121 (a) (2006).
79. Pompe, T., Keller, K., Mothes, G., Nitschke, M., Teese, M., Zimmermann, R., and Werner, C. Surface modification of poly(hydroxybutyrate) films to control cell-matrix adhesion. *Biomaterials*, **28**(1), 28–37 (2007).
80. Deng, Y., Lin, X. S., Zheng, Z., Deng, J. G., Chen, J. C., Ma, H., and Chen, G. Q. Poly(hydroxybutyrate-co-hydroxyhexanoate) promoted production of extracellular matrix of articular cartilage chondrocytes in vitro. *Biomaterials*, **24**(23), 4273–4281 (2003).
81. Zheng, Z., Bei, F. F., Tian, H. L., and Chen, G. Q. Effects of crystallization of polyhydroxyalkanoate blend on surface physico-chemical properties and interactions with rabbit articular cartilage chondrocytes. *Biomaterials*, **26**, 3537–3548 (2005).
82. Qu, X. H., Wu, Q., Liang, J., Zou, B., and Chen, G. Q. Effect of 3-hydroxyhexanoate content in poly(3-hydroxybutyrate-co-3-hydroxyhexanoate) on in vitro growth and differentiation of smooth muscle cells. *Biomaterials*, **27**(15), 2944–2950 (b) (May, 2006).
83. Shishatskaya, E. I. and Volova, T. G. A comparative investigation of biodegradable polyhydroxyalkanoate films as matrices for in vitro cell cultures. *J. Mater. Sci. Mater. M.*, **15**, 915–923 (2004).
84. Fischer, D., Li, Y., Ahlemeyer, B., Kriglstein, J., and Kissel, T. In vitro cytotoxicity testing of polycations influence of polymer structure on cell viability and hemolysis. *Biomaterials*, **24**(7), 1121–1131 (2003).
85. Nitschke, M., Schmack, G., Janke, A., Simon, F., Pleul, D., and Werner, C. Low pressure plasma treatment of poly(3-hydroxybutyrate) toward tailored polymer surfaces for tissue engineering scaffolds. *J. Biomed. Mater. Res.*, **59**(4), 632–638 (2002).
86. Chanvel-Lesrat, D. J., Pellen-Mussi, P., Auroy, P., and Bonnaure-Mallet, M. Evaluation of the in vitro biocompatibility of various elastomers. *Biomaterials*, **20**, 291–299 (1999).
87. Boyan, B. D., Hummert, T. W., Dean, D. D., and Schwartz, Z. Role of material surfaces in regulating bone and cartilage cell response. *Biomaterials*, **17**, 137–146 (1996).
88. Bowers, K. T., Keller, J. C., Randolph, B. A., Wick, D. G., and Michaels, C. M. Optimization of surface micromorphology for enhanced osteoblasts responses in vitro. *Int. J. Oral. Max. Impl.*, **7**, 302–310 (1992).
89. Cochran, D., Simpson, J., Weber, H., and Buser, D. Attachment and growth of periodontal cells on smooth and rough titanium. *Int. J. Oral. Max. Impl.*, **9**, 289–297 (1994).
90. Sevastianov, V. I., Perova, N. V., Shishatskaya, E. I., Kalacheva, G. S., and Volova, T. G. Production of purified polyhydroxyalkanoates (PHAs) for applications in contact with blood. *J. Biomater. Sci. Polym. Ed.*, **14**, 1029–1042 (2003).
91. Seebach, D., Brunner, A., Burger, H. M., Schneider, J., and Reusch, R. N. Isolation and 1H-NMR spectroscopic identification of poly(3-hydroxybutanoate) from prokaryotic and eukaryotic organisms. Determination of the absolute configuration (R) of the monomeric unit

3-hydroxybutanoic acid from Escherichia coli and spinach. *Eur. J. Biochem.*, **224**(2), 317–328 (1994).

92. Reusch, R. N. Poly-β-hydroxybutryate/calcium polyphosphate complexes in eukaryotic membranes. *Proc. Soc. Exp. Biol. Med.*, **191**, 377–381 (1989).
93. Reusch, R. N. Biological complexes of poly-β-hydroxybutyrate. *FEMS Microbiol. Rev.*, **103**, 119–130 (1992).
94. Reusch, R. N. Low molecular weight complexed poly(3-hydroxybutyrate) a dynamic and versatile molecule in vivo. *Can. J. Microbiol.*, **41**(1), 50–54 (1995).
95. Müller, H. M. and Seebach, D. Polyhydroxyalkanoates: a fifth class of physiologically important organic biopolymers. *Angew Chemie*, **32**, 477–502 (1994).
96. Huang, R. and Reusch, R. N. Poly(3-hydroxybutyrate) is associated with specific proteins in the cytoplasm and membranes of Escherichia coli. *J. Biol. Chem.*, **271**, 22196–22201 (1996).
97. Reusch, R. N., Bryant, E. M., and Henry, D. N. Increased poly-(R)-3-hydroxybutyrate concentrations in streptozotocin (STZ) diabetic rats. *Acta Diabetol.*, **40**(2), 91–94 (2003).
98. Reusch, R. N., Sparrow, A. W., and Gardiner, J. Transport of poly-β-hydroxybutyrate in human plasma. *Biochim. Biophys. Acta*, **1123**, 33–40 (1992).
99. Reusch, R. N., Huang, R., and Kosk-Kosicka, D. Novel components and enzymatic activities of the human erythrocyte plasma membrane calcium pump. *FEBS Lett.*, **412**(3), 592–596 (1997).
100. Pavlov, E., Zakharian, E., Bladen, C., Diao, C. T. M., Grimbly, C., Reusch, R. N., and French, R. J. A large, voltage-dependent channel, isolated from mitochondria by water-free chloroform extraction. *Biophysical Journal*, **88**, 2614–2625 (2005).
101. Theodorou, M. C., Panagiotidis, C. A., Panagiotidis, C. H., Pantazaki, A. A., and Kyriakidis, D. A. Involvement of the AtoS-AtoC signal transduction system in poly-(R)-3-hydroxybutyrate biosynthesis in Escherichia coli. *Biochim. Biophys. Acta.*, **1760**(6), 896–906 (2006).
102. Wiggam, M. I., O'Kane, M. J., Harper, R., Atkinson, A. B., Hadden, D. R., Trimble, E. R., and Bell, P. M. Treatment of diabetic ketoacidosis using normalization of blood 3-hydroxy-butyrate concentration as the endpoint of emergency management. *Diabetes Care*, **20**, 1347–1352 (1997).
103. Larsen, T. and Nielsen, N. I. Fluorometric determination of beta-hydroxybutyrate in milk and blood plasma. *J. Dairy Sci.*, **88**(6), 2004–2009 (2005).
104. Agrawal, C. M. and Athanasiou, K. A. Technique to control pH in vicinity of biodegrading PLA-PGA implants. *J. Biomed. Mater. Res.*, **38**(2), 105–114 (1997).
105. Ignatius, A. A. and Claes, L. E. In vitro biocompatibility of bioresorbable polymers poly(l, dl-lactide) and poly(l-lactide-co-glycolide). *Biomatrials*, **17**(8), 831–839 (1996).
106. Rihova, B. Biocompatibility of biomaterials hemocompatibility, immunocompatibility and biocompatibility of solid polymeric materials and soluble targetable polymeric carriers. *Adv. Drug Delivery Rev.*, **21**, 157–176 (1996).
107. Ceonzo, K., Gaynor, A., Shaffer, L., Kojima, K., Vacanti, C. A., and Stahl, G. L. Polyglycolic acid-induced inflammation: role of hydrolysis and resulting complement activation. *Tissue Eng.*, **12**(2), 301–308 (2006).
108. M. Chasin and R. Langer (Eds.). *Biodegradable Polymers as Drug Delivery Systems.* Marcel Dekker, New York (1990).
109. Johnson, O. L. and Tracy, M. A. Peptide and protein drug delivery. E. Mathiowitz (Ed.). In *Encyclopedia of Controlled Drug Delivery Vol 2.* Hoboken, John Wiley and Sons. New Jersey 816–832 (1999).
110. Jain, R. A. The manufacturing techniques of various drug loaded biodegradable poly(lactide-co-glycolide) (PLGA) devices. *Biomaterials*, **21**, 2475–2490 (2000).
111. Gursel, I. and Hasirci, V. Properties and drug release behavior of poly(3-hydroxybutyric acid) and various poly(3-hydroxybutyrate-hydroxyvalerate) copolymer microcapsules. *J. Microencapsul.*, **12**(2), 185–193 (1995).
112. Li, J., Li, X., Ni, X., Wang, X., Li, H., and Leong, K. W. Self-assembled supramolecular hydrogels formed by biodegradable PEO–PHB–PEO triblock copolymers and a-cyclodextrin for controlled drug delivery. *Biomaterials*, **27**(22), 4132–4140 (2006).

113. Akhtar, S., Pouton, C. W., and Notarianni, L. J. Crystallization behavior and drug release from bacterial polyhydroxyalkanoates. *Polymer*, **33**(1), 117–126 (1992).
114. Akhtar, S., Pouton, C. W., and Notarianni, L. J. The influence of crystalline morphology and copolymer composition on drug release from solution cast and melting processed P(HB-HV) copolymer matrices. *J. Controlled Release*, **17**, 225–234 (1991).
115. Korsatko, W., Wabnegg, B., Tillian, H. M., Braunegg, G., and Lafferty, R. M. Poly-D-hydroxybutyric acid-a biologically degradable vehicle to regard release of a drug. *Pharm. Ind.*, **45**, 1004–1007 (1983).
116. Korsatko, W., Wabnegg, B., Tillian, H. M., Egger, G., Pfragner, R., and Walser, V. Poly D(-)-3-hydroxybutyric acid (poly-HBA)-a biodegradable former for long-term medication dosage. 3. Studies on compatibility of poly-HBA implantation tablets in tissue culture and animals. *Pharm. Ind.*, **46**, 952–954 (1984).
117. Kassab, A. C., Piskin, E., Bilgic, S., Denkbas, E. B., and Xu, K. Embolization with polyhydroxybutyrate (PHB) micromerspheres in vivo studies. *J. Bioact. Compat. Polym.*, **14**, 291–303 (1999).
118. Kassab, A. C., Xu, K., Denkbas, E. B., Dou, Y., Zhao, S., and Piskin, E. Rifampicin carrying polyhydroxybutyrate microspheres as a potential chemoembolization agent. *J. Biomater. Sci. Polym. Ed.*, **8**, 947–961 (1997).
119. Sendil, D., Gursel, I., Wise, D. L., and Hasirci, V. Antibiotic release from biodegradable PHBV microparticles. *J. Control. Release*, **59**, 207–17 (1999).
120. Gursel, I., Yagmurlu, F., Korkusuz, F., and Hasirci, V. In vitro antibiotic release from poly(3-hydroxybutyrate-co-3-hydroxyvalerate) rods. *J. Microencapsul.*, **19**, 153–164 (2002).
121. Turesin, F., Gursel, I., and Hasirci, V. Biodegradable polyhydroxyalkanoate implants for osteomyelitis therapy in vitro antibiotic release. *J. Biomater. Sci. Polym. Ed.*, **12**, 195–207 (2001).
122. Turesin, F., Gumusyazici, Z., Kok, F. M., Gursel, I., Alaeddinoglu, N. G., and Hasirci, V. Biosynthesis of polyhydroxybutyrate and its copolymers and their use in controlled drug release. *Turk. J. Med. Sci.*, **30**, 535–541 (2000).
123. Gursel, I., Korkusuz, F., Turesin, F., Alaeddinoglu, N. G., and Hasirci, V. In vivo application of biodegradable controlled antibiotic release systems for the treatment of implant-related osteomyelitis. *Biomaterials*, **22**(1), 73–80 (2001).
124. Korkusuz, F., Korkusuz, P., Eksioglu, F., Gursel, I., and Hasirci, V. In vivo response to biodegradable controlled antibiotic release systems. *J. Biomed. Mater. Res.*, **55**(2), 217–228 (2001).
125. Yagmurlu, M. F., Korkusuz, F., Gursel, I., Korkusuz, P., Ors, U., and Hasirci, V. Sulbactam-cefoperazone polyhydroxybutyrate-co-hydroxyvalerate (PHBV) local antibiotic delivery system in vivo effectiveness and biocompatibility in the treatment of implantrelated experimental osteomyelitis. *J. Biomed. Mater. Res.*, **46**, 494–503 (1999).
126. Khang, G., Kim, S. W., Cho, J. C., Rhee, J. M., Yoon, S. C., and Lee, H. B. Preparation and characterization of poly(3-hydroxybutyrate-co-3-hydroxyvalerate) microspheres for the sustained release of 5-fluorouracil. *Biomed. Mater. Eng.*, **11**, 89–103 (2001).
127. Bonartsev, A. P., Bonartseva, G. A., Makhina, T. K., Mashkina, V. L., Luchinina, E. S., Livshits, V. A., Boskhomdzhiev, A. P., Markin, V. S., and Iordanskii, A. L. New poly-(3-hydroxybutyrate)-based systems for controlled release of dipyridamole and indomethacin. *Prikl. Biokhim. Mikrobiol.*, **42**(6), 710–715 (2006).
128. Salman, M. A., Sahin, A., Onur, M. A., Oge, K., Kassab, A., and Aypar, U. Tramadol encapsulated into polyhydroxybutyrate microspheres in vitro release and epidural analgesic effect in rats. *Acta Anaesthesiol. Scand.*, **47**, 1006–1012 (2003).
129. Bonartsev, A. P., Livshits, V. A., Makhina, T. A., Myshkina, V. L., Bonartseva, G. A., and Iordanskii, A. L. Controlled release profiles of dipyridamole from biodegradable microspheres on the base of poly(3-hydroxybutyrate). *Express Polymer Letters*, **1**(12), 797–803 (b) (2007).
130. Bonartsev,A. P., Postnikov, A. B., Myshkina, V. L., Artemieva, M. M., and Medvedeva, N. A. A new system of nitric oxide donor prolonged delivery on basis of controlled-release polymer, polyhydroxybutyrate. *American Journal of Hypertension*, **18**(5A), p.A (2005).

131. Bonartsev, A. P., Postnikov, A. B., Mahina, T. K., Myshkina, V. L., Voinova, V. V., Boskhomdzhiev, A. P., Livshits, V. A., Bonartseva, G. A., and Iorganskii, A. L. A new in vivo model of prolonged local nitric oxide action on arteries on basis of biocompatible polymer. *The Journal of Clinical Hypertension Suppl. A.*, **9**(5), A152 (2007) (c).
132. Pouton, C. W. and Akhtar, S. Biosynthetic polyhydroxyalkanoates and their potential in drug delivery. *Adv. Drug Deliver. Rev.*, **18**, 133–162 (1996).
133. Kots, A. Y., Grafov, M. A., Khropov, Y. V., Betin, V. L., Belushkina, N. N., Busygina, O. G., Yazykova, M. Y., Ovchinnikov, I. V., Kulikov, A. S., Makhova, N. N., Medvedeva, N. A., Bulargina, T. V., and Severina, I. S. Vasorelaxant and antiplatelet activity of 4,7-dimethyl-1,2,5-oxadiazolo[3,4-d]pyridazine 1,5,6-trioxide: role of soluble guanylate cyclase, nitric oxide and thiols. *Br. J. Pharmacol.*, **129**(6), 1163–1177 (2000).

17 On Polymer Nanocomposites

A. A. Olkhov, D. J. Liaw, G. V. Fetisov, M. A. Goldschtrakh,N. N. Kononov, A. A. Krutikova, P. A. Storozhenko, G. E. Zaikov, and A. A. Ischenko

CONTENTS

17.1 INTRODUCTION

In recent years, considerable efforts have been devoted for search new functional nanocomposite materials with unique properties that are lacking in their traditional analogues. Control of these properties is an important fundamental problem. The use of nanocrystals as one of the elements of a polymer composite opens up new possibilities for targeted modification of its optical properties because of a strong dependence of the electronic structure of nanocrystals on their sizes and geometric shapes. An increase in the number of nanocrystals in the bulk of composites is expected to enhance long-range correlation effects on their properties. Among the known nanocrystals, nanocrystalline silicon (nc-Si) attracts high attention due to its extraordinary optoelectronic properties and manifestation of quantum size effects. Therefore, it is widely used for designing new generation functional materials for nanoelectronics and information technologies. The use of nc-Si in polymer composites calls for a knowledge of the processes of its interaction with polymeric media. Solid nanoparticles can be combined into aggregates (clusters), and when the percolation threshold is achieved, a continuous cluster is formed.

An orderly arrangement of interacting nanocrystals in a long-range potential minimum leads to formation of periodic structures. Because of the well-developed interface, an important role in such systems belongs to adsorption processes, which are determined by the structure of the nanocrystal surface. In a polymer medium, nanocrystals are surrounded by an adsorption layer consisting of polymer, which may change the electronic properties of the nanocrystals. The structure of the adsorption layer has an effect on the processes of self-organization of solid-phase particles, as well as on the size, shape, and optical properties of resulting aggregates. According to data obtained for metallic [1] and semiconducting [2] clusters, aggregation and adsorption in three-phase systems with nanocrystals have an effect on the optical properties of the whole system. In this context, it is important to reveal the structural features of systems containing nanocrystals, characterizing aggregation and adsorption processes in these systems, which will make it possible to establish a correlation between the structural and the optical properties of functional nanocomposite systems.

Silicon nanoclusters embedded in various transparent media are a new interesting object for physico-chemical investigation. For example, for particles smaller than 4 nm in size, quantum size effects become significant. It makes possible to control the luminescence and absorption characteristics of materials based on such particles using of these effects [3, 4]. For nanoparticles about 10 nm in size or larger (containing $\sim 10^4$ Si atoms), the absorption characteristics in the UV and visible ranges are determined in many respects by properties typical of massive crystalline or amorphous silicon samples. These characteristics depend on a number of factors: the presence of structural defects and impurities, the phase state, and so on [5, 6]. For effective practical application and creation on a basis nc-Si the new polymeric materials possessing useful properties: sun protection films [7] and the coverings [8] photoluminescent and electroluminescent composites [9, 10], stable to light dye [11], embedding of these nanosized particles in polymeric matrixes becomes an important synthetic problem.

The method of manufacture of silicon nanoparticles in the form of a powder by plasma chemical deposition, which was used in this study, makes possible to vary the chemical composition of their surface layers. As a result, another possibility of controlling their spectral characteristics arises, which is absent in conventional methods of manufacture of nc-Si in solid matrices (for example, in α–SiO_2) by implantation of charged silicon particles [5] or radio frequency deposition of silicon [2]. Polymer composites based on silicon nanopowders are a new object for comprehensive spectral investigation. At the same time, detailed spectral analysis has been performed for silicon nanopowder prepared by laser induced decomposition of gaseous SiH_4 (see, for example, [6, 12]. It is of interest to consider the possibility of designing new effective UV protectors based on polymer containing silicon nanoparticles [13]. An advantage of this nanocomposite in comparison with other known UV protectors is its environmental safety, that is, ability to hinder the formation of biologically harmful compounds during UV-induced degradation of components of commercial materials. In addition, changing the size distribution of nanoparticles and their concentration in a polymer and correspondingly modifying the state of their surface, one can deliberately change the spectral characteristics of nanocomposite as a whole. In this case, it is necessary to minimize the transmission in the wavelength range below 400 nm (which determines the properties of UV-protectors [13]) by changing the characteristics of the silicon powder.

17.2 OBJECTS OF RESEARCH

In this study, the possibilities of using polymers containing silicon nanoparticles as effective UV protectors are considered. First, the structure of nc-Si obtained under different conditions and its aggregates, their adsorption and optical properties was studied in order to find ways of control the UV spectral characteristics of multiphase polymer composites containing nanocrystalline silicon. Also, the purpose of this work was to investigate the effect of the concentration of silicon nanoparticles embedded in polymer matrix and the methods of preparation of these nanoparticles on the spectral characteristics of such nanocomposites. On the basis of the data obtained, recommendations for designing UV protectors based on these nanocomposites were formulated.

The nc-Si consists of core shell nanoparticles in which the core is crystalline silicon coated with a shell formed in the course of passivation of nc-Si with oxygen and/or nitrogen. The nc-Si samples were synthesized by an original procedure in argon plasma in a closed gas loop. To do this, we used a plasma vaporizer/condenser operating in a low frequency arc discharge. A special consideration was given to the formation of a nanocrystalline core of specified size. The initial reagent was a silicon powder, which was fed into a reactor with a gas flow from a dosing pump. In the reactor, the powder vaporized at 7,000–10,000°C. At the outlet of the high temperature plasma zone, the resulting gas vapor mixture was sharply cooled by gas jets, which resulted in condensation of silicon vapor to form an aerosol. The synthesis of nc-Si in a low frequency arc discharge was described [3].

The microstructure of nc-Si was studied by transmission electron microscopy (TEM) on a Philips NED microscope. X-ray powder diffraction analysis was carried out on a Shimadzu Lab XRD-6000 diffractometer. The degree of crystallinity of nc-Si was calculated from the integrated intensity of the most characteristic peak at 2θ = 28°. Low temperature adsorption isotherms at 77.3K were measured with a Gravimat-4303 automated vacuum adsorption apparatus. The FTIR spectra were recorded on in the region of 400–5,000 cm^{-1} with resolution of about 1 cm^{-1}.

Three samples of nc-Si powders with specific surfaces of 55, 60, and 110 m^2/g were studied. The D values for these samples calculated by Equation (2) are 1.71, 1.85, and 1.95 respectively, that is, they are lower than the limiting values for rough objects. The corresponding D values calculated by Equation (3) are 2.57, 2.62, and 2.65 respectively. Hence, the adsorption of nitrogen on nc-Si at 77.3K is determined by capillary forces acting at the liquid gas interface. Thus, in argon plasma with addition of oxygen or nitrogen, ultra disperse silicon particles are formed, which consist of a crystalline core coated with a silicon oxide or oxynitride shell. This shell prevents the degradation or uncontrollable transformation of the electronic properties of nc-Si upon its integration into polymer media. Solid structural elements (threads or nanowires) are structurally similar, which stimulates self-organization leading to fractal clusters. The surface fractal dimension of the clusters determined from the nitrogen adsorption isotherm at 77.3K is a structurally sensitive parameter, which characterizes both the structure of clusters and the morphology of particles and aggregates of nc-Si.

As the origin materials for preparation film nanocomposites served polyethylene of low density (LDPE) marks 10803-020 and ultradisperse crystal silicon. Silicon powders have been received by a method plazmochemical recondensation of coarse-crystalline silicon in nanocrystalline powder. Synthesis nc-Si was carried out in argon plasma in the closed gas cycle in the plasma evaporator the condenser working in the arc low frequency category. After particle synthesis nc-Si were exposed microcapsulating at which on their surfaces the protective cover from SiO_2, protecting a powder from atmospheric influence and doing it steady was created at storage. In the given work powders of silicon from two parties were used: nc-Si-36 with a specific surface of particles ~36 m^2/g and nc-Si-97 with a specific surface ~97 m^2/g.

Preliminary mixture of polyethylene with a powder nc-Si firms "Brabender" (Germany) carried out by means of closed hummer chambers at temperature 135 ± 5°C, within 10 min and speed of rotation of a rotor of 100 min^{-1}. Two compositions LDPE + nc-Si have been prepared: (1) composition PE + 0.5% nc-Si-97 on a basis nc-Si-97, containing 0.5 weights silicon %, (2) composition PE + 1% nc-Si-36 on a basis nc-Si-36, containing 1.0 weights silicon %.

Formation of films by thickness 85 ± 5 micron was spent on semiindustrial extrusion unit ARP-20-150 (Russia) for producing the sleeve film. The temperature was 120–190°C on zones extruder and extrusion die. The speed of auger was 120 min^{-1}. Technological parameters of the nanocomposites choose, proceeding from conditions of thermostability and the characteristic viscosity recommended for processing polymer melting.

17.3 EXPERIMENTAL METHODS

Mechanical properties and an optical transparency of polymer films, their phase structure and crystallinity, and also communication of mechanical and optical properties with a microstructure of polyethylene and granulometric structure of modifying powders nc-Si were observed.

Physico-mechanical properties of films at a stretching (extrusion) measured in a direction by means of universal tensile machine EZ-40 (Germany) in accordance with Russian State Standard GOST-14236-71. Tests are spent on rectangular samples in width of 10 mm and a working site of 50 mm. The speed of movement of a clip was 240 mm/min. The five parallel samples were tested.

Optical transparency of films was estimated on absorption spectra. Spectra of absorption of the obtained films were measured on spectrophotometer SF-104 (Russia) in a range of wavelengths 200–800 nm. Samples of films of polyethylene and composite films PE + 0.5% nc-Si-36 and PE + 1% nc-Si-36 in the size 3 × 3 cm were investigated. The special holder was used for maintenance uniform a film tension.

X-ray diffraction analysis by wide-angle scattering of monochromatic X-rays data was applied for research phase structure of materials, degree of crystallinity of a polymeric matrix, the size of single crystal blocks in powders nc-Si and in a polymeric matrix, and also functions of density of distribution of the size crystalline particles in initial powders nc-Si.

X-ray diffraction measurements were observed on Guinier diffractometer: chamber G670 Huber [14] with bent Ge (111) monochromator of a primary beam which are cutting out line $K\alpha_1$ (length of wave $\lambda = 1.5405981$ Å) characteristic radiation of x-ray tube with the copper anode. The diffraction picture in a range of corners 2θ from 3° to 100° was registered by the plate with optical memory (IP detector) of the camera bent on a circle. Measurements were spent on original powders nc-Si-36 and nc-Si-97, on the pure film LDPE further marked as PE, and on composite films PE + 0.5% nc-Si-97 and PE + 1.0% nc-Si-36. For elimination of tool distortions effect diffractogram standard SRM660a NIST from the crystal powder LaB_6 certificated for these purposes by Institute of standards of the USA was measured. Further it was used as diffractometer tool function.

Samples of initial powders nc-Si-36 and nc-Si-97 for x-ray diffraction measurements were prepared by drawing of a thin layer of a powder on a substrate from a special film in the thickness six microns (MYLAR, Chemplex Industries Inc., Cat. No: 250, Lot No: 011671). Film samples LDPE and its composites were established in the diffractometer holder without any substrate, but for minimization of structure effect two layers of a film focused by directions extrusion perpendicular each other were used.

Phase analysis and granulometric analysis was spent by interpretation of the X-ray diffraction data. For these purposes the two different full crest analysis methods [15, 16] were applied: (1) method of approximation of a profile diffractogram using analytical functions, polynoms, and splines with diffractogram decomposition on making parts, (2) method of diffractogram modeling on the basis of physical principles of scattering of X-rays. The package of computer programs WinXPOW was applied to approximation and profile decomposition diffractogram ver. 2.02 (Stoe, Germany) [17], and diffractogram modeling at the analysis of distribution of particles in the sizes was spent by means of program PM2K (version 2009) [18].

17.4 DISCUSSION AND RESULTS

The results of mechanical tests of the prepared materials are presented to Table 1 from which it is visible that additives of particles nc-Si have improved mechanical characteristics of polyethylene.

TABLE 1 Mechanical characteristics of nanocomposite films based of LDPE and nc-Si.

Sample	Tensile strength, kg/cm²	Relative elongation at break, %
PE	100 ± 12	200 – 450
PE + 1% ncSi-36	122 ± 12	250 – 390
PE + 0.5% ncSi-97	118 ± 12	380 – 500

The results presented in the table show that additives of powders of silicon raise mechanical characteristics of films, and the effect of improvement of mechanical properties is more expressed in case of composite PE + 0.5% nc-Si-97 at which in comparison with pure polyethylene relative elongation at break has essentially grown.

Transmittance spectra of the investigated films are shown on Figure 1.

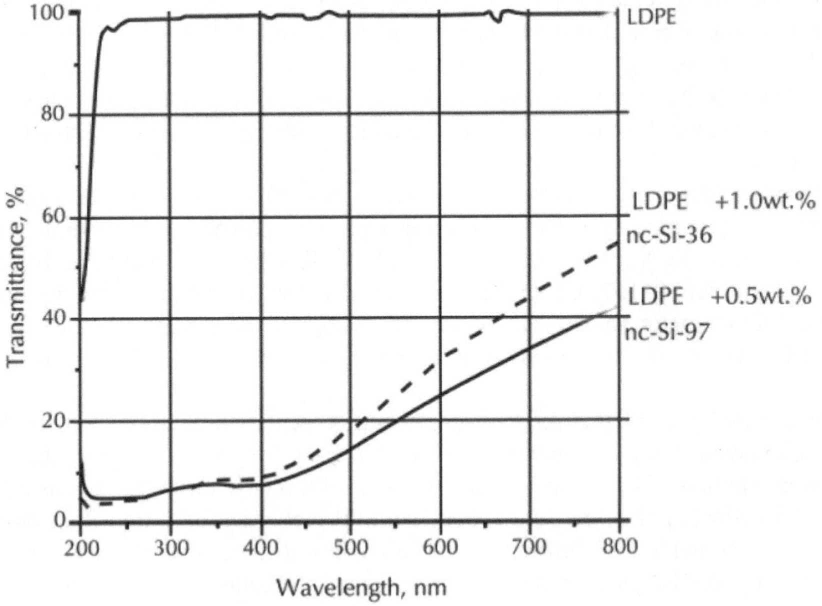

FIGURE 1 Transmittance spectra of the investigated films LDPE and nanocomposite films PE + 0.5% nc-Si-97 and PE + 1.0% nc-Si-36.

It is visible that additives of powders nc-Si reduce a transparency of films in all investigated range of wavelengths, but especially strong decrease transmittance (almost in 20 times) is observed in a range of lengths of waves of 220–400 nm, that is in UV areas.

The wide-angle scattering of X-rays data were used for the observing phase structure of materials and their component. Measured x-ray diffractograms of initial powders nc-Si-36 and nc-Si-97 on intensity and Bragg peaks position completely corresponded to a phase of pure crystal silicon (a cubic elementary cell of type of diamond–spatial group $Fd\bar{3}m$, cell parameter $a_{Si} = 0.5435$ nm).

For the present research granulometric structure of initial powders nc-Si is of interest. Density function of particle size in a powder was restored on X-ray diffractogram a powder by means of computer program PM2K [18] in which the method [19] modellings of a full profile diffractogram based on the theory of physical processes of diffraction of X-rays is realized. Modeling was spent in the assumption of the spherical form of crystalline particles and logarithmically normal distributions of their sizes. Deformation effects from flat and linear defects of a crystal lattice were considered. Received function of density of distribution of the size crystalline particles for initial powders nc-Si are represented graphically on Figure 2, in the signature to which statistical parameters of the found distributions are resulted. These distributions are characterized by such important parameters, as *Mo(d)*–position of maximum (a distribution mode), $<d>_V$–average size of crystalline particles based on volume of the sample (the average arithmetic size) and *Me(d)*–the median of distribution defining the size *d*, specifying that particles with diameters less than this size make half of volume of a powder.

The results represented on Figure 2, show that initial powders nc-Si in the structure has particles with the sizes less than 10 nm which especially effectively absorb UV radiation. The both powders modes of density function of particle size are very close, but median of density function of particle size of a powder nc-Si-36 it is essential more than at a powder nc-Si-97. It suggests that the number of crystalline particles with diameters is less 10 nm in unit of volume of a powder nc-Si-36 much less than in unit of volume of a powder nc-Si-97. As a part of a powder nc-Si-36 it is a lot of particles with a diameter more than 100 nm and even there are particles more largely 300 nm whereas the sizes of particles in a powder nc-Si-97 do not exceed 150 nm and the basic part of crystalline particles has diameter less than 100 nm.

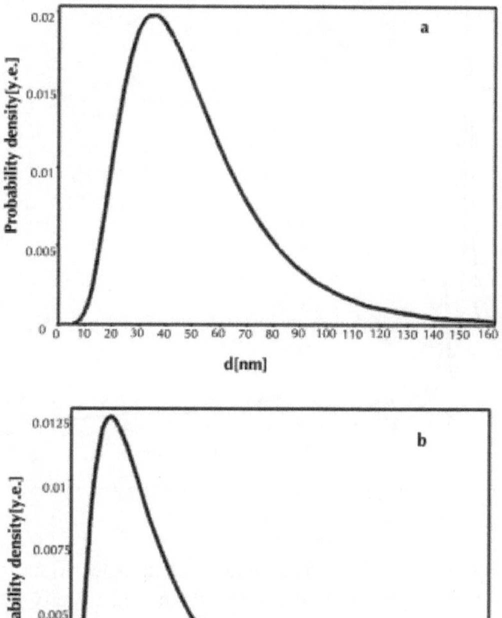

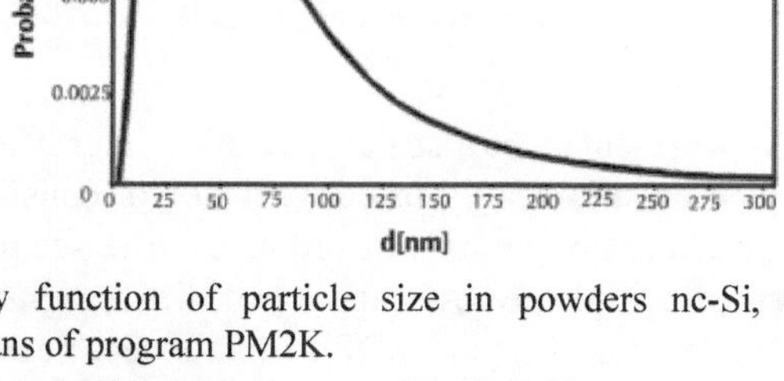

FIGURE 2 Density function of particle size in powders nc-Si, received from x-ray diffractogram by means of program PM2K.

(a) – nc-Si-97	Mo(d) = 35 nm	Me(d) = 45 nm	$<d>_V = 51$ nm;
(б) – nc-Si-36	Mo(d) = 30 nm	Me(d) = 54 nm	$<d>_V = 76$ nm.

The phase structure of the obtained films was estimated on wide-angle scattering diffractogram only qualitatively. Complexity of diffraction pictures of scattering and structure do not poses the quantitative phase analysis of polymeric films [20]. At the phase analysis of polymers often it is necessary to be content with the comparative qualitative analysis which allows watching evolution of structure depending on certain parameters of technology of production. Measured wide-angle X-rays scattering diffractograms of investigated films are shown on Figure 3. Diffractograms has a typical form for polymers. As a rule, polymers are the two-phase systems consisting of an amorphous phase and areas with distant order, conditionally named crystals. Their diffractograms represent [20] superposition of intensity of scattering by the amorphous phase which is looking like wide halo on the small-angle area (in this case in area 2θ between 10° and 30°), and intensity Bragg peaks scattering by a crystal phase.

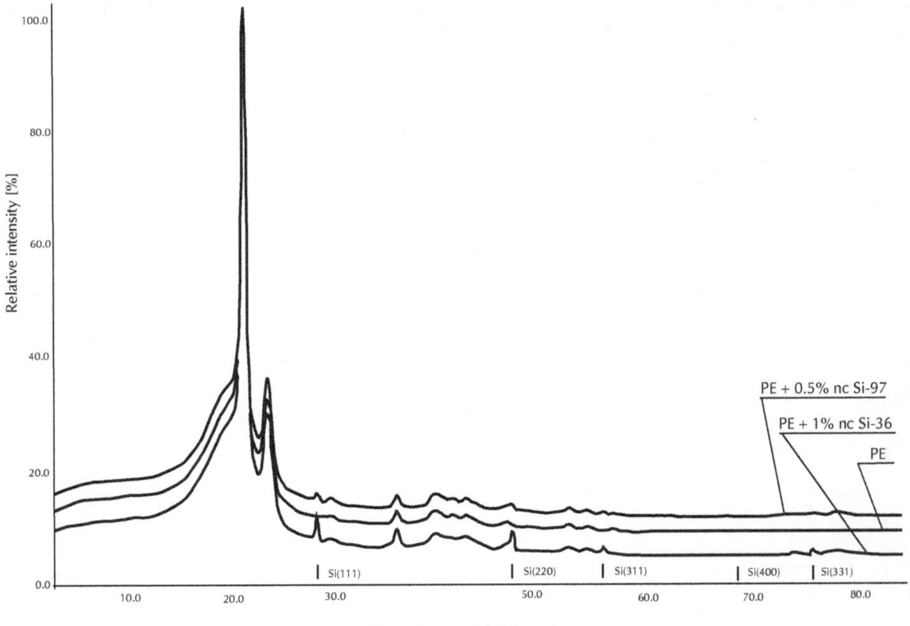

FIGURE 3 Diffractograms of the investigated composite films in comparison with diffractogram of pure polyethylene. Below vertical strokes specify reference positions of diffraction lines of silicon with their interference indexes (hkl).

Data on Figure 3 is presented in a scale of relative intensities (intensity of the highest peak is accepted equal 100%). For convenience of consideration curves are represented with displacement on an axis of ordinates. The scattering plots without displacement represented completely overlapping of diffractogram profiles of composite films with diffractogram of a pure LDPE film, except peaks of crystal silicon which were not present on PE diffractogram. It testifies that additives of powders nc-Si practically have not changed crystal structure of polymer.

The peaks of crystal silicon are well distinguishable on diffractograms of films with silicon (the reference positions with Miller's corresponding indexes are pointed). Heights of the peaks of silicon with the same name (i.e. peaks with identical indexes) on diffractograms of the composite films PE + 0.5% nc-Si-97 and PE + 1.0% nc-Si-36 differ approximately twice that corresponds to a parity of mass concentration Si set at their manufacturing.

Degree of crystallinity of polymer films (a volume fraction of the crystal ordered areas in a material) in this research was defined by diffractograms Figure 3 for a series of samples only semi-quantitative (more/less). The essence of the method of crystallinity definition consists in analytical division of a diffractogram profile on the Bragg peaks from crystal areas and diffusion peak of an amorphous phase [20], as is shown in Figure 4.

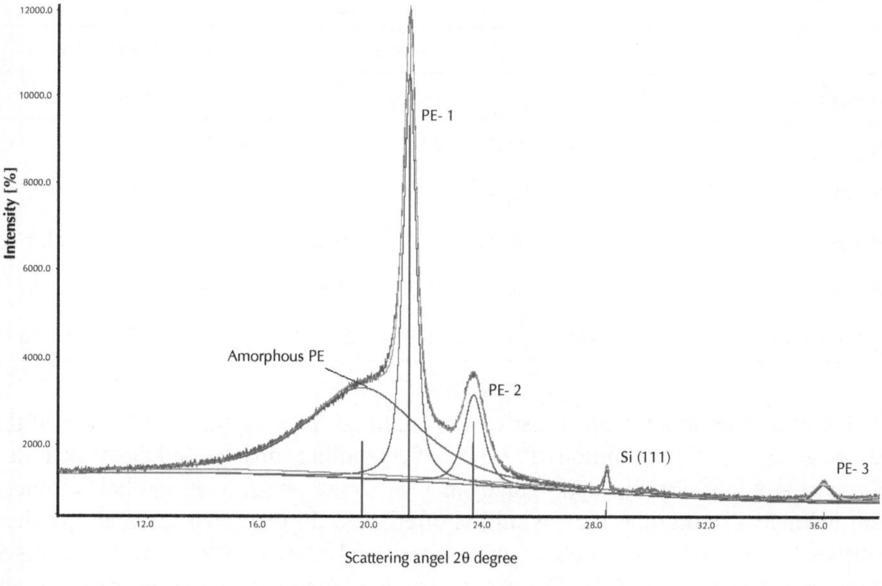

FIGURE 4 Diffractogram decomposition on separate peaks and a background by means of approximation of a full profile by analytical functions on an example of the data from sample PE+1% nc-Si-36 (Figure 3). PE-n designate Bragg peaks of crystal polyethylene with serial numbers n from left to right. Si (111)–Bragg silicon peak nc-Si-36. Vertical strokes specify positions of maxima of peaks.

Peaks profiles of including peak of an amorphous phase, were approximated by function pseudo Voigt, a background four order polynoms of Chebysheva. The nonlinear method of the least squares minimized a difference between intensity of points experimental and approximating curves. The width and height of approximating functions, positions of their maxima, and the integrated areas, and also background parameters were thus specified. The relation of integrated intensity of a scattering profile by an amorphous phase to fully integrated intensity of scattering by all phases except for particles of crystal silicon gives a share of amorphy of the sample, and crystallinity degree turns out as a difference between unit and an amorphy fraction.

It was supposed that one technology of film obtaining allowed an identical structure. It proved to be true by coincidence relative intensities of all peaks on diffractograms Figure 3, and samples consists only crystal and amorphous phases of the same chemical compound. Therefore, received values of degree of crystallinity should reflect correctly a tendency of its change at modification polyethylene by powders nc-Si though because of a structure of films they can quantitatively differ considerably from the valid concentration of crystal areas in the given material. The found values of degree of crystallinity are represented in Table 2.

TABLE 2 Characteristics of the ordered (crystal) areas in polyethylene and its composites with nc-Si.

PE			**PE + 1% nc-Si-36**			**PE + 0.5% nc-Si-97**		
Crystallinity	**46%**			**47,5%**			**48%**	
2θ [°]	d [E]	ε	2θ [°]	d [E]	ε	2θ [°]	d [E]	e
21.274	276	8.9	21.285	229	7.7	21.282	220	7.9
23.566	151	12.8	23.582	128	11.2	23.567	123	11.6
36.038	191	6.8	36.035	165	5.8	36.038	162	5.8
Average values	206	9.5×10^{-3}		174	8.2×10^{-3}		168	8.4×10^{-3}

One more important characteristic of crystallinity of polymer is the size *d* of the ordered areas in it. For definition of the size of crystalline particles and their maximum deformation ε in X-ray diffraction analysis [21] Bragg peaks width on half of maximum intensity (Bragg lines half width) is often used. In the given research the sizes of crystalline particles in a polyethylene matrix calculated on three well expressed diffractogram peaks Figure 3. The peaks of polyethylene located at corners 2 θ approximately equal 21.28°, 23.57°, and 36.03° (peaks PE-1, PE-2 and PE-3 on Figure 4 see) were used. The ordered areas size *d* and the maximum relative deformation ε of their lattice were calculated by the joint decision of the equations of Sherrera and Wilson [21] with use of half width of the peaks defined as a result of approximation by analytical functions, and taking into account experimentally measured diffractometer tool function. Calculations were spent by means of program *WinX^{POW}* size/strain. Received *d* and ε, and also their average values for investigated films are presented in Table 2. The updated positions of maxima of diffraction peaks used at calculations are specified in the table.

The offered technology allowed the obtaining of films LDPE and composite films LDPE + 1% nc-Si-36 and LDPE + 0.5% nc-Si-97 an identical thickness (85 microns). Thus concentration of modifying additives nc-Si in composite films corresponded to the set structure that is confirmed by the X-ray phase analysis.

By direct measurements it is established that additives of powders nc-Si have reduced a polyethylene transparency in all investigated range of lengths of waves, but especially strong transmittance decrease (almost in 20 times) is observed in a range of lengths of waves of 220–400 nm, that is in UV areas. Especially strongly effect of suppression UV of radiation is expressed in LDPE film + 0.5% nc-Si-97 though concentration of an additive of silicon in this material is less. It is possible to explain this fact to that according to experimentally received function of density of distribution of the size the quantity of particles with the sizes is less 10 nm on volume/weight unit in a powder nc-Si-97 more than in a powder nc-Si-36.

Direct measurements define mechanical characteristics of the received films–durability at a stretching and relative lengthening at disrupture (Table 1). The received results show that additives of powders of silicon raise durability of films approximately on 20% in comparison with pure polyethylene. Composite films in comparison with pure polyethylene also have higher lengthening at disrupture; especially this improvement is expressed in case of composite PE + 0.5% nc-Si-97. Observable improvement of mechanical properties correlates with degree of crystallinity of films and the average sizes of crystal blocks in them (Table 2). By results of the X-ray analysis the highest crystallinity at LDPE film + 0.5% nc-Si-97, and at it the smallest size the crystal ordered areas that should promote durability and plasticity increase.

KEYWORDS

- **Amorphous phase**
- **Nanocrystalline silicon**
- **Nanocrystals**
- **Nanoparticles**
- **Transmission electron microscopy**

ACKNOWLEDGMENT

This work is supported by grants RFBR № 10-02-92000 and RFBR № 11-02-00868 also by grants FCP "Scientific and scientific and pedagogical shots of innovative Russia", contract № 2353 from 17.11.09 and contract № 2352 from 13.11.09.

REFERENCES

1. Karpov, S. V. and Slabko, V. V. *Optical and Photophysical Properties of Fractally Structured Metal Sols*. Sib. Otd. Ross. Akad. Nauk, Novosibirsk (2003).
2. Varfolomeev, A. E., Volkov, A. V., Godovskii, D. Yu., et al. *Pis'ma Zh. Eksp. Teor. Fiz.*, **62**, 344 (1995).
3. Delerue, C., Allan, G., and Lannoo, M. *J. Lumin.*, **80**, 65 (1999).
4. Soni, R. K., Fonseca, L. F., Resto, O., et al. *J. Lumin.*, **83–84**, 187 (1999).
5. Altman, I. S., Lee, D., Chung, J. D., et al. *Phys. Rev. B Condens. Matter Mater. Phys.* **63**, 161402 (2001).
6. Knief, S. and von Niessen, W. *Phys. Rev. B: Condens. Matter Mater. Phys.*, **59**, 12940 (1999).
7. Olkhov, A. A., Goldschtrakh, M. A., and Ischenko, A. A. RU Patent № 2009145013 (2009).

8. Bagratashvili, V. N., Tutorskii, I. A., Belogorokhov, A. I. et al. Reports of Academy of Sciences. *Physical Chemistry*, **405**, 360 (2005).
9. V. Kumar. (Ed.). *Nanosilicon*. Elsevier Ltd., **13**, 368 (2008).
10. *Nanostructured Materials Processing, Properties, and Applications*. Carl C. Koch. (Ed.) William Andrew Publishing, New York p.752 (2009).
11. Ischenko, A. A., Dorofeev, S. G., and Kononov, N. N., et al. RU Patent №2009146715 (2009).
12. Kuzmin, G. P., Karasev, M. E., Khokhlov, E. M., et al. *Laser Phys*. **10**, 939 (2000).
13. Beckman, J. and Ischenko, A. A. RU Patent №. 2 227 015 (2003).
14. Stehl, K. The Huber G670 imaging-plate Guinier camera tested on beamline I711 at the MAX II synchrotron. *J. Appl. Cryst.*, **33**, 394–396 (2000).
15. Fetisov, G. V. *The X-ray phase analysis*. **11**, 153–184. *Analytical chemistry and physical and chemical methods of the analysis*. T. 2./Red. A.A. Ischenko. M. ITc Academy, p.416 (2010).
16. Scardi, P. and Leoni, M. Line profile analysis pattern modeling versus profile fitting. *J. Appl. Cryst.*, **39**, 24–31 (2006).
17. WINX^{POW} Version 1.06. STOE & CIE GmbH Darmstadt/Germany (1999).
18. Leoni, M., Confente, T., and Scardi, P. PM2K a flexible program implementing Whole Powder Pattern Modeling. *Z. Kristallogr. Suppl.*, **23**, 249–254 (2006).
19. Scardi, P. Recent advancements in whole powder pattern modeling. *Z. Kristallogr. Suppl.*, **27**, 101–111 (2008).
20. Strbeck, N. *X-ray scattering of soft matter*. Springer-Verlag Berlin Heidelberg. **20**, 238 (2007).
21. Iveronova, V. I. and Revkevich, U. P *The theory of scattering of X-rays*. MGU, Mascow p. 278 (1978).

18 On Morphology of Polyhydroxybutyrate

A. A. Olkhov, A. L. Iordanskii, M. M. Feldshtein, and G. E. Zaikov

CONTENTS

18.1 INTRODUCTION

It is in process installed, that to growth of polarity of dissolvent there is a perfection of a crystal structure and decrease of a share of amorphous phase polyhydroxybutyrate (PHB). It is shown, that a share of a defective crystal phase in films PHB directly proportional to magnitude of cohesive energy of dissolvent.

To giving products from polymeric materials of a necessary complex of operational properties apply various aspects of chemical, physical, or physical and chemical modification at a stage of preparation of raw materials or in the course of product moulding. Thus, there is a radical structural change of a polymeric material at various levels of the structural organization [1].

In this connection there is a necessity for interconnection determination between factors of affecting and change of parameters of structure of a polymeric material. Determination of such interconnection allows changing targeted properties of a polymeric system depending on a range of application of the yielded product.

As a rule, the majority of biopolymers of a bacterial origin is very sensitive to elevated temperature, therefore, moulding of products from them carry out from a solution.

It is especially important for products of medical appointment, for example, matrixes for controllable liberation of medical products as medicine introduction in polymer is carried out in the core through a solution [2].

One of the progressive polymers applied in medicine, is poly(3-hydroxybutyrate) (PHB). This bacterial biopolymer possesses high crystallinity [3].

Therefore, changing degree of crystallinity of polymer at a stage of formation of films from a solution, it is possible to control over a wide range their transport properties [4].

The purpose of the yielded work consisted in definition of influence of dissolvent of various polarities on formation of structure PHB of films.

18.2 OBJECT AND RESEARCH METHOD

As subjects of inquiry served weed PHB Lot M-0997 (Germany, "Biomer"), M_h = 2,50,000, Solvents was chemically pure dioxane, chloroform, formic acid, and dichloroethane. The PHB dissolved in the specified solvents at temperature 90°C and jolted on a glass surface. For residual solvent removal the generated films subjected to heat treatment in pressure (0, 2 bar) at temperature 110°C in a current of 30 min.

Research of structure of films made method differential scanning calorimeter (DSC) at speed of scanning of 20 wasps/min by mean of a differential scanning calorimeter of firm "Mettler" (USA).

18.3 DISCUSSION AND RESULTS

In Figure 1 typical thermograms of heating PHB of films are presented. From the analysis of thermograms follows, that after machining by dissolvents initial structure PHB is broken: there is the amorphous phase characterized by obviously expressed transferring of glass transition and there is a fine-crystalline modification or areas with the broken crystallinity. Emersion of low temperature peak of fusion testifies to it.

In Table 1 thermal parametres of films on the basis of PHB and parametres of dissolvents are presented.

From the table it is visible, that to growth of electrical dipole moment of dissolvent there is a perfection (streamlining) of crystal phase PHB. Growth of melting point of polymer testifies to it, see Figure 2.

Simultaneously with it heat capacity PHB decreases. It can be connected with decrease of a share of an amorphous phase of polymer.

Molecule PHB contains an ester group in the main circuit, capable to enter interacting with polar molecules of dissolvent [5].

The change of conformation of macromolecules in crystal and amorphous regions of polymer under the influence of dissolvent molecules is thus possible. The structure of crystallites thereof is stabilized, and macromolecules in an amorphous phase lose flexibility. The yielded effects have been revealed by the authors [6], investigating agency of molecules of water on films from PHB. They have installed, that under the influence of polar molecules of water communication between links of the next chains, result inning stabilization of a crystal structure of polymer is realized. Formed on boundary line of crystallites the grid of hydrogen bridges promotes more strongly pronounced orientation of crystallites in a polymer matrix.

In an observed number of the dissolvents, having electrical dipole moment more low or above (dichloroethane), than at water (1,84 D) [7], the described phenomena can proceed to a lesser degree owing to presence in films PHB of crystallites with the defective structure, having the downgraded melting point, see Figure 1.

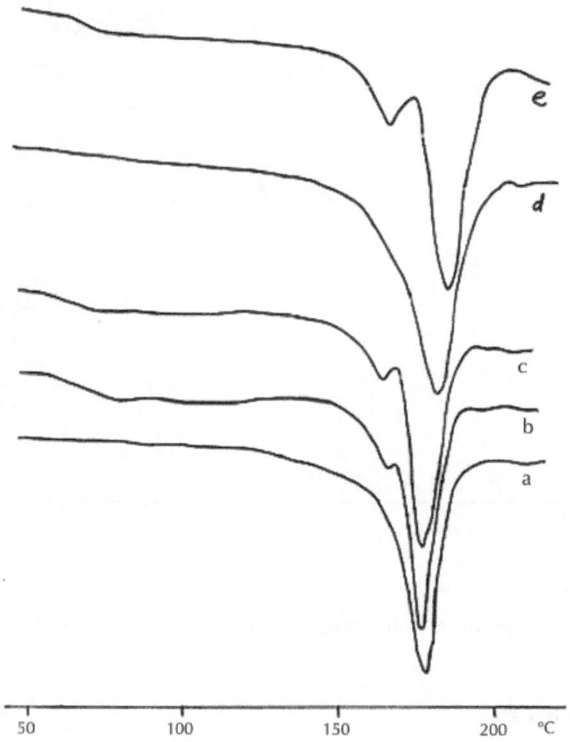

FIGURE 1 Thermograms of heating of films on the basis of PHB.
a = initial PHB
b = PHB from dioxane
c = PHB from chloroform
d = PHB from formic acid
e = PHB from dichloroethane

TABLE 1 Thermal characteristics PHB films from various solvents.

Solvents	ΔG_m, J/g	T_m, °C	T_g, °C	ΔC_p, J/g K	μ, D	$e^{1/2}$
	Polymer films parameters				Solvents characteristics	
dioxane	– 0,076	174,3	63,0	0,39	0,45	10,05
chloroform	– 0,140	175,6	53,9	0,24	1,15	9,3
formic acid	– 0,106	176,2	–	–	1,4	13,5
dichloro-ethane	– 0,173	178,5	74,0	0,1	2,06	9,0

ΔG_m, J/g - Gibbs energy melting; T_m, °C - melting temperature; T_g, °C - glass-transition temperature; ΔC_p, J/g K - specific heat; μ, D - dipole moment - [9, 10]
$e^{1/2}$ - cohesions energy density - [8].

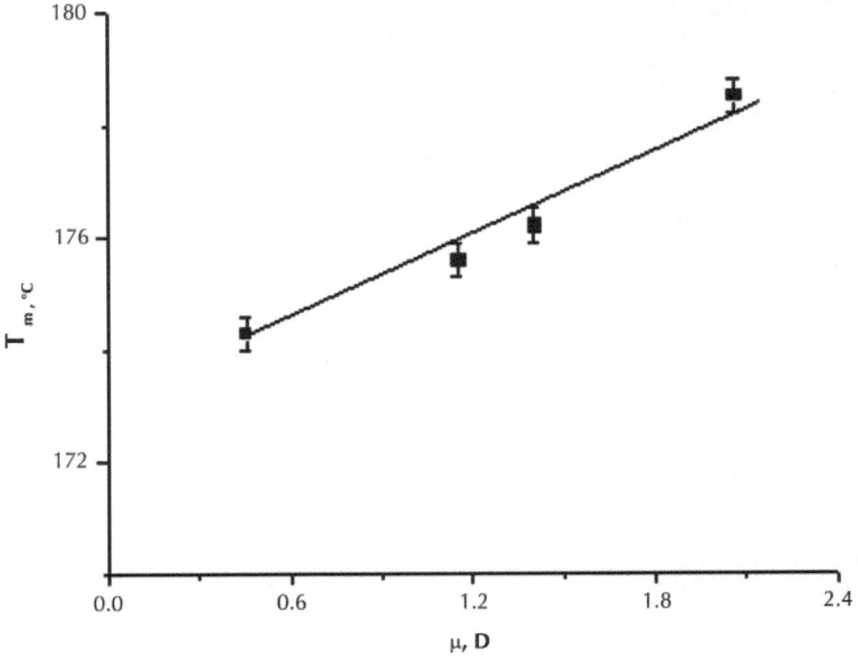

FIGURE 2 Dependence of melting point PHB (T_m) from electrical dipole moment of dissolvents (μ).

18.4 CONCLUSION

Summarizing all told, it is possible to conclude that, with increase in polarity of dissolvent its ability to enter interacting with polar groups PHB increases. As a result of this interacting there is a perfection of crystallites and consolidation of an amorphous phase of polymer in the course of film formation.

Inprocess also it has been installed, that character change energy of Gibbs fusion of crystal phase PHB and magnitude of cohesive energy density of molecules of dissolvent are practically identical in certain sequence, see Table 1.

It is known, that magnitude of cohesive energy density characterizes intermolecular interaction extent in substance [8].

From Table 1, it follows that with decrease of extent of intermolecular interaction in dissolvent were increase the disorder in a crystal phase of films PHB, characterized by decrease of Gibbs energy of fusion.

In other words, the intermolecular interaction in dissolvent, the mobility of its molecules and, hence, speed of their transferring in a vaporous phase (transpiration) there is less. Owing to various speed change the phase of molecules of observed dissolvents, formation of films PHB proceeds differently. It is expressed in extent of completeness of process of crystallization and parametres of a crystal structure of polymer.

On the basis of the made researches, it is possible to formulate following leading-outs:

(1) To increase in polarity of dissolvent occur perfection of a crystal structure and the share of amorphous phase PHB decreases.
(2) With decrease of extent of intermolecular interaction in dissolvent the share of a defective crystal phase in PHB films grows.

KEYWORDS

- **Amorphous phase**
- **Biopolymers**
- **Gibbs fusion**
- **Polyhydroxybutyrate**
- **Vaporous phase**

ACKNOWLEDGMENTS

The authors are grateful to U. J. Hanggi (Biomer company, Germany) for kindly providing the PHB sample for investigation.

REFERENCES

1. Kryzhanovsky, V. K. and Burlov, V. V. *The applied physics of polymeric materials*. SPb. SPbSTI (technical university), (2001).
2. *Biopolymers*. J. M Imanisi.(Ed.). the World, Moscow (1988).
3. Seebach, D., Brunner, A., Bachmann, B. M., et al. *Biopolymers and Oligimers of (R)-3-hydroxyalkanoic Acids*. Ernst Scherring Research Foundation, Berlin (1995).
4. Iordanskii, A. L., Kamaev, P. P., and Hanggi, U. J. *J. of Appl. Polym. Sci.*, **76**(6), 475 (2000).
5. Yokoushi, M., Chatani, Y., Tadokoro, H, et al. *Polymer*, 14(6), 267 (1973).
6. Iordanskii, A. L., Kamaev, P. P., and Zaikov, G. E. *Oxidation Communications*, **21**(3), 305 (1998).
7. The short directory of physical and chemical magnitudes. *Chemistry*, Leningrad (1983).
8. Herausgegeben von Holzmuller, W. and Altenburg, K. Physik der Kunststoffe (Ed.). Akademie - Verlag, Berlin (1961).
9. The chemical's directory. B. P. M. Nikolsky (Ed.). *Chemistry*, (1961–1968).
10. The short chemical encyclopedia. M: the Soviet encyclopaedia, (1961).

19 On the Influence of Melafen - Plant Growth Regulator

O. M. Alekseeva and G. E. Zaikov

CONTENTS

19.1 INTRODUCTION

The effects of plant growth regulator Melafen to the some animal cell metabolic pathways were investigated at our work. Earlier literature data recognized that the fluctuations of animal cellular volume are correlated with some Ca^{2+} dependent cellular metabolic pathways. The cellular volume was registered by spectral method – the first light scattering, of diluted cellular suspensions that were incubated under the Melafen aqua solutions at wide concentration region (10^{-21}-10^{-3} M). The Ehrlich ascetic carcinoma (EAC) cells, that have the metabotropic purinoreceptors P2Y at its surface, showed two maximums at light scattering kinetic curve that correlated with two maximums of volumes increasing after the one addition of adenosine triphosphate (ATP). The leukocytes, which have the Ca^{2+} channel former purinoreceptors P2X at its surface, showed that the lag phase at light scattering kinetic curve before the cellular volume increasing after the ATP addition was in dependence of Melafen concentration. The thymocytes, which have the Ca^{2+} channel former purinoreceptors P2Z (and other too) at its surface, showed two maximums at light scattering kinetic curve after the one addition of ATP also. But the second answer that correlated with the CRAC activation was in bimodal dependence of Melafen concentration. We conclude that the three main types of purinoreceptors: metabotropic P2Y and two Ca^{2+} channel formers P2X and P2Z were under the direct Melafen actions. And CRAC - CIF-activated Ca^{2+} store regulated Ca^{2+} channel, was under the indirect Melafen actions. Thus, the three first points of purine dependent Ca^{2+} transduction pathways—three types of receptors, are under the direct Melafen influence.

This investigation deals with the of melamine salt of bis (oximethyl) phosphinic acid – Melafen, that was synthesized by works of laboratory of A. E. Arbuzov Institution of Organic and Physical Chemistry [1]. Researsher tested the influence this hydrophilic substance at the wide concentration range (10^{-21}–10^{-3} M) on the functional properties of animal cells.

The animal cells must have a number of the variable targets for hydrophilic substance action.

If we deal with the hydrophobic substance, we must notes the certain points. The hydrophobe substances go into the membrane directly or to the hydrophobe pockets into the protein molecules, and are incorporated there, and then the substances are dissipated among hydrophobic phase, or formed its own phase in the cellular compartments [2]. Its action has a negligible specificity without certain targets almost always. The hydrophobe substances have the possibility to show its specificity at protein hydrophobe pocket only. It can influence to the micro viscosity of lipid and protein membrane components [3]. Its actions to the protein structure and functions may be mediated by hydrophobic phase changing or immediate by hydrophobic targets bindings.

On the contrary, the hydrophilic substance, as the Melafen, may have the contacts with any charged or polar surfaces. Thus, the molecules of Melafen may have their interactions with a number nonspecific targets. But the incorporation to the membrane and dissipate among the lipid molecules does not happen. The targets are unknown, but the aftermath's actions of Melafen were known. We will describe the overall results of Melafen actions.

From literary data it is known that the hydrophilic Melafen change the fatty acid composition and lipid and protein microviscosity of cellular microsome and mitochondrial membranes of vegetable [2]. Low concentrations (from 4×10^{-12} to 2×10^{-7} M) of Melafen changed the structural characteristics of plant and animal cell membranes. The Melafen changes the microviscosity of free bilayer lipids and annular lipids bounded with protein clusters with different effective concentrations for plant and animal membranes. The Melafen decreased the level of lipid peroxide oxidation (LPO) in biological membranes under bed environment conditions also. The Melafen concentrations that affect to the microviscosity of free and annular lipids, decreases the intensity of LPO processes too [2]. On a foundation of the research conducted, authors assumed that high physiological activity of Melafen is linked to its actions to the physical and chemical state of biological membranes, resulting in change of lipid-protein interaction, influencing the activity by membrane-associated enzymes and channels. The Melafen increases the effectiveness of energy metabolism of plant cells [3]. The exclusive influence of Melafen was shown on electron transport in respiratory chain of mitochondria [4], and stress resistance of vegetables and cereal corn in bed environment conditions as result [5]. The seed treatment by the Melafen or Pyrafen (Melafen analog) reduces the intensity of lipid peroxidation and strengthens the mitochondria energy of 6 day pea seedlings, which have undergone stress in conditions of moistening shortage and moderate cooling [6]. The Melafen changes the fatty acid composition of mitochondria greatly in the presence of its effective concentrations [7]. These are the reason why the crop yield much rose.10 The influences to the animal microsomes and mitochondrial were showed also [4–8].

The laboratory, investigated the possibility of Melafen influence to the first targets at animal objects--the protein in blood vascular system, cellular membranes, and its components. We found that Melafen under the large concentrations changes greatly (may be loosed) the quarter structure of bovine serum albumin (BSA) – the main soluble protein in blood vascular system. The intrinsic fluorescence of two tryptophanils that are contained at the BSA molecule was quenched because the tryptophanils became access to the main quencher – H_2O [11].

But to the integral membrane bounded proteins Melafen influences were negligible even under the large concentrations. Thus, the organization of protein microdomains of erythrocyte ghost did not change, that we registered with aid of differential scanning microcalorymetry (DSK). The thermo induced parameters of usual cytoskeleton protein components of cellular membrane—spektrin, ankyrin, actin, demantin, fragments of ion channels and other, stood unchanged under the Melafen presence [12].

However, the thermo induced parameters of lipids microdomains at membranes multulammelar liposomes, formed by individual neutral saturated phospholipids dimyristoilphosphatidylcholine (DMPC), were changed greatly under the Melafen influence at the wide concentration diapason. Having the using of the specific method of membrane extraction from steady state - the different rates of heat supplied to the cell with liposomes suspension, received the glaring picture of dependence of thermally induced parameters--enthalpies, maximum temperatures thermally induced transition and cooperatives transition, under Melafen concentrations in the range 10^{-17}–10^{-3} M. The main extreme was under the 10^{-14}–10^{-8} M of Melafen for all rates (1 °C/min, 0, 5 °C/min; 0, 25 °C/min; 0, 125 °C/min) of heat supplied to the cell with liposomes suspension [13].

But the reciprocal location and density of packaging of membranes at multulammelar liposomes did not change under the Melafen influence at the wide concentration diapason. At this case, the membranes of multulammelar liposomes were formed by egg lecithin that is the mix of natural saturated and no saturated, neutral and charged phospholipids [14]. At that type of membrane, where the belayer structure is reinforced by balancing on charge and by the location of length of fatty acids tails at nature phospholipids mix, the membrane thickness from data of X-ray diffraction method does not change under the Melafen presence [15]. The microdomains organizations, concerning in such mixture, we say nothing. Since the application of differential scanning calorimetry for lipids mixture have been hampered.

The structure (and functions and fate may be) of native cells were under Melafen influenced too. The small doses of 10^{-11}–10^{-5} M Melafen influence *in vivo* on the morphology of the erythrocytes in mice. We obtained the decrease of height, area, and volume of the atomic force microscopy (AFM) image of red blood cells, that registered by AFM [16].

The low doses of 10^{-12}–10^{-5} M Melafen influence to the fate of animal tumor cells *in vitro*. The content of protein "labels of apoptosis"--protein regulator p53 (increase) and antiapoptosis protein Bcl-2 (decrease), that were showed by immunoblotting methods at the EAC cells. This fact indicated that the apoptosis was developed under the 1, 5 hr of Melafen action. And similar dozes of Melafen *in vivo* suppressed the growth of Luis carcinoma [17].

Thus, Melafen looses the structure of the soluble protein, changes the microdomaine organization of DMPC membrane and does not change the structural organizations of ghost and egg lecithin membranes. But Melafen global change the erythrocyte morphology and delayed the rate of growth of solid Luis carcinoma.

Taking into account, data obtained the A. E. Arbuzov Institution of Organic and Physical Chemistry, about formation by the Melafen in aqueous solutions of supramolecular structures involving of water molecules [18]. It can be assumed that just such the structures change the microdomaines organization in attachable delicately organized and labile structures. This assumes may be real only for interactions between biological objects with Melafen in aqua solutions. But in the cellular interior there is not the sufficient amount of free waters molecules for such supramolecular structures formation. It is hard to suppose, that the linkage with Melafen will find stronger, than with nature cellular chelating agents of water. To withdraw, the water molecules from the coats of cell components in the presence of Melafen are not likely. On the contrary, to structure the bulk water and the neared membrane water layers it is quite likely.

However, at present chapter, we are emphasized the attention to the Melafen aqua solutions actions at the animal cellular level *in vitro*.

This is why, we tried to clear up the most possible specific targets for Melafen on animal's cells surface. And several types of easily emitted cells of different etiology or different origin were having picked up. The main task of work was the investigated the Melafen action to the three types of cellular plasmalemmal receptors – purinoreceptors P2Y, P2X, P2Z that present at the three types of cells—EAK cells, thymocytes, and lymphocytes.

The Melafen is a melamine salt of bis (hydroximethyl) phosphinic acid. It is a hydrophilic poly function substance with multi targets for its actions (Figure 1). There are the phosphoric, hydroximethyl groups, and nitrogen contained structures at Melafen molecule, potentially pointed to the certain targets at the biological cells. The pretreatment of crop seeds by aqueous solutions of Melafen at concentration 10^{-11}–10^{-9} M increased the yield of plant production by 11% or more due to the increasing of plants stress tolerance under the bed environment [10]. But the increasing of the concentration of Melafen to 10^{-7} M and higher inhibits the processes of plants growth completely [19]. Therefore, studies were carried out at a wide range of concentrations (10^{-21}–10^{-2} M). The main purpose was to determine how the aqueous solutions of Melafen in a wide range of concentrations influence to the function of animal cells *in vitro* only.

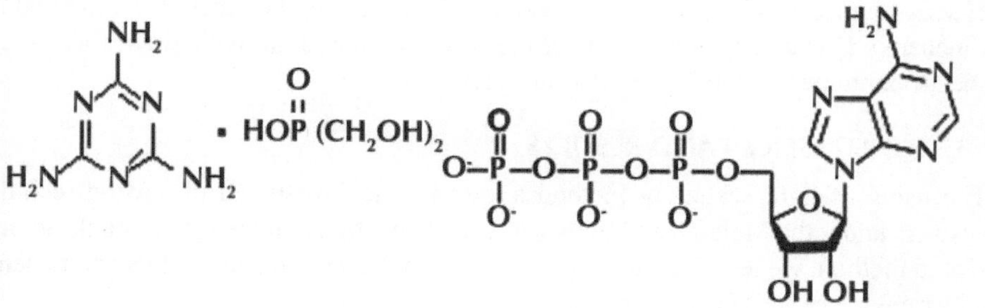

FIGURE 1 Structural formula of Melafen and ATP.

It can be assumed that some fragmental structural similarity of Melafen - organophosphorus plant growth regulator, and ATP molecules (Figure 1 and 2) may define the binding with similar active sites. As a result, we may observe the both substances actions on purine receptors. But the vectors of consequences of Melafen and ATP molecules actions are opposed to each. Also we must note that the global activating influence of ATP applications to the Ca^{2+} signal transductions at the EAK cells at the 7–8 days of development that was described by Zamai [20] and Zinchenko [21]. The main metabolic pathways points are indicated and at this case [22]. It deal with the testing of Melafen actions to the ATP-binding purinoreceptors at the base of some similarity of ATP and Melafen molecular construction.

19.2 EXPERIMENTAL PART

19.2.1 Materials

The ATP ("Bochringer"Germany), HENKS (138 mM NaCl, 5, 4 mM KCl, 1, 2 mM $CaCl_2$, 0,4 mM KH_2PO_4, 0, 8 mM $MgSO_4$, 0, 3mM Na_2HPO_4, and 5, 6 mM D-glucose и 10 mM HEPES pH 7, 2), cytrat Na (PAN EKO Russia), DMPC dimyristoilphosphatidylcholine ("Sigma"), NaCl, KCl, $CaCl_2$, KH_2PO_4, $MgSO_4$, Na_2HPO_4, HEPES ("Sigma), A23187 (Sigma).

19.2.2 Methods

The cells EAK received by methods [21, 23]. The EAK induced at pubescent white mice of males of NMRI introduction intraperitoneally on 10^6 cells of diploid strain of EAC. The cells EAK insulated from mice on 7-e days after the transplantation.

Thymocytes insulated by the pulping through the capron net from thymus the white Wister rat. The cells were washed out three times centrifugation when 800 rate/min. 10 min on medium of HENKS when 4°C. The pellet resuspended on medium of HENKS in concentration about $1–5\times10^8$ cells/ml.

The pooled fraction of lymphocytes and platelets received after spontaneous deposition erythrocytes from the blood of white Wister rat in the presence of medium of HENKS (1: 1) and 5% citrate Na. The cell viability was estimated on cells overtone 0, 04% trypan blue and compiled in all experiences not less than 95%.

The registration of light diffusion by dilute suspension of EAK cell, of thymocytes, of lymphocytes and platelets, was held by the method Cornet [23], modified by Zinchenko [21] at a right angle under the wavelength 510 nm with aid of fluorescent spectrophotometer "MPF-44B" PerkinElmer.

19.3 DISCUSSION AND RESULTS

The using of light scattering method allowed us to investigate the overall cellular answers under the Melafen additions without any artificial messengers. Thanks to presented method, we tested the action of Melafen to the animal cells and its components under a wide range of concentrations.

The three cellular objects were used, as the dilute cells cellular suspension: EAC cells - transformed cells with uncontrolled growth, and normal cells thymocytes and lymphocytes (the white fraction of blood without the platelets and erythrocytes).

First EAC cells, as a good model of cells with the complete cellular transduction system. The active P2Y purinoreceptors are presented at the cellular plasmalemma surface at the 7–8 days of carcinoma growth [20]. Thus the ATP or ADP additions initiated the Ca^{2+} signal transduction (Figure 2). The ATP is the first messenger that deals with the extracellular signal transductions pathways. The ATP is released to the extracellular space. At least two subtypes of receptors for extracellular ATP are currently known: the G-protein–coupled P2Y receptors, which are methabotrophic receptors, and the ATP-gated cation channels classified as P2X receptors (and its subspecy – P2Z receptors). The EAC cells gave a typical cellular response to a signal (ATP-addition). It sent the signal from the cell surface to inside the endoplasmic reticulum (ER) $InsP_3$-receptor, and backward to the CRAC at the cell surface.

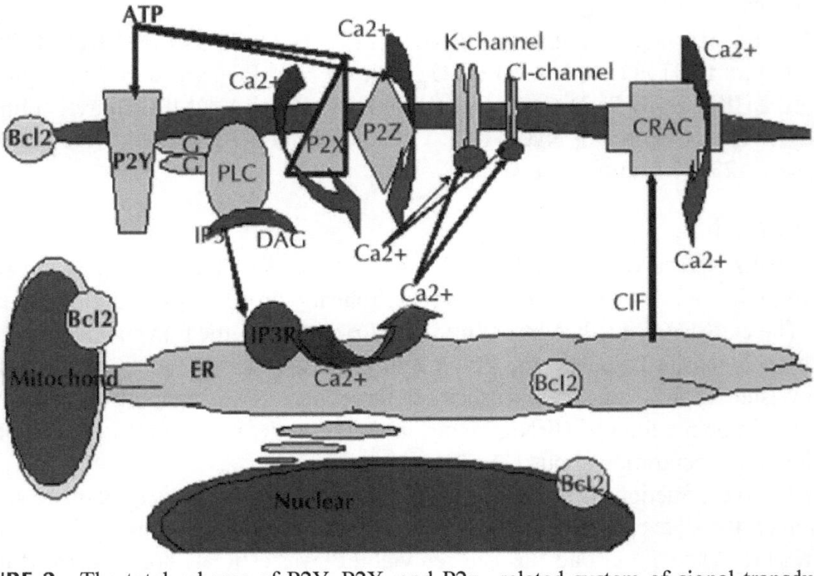

FIGURE 2 The total scheme of P2Y, P2X, and P2z –related system of signal transduction (scheme was modified from Alekseeva O.M. 2010).

At the Figure 2 shows scheme of the purine-dependent Ca^{2+} transduction at the cells. The metabotropic purinoreceptors P2Y throughout the G-proteins activate PLC that produces $InsP_3$, then $InsP_3$ bounds with $InsP_3$-receptors, that release Ca^{2+} from ER. And retrograde signal CIF (related to ER Ca^{2+} store depletion) go to the cellular surface, and the large flow of Ca^{2+} introduced to the cell throughout the Ca^{2+} release-activated Ca^{2+} channels (CRAC) [24]

So, after ATP addition occur two large increasing of intracellular Ca^{2+} concentrations. As a result Ca^{2+}-dependent K^+ и Cl^- channels are activated. The volume of cells changed, because these channels regulate the cell volume. The compensators H_2O flows occurred to the cellular interiors. And cellular volumes are increased greatly and bitterly. Than Ca^{2+} concentrations are decreased quickly, Ca^{2+} is attenuated by mitochondrial, or ER or it is removed from the cell. After than Ca^{2+} concentration in the cell increases smoothly again. Ca^{2+} dependent K^+ и Cl^- channels are activated again. The cellular volume increase slowly, too. Thus, the light scattering of dilute suspension of EAC cells were changed too twice. There are two large increasing of intracellular Ca^{2+} concentrations, as the result we obtain two maximums at the light scattering kinetic curves.

We recorded of right angle light scattering of dilute suspension of EAC cells with PerkinElmer-44B spectrophotometer at a wavelength of 510 nm. The control samples showed the bimodal cellular responses (Figure 3). Urgently after the first addition of ATP to the cell suspension the big maximum appeared quickly on the kinetic curve (first peak). Then it bust momentary. After that the low plate is appeared. And than the slow rise up is developing to plate, that is equal or above to the reference levels (second peak). When we repeated the addition of ATP, the picture was repeated. Melafen inhibited the response development by the dozes dependent manner (Figure 3).

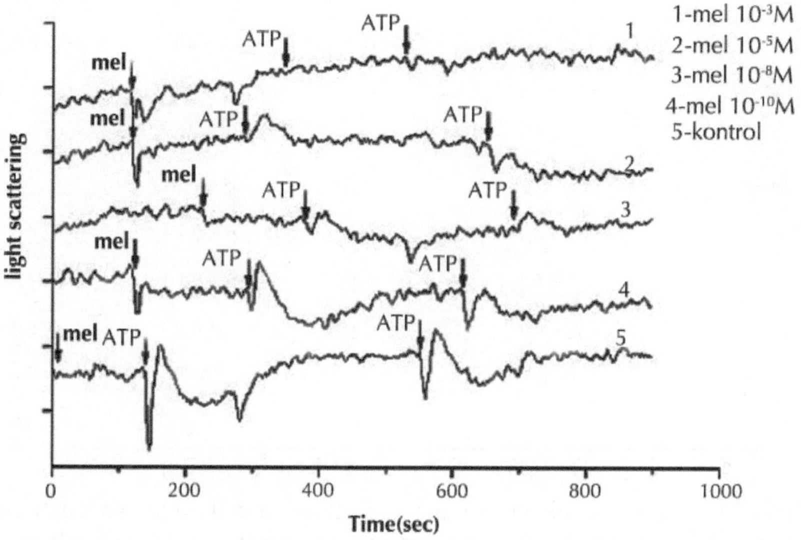

FIGURE 3 The influences of Melafen to the first and second cellular answer of EAC cells. Kinetic curves of light scattering of the cell suspension under the wide concentrations region of Melafen.

But Melafen has a bidirectional effect to the EAC cells (Table 1). We founded that at super low concentrations Melafen (10^{-12} and 10^{-13} M) stimulated the signal transduction, increasing the Ca^{2+} releasing from intracellular Ca^{2+} store (the first peak). But under the bigger concentrations the Melafen actions changed its vector and the depressing of the overall cell responses were began. At the case, when the Melafen concentrations were increased, the amplitudes of first and second extremes were decreased. The second cell response – the Ca^{2+} entering through the plasma membrane (second peak) was not activated by Melafen (under the anywhere concentrations), and it shown the bigger sensitivity to the Melafen action. It was inhibited by the Melafen at the smaller concentrations. Thus, Melafen 10^{-7} M decreased by 50% the first extreme and by 70% – the second extreme. Melafen 10^{-4} M really eliminated the overall cell response fully; it inhibited the purine-dependent Ca^{2+} transduction – both peaks. Hence, the carcinoma cells that characterized by uncontrolled cell division, can be depressed by such doses of Melafen, which are harmless for erythrocytes that we obtain earlier [16].

TABLE 1 The influences of Melafen to the first (A) and second (B) cellular answer of EAC cells at the wide concentrations diapason.

The melafen influences to EAC cell bimodal responses (1 peak and 2 peak) under the ATP adding				
Sample +melafen	**1 peac (rel.un.)**	**D (%)**	**1 peac (rel.un.)**	**D (%)**
control	27,5 ± 0,1	-	20 ± 0,1	-
$+10^{-13}$ M	35 ± 0,1	+27 ± 0,01	19 ± 0,1	-5 ± 0,01
+ 10^{-12} M	33 ± 0,1	+20 ± 0,01	19 ± 0,1	-5 ± 0,01
$+10^{-11}$ M	27,5 ± 0,1	0	13 ± 0,1	-35 ± 0,01
$+10^{-10}$ M	25 ± 0,1	0	13 ± 0,1	-35 ± 0,01
$+10^{-9}$ M	22 ± 0,1	-20 ± 0,01	10 ± 0,1	-50 ± 0,01
$+10^{-8}$ M	16 ± 0,01	7 ± 0,1	-65 ± 0,01	
$+10^{-7}$ M	13 ± 0,1	-53 ± 0,01	6 ± 0,1	-70 ± 0,01
$+10^{-6}$ M	12 ± 0,1	-56 ± 0,01	3 ± 0,1	-85 ± 0,01
$+10^{-5}$ M	11 ± 0,1	-60 ± 0,01	2 ± 0,1	-85 ± 0,01
$+10^{-4}$ M	7 ± 0,1	-74,5 ± 0,01	0	-100
$+10^{-3}$ M	0	-100	0	-100

Thus, Melafen influence on two targets surfaces of cells simultaneously on purinoreceptors PY2 and on CRAC, considerably reducing their activity in plants growth stimulated doses. Really, Melafen changes the functions of surface receptors and intracellular signal transduction.

It will be interesting to test the Melafen influence to the overall cellular answer of the normal cells that may be activated by ATP additions. Because that we recorded of right angle of light scattering of dilute suspension thymocytes and leucocytes. These cells have ranking among P2Y receptors have another types of purinoreceptors non-methabotrophic channel formers P2X (at leukocytes) [25] and its modification P2Z (at thymocytes) [26]. The compositions of groups of P2 receptors at thymocytes are showing both P2Z and P2X receptor activation characteristics are in depending on the stage of cellular growth. The additions of ATP to the thymocytes suspension caused the two-phase of change of cell volume (Figure 4).

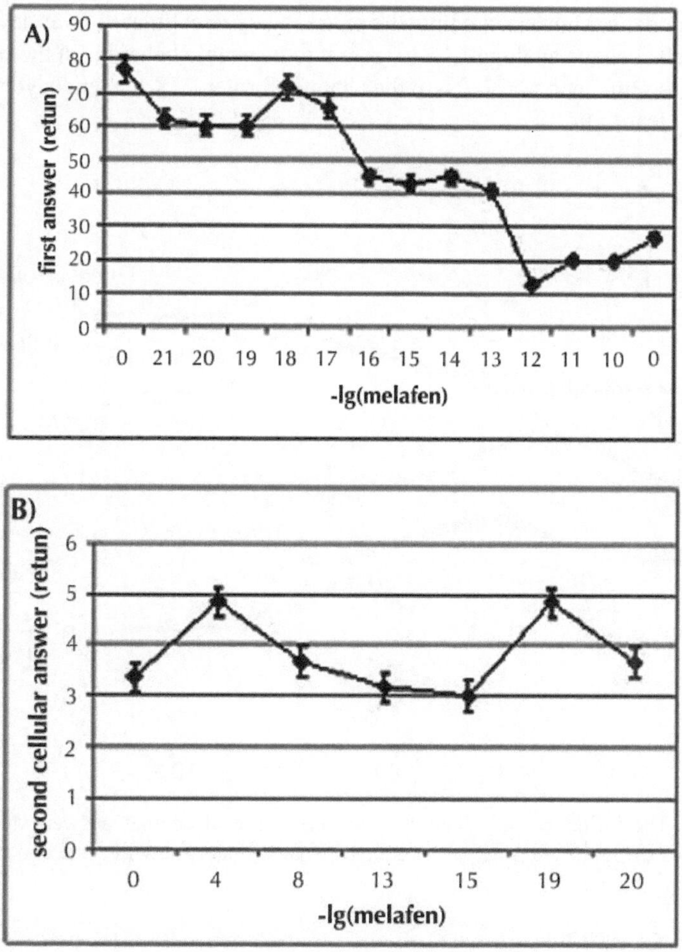

FIGURE 4 The influence of Melafen to the first (A) and second (B) cellular answers of cell suspension of rat thymocytes under the wide concentrations region of Melafen. The amplitude of light scattering risings after the ATP additions to the cell suspension in the dependence of Melafen concentrations.

Melafen influences on both phases. The questions, what certain points of Ca^{2+}-transduction at rat thymocytes are involved to Melafen effects, do not closed. We may conclude that P2X and P2Z Ca^{2+} channel formed receptors are susceptible to Melafen influence under a wide region of concentrations.

The additions of ATP to leucocytes suspension caused releasing of Ca^{2+} from intracellular stores through activating of metabotropic P2Y purine receptors (at the medium of measurement Ca^{2+} do not present). Thus, eliminated the possibility of introducing to the cell interior the extra cellular Ca^{2+}. We used the measurements medium without the Ca^{2+} ions. At this case the non-methabotrophic P2X and P2Z channel formers were silent structure. And as itself will lead the P2Y methabotrophic receptor. What activate the Ca^{2+} ions releasing from intracellular Ca^{2+} stores (endoplasmic reticulum). The Melafen attendance shortens the time the phase lag up to cellular answer development. The EAC cells behave analogously. Its go out to the stable behavior in the presence of Melafen faster. Impact on Melafen to the channel former P2X the leucocytes receptors are coming clear to up.

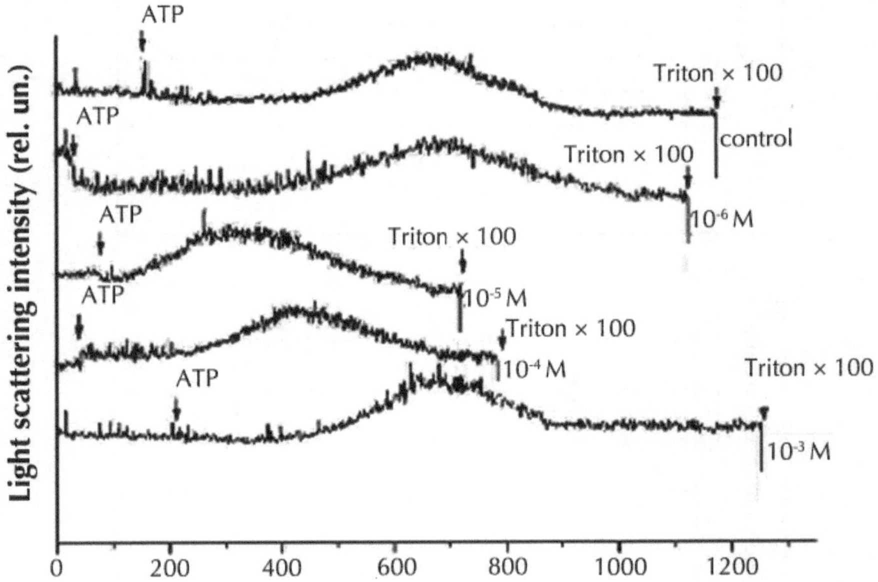

FIGURE 5 The influences of Melafen to the lag phase of cellular answers of leucocytes. Kinetic curves of light scattering of the cell suspension under the wide concentrations region of Melafen.

19.4 CONCLUSION

The Melafen actions to the three types of animal cells have the three global targets P2Y, P2X, P2Z – purinoreceptors. Plant growth regulator has the negative actions to the animal cells of different origin. And its actions were present under the low and ultra low concentrations.

KEYWORDS

- **Ehrlich ascetic carcinoma**
- **Hydrophobic phase**
- **Lipid peroxide oxidation**
- **Melafen**
- **Plasma membrane**
- **Spectral method**

REFERENCES

1. Fattachov, S. G., Reznik, V. S., and Konovalov, A. I. In set of articles. Reports of 13th international conference on chemistry of phosphorus compounds. S. 80.2. Melamine Salt of Bis(hydroxymethyl)phosphinic Acid. (Melaphene) As a New Generation Regulator of Plant growth regulator. S. Petersburg, (2002).
2. Alekseeva, O. M., Narimanova, R. A., Yagolnik, E. A., and Kim, Yu. A.. "*Phenosan influence and its hybrid derivative on membrane components.*" Bashkir State University All-Russian Conference Biostimulators in medicine and agriculture Ufa pages 15–19 article in collection (March 15, 2011).
3. Vekshina, O. M., Fatkullina, L. D, Kim, Yu. A., and Burlakova, E. B. The changes of structure and functions of erythrocyte membranes and cells of ascetic Ehrlich carcinoma when action of hybrid antioxidant of rising generation ICHFAN-10. *Bulletin of experimental biology and medicine*, (4), p. 402–406 (2007).
4. Zhigacheva, I. V., Fatkullina, L. D., Burlakova, E. B., Shugaev, A. G., Generozova, I. P., Fattakhov, S. G., and Konovalov, A. I. Effects of the Organophosphorus Compound "Melaphen – Plant Growth Regulator – on Structural Characteristics of Membranesof Plant and Animal Origin. *Biol. Membrane*, **25**(2), 128–134 (2008).
5. Zhigacheva, I. V., Fatkullina, L. D., Shugaev, A. G., Fattakhov, S. G.,, Reznik, V. S., and Konovalov, A. I. Melafen and energy status of cells of plant and animal origin. *Dokl Biochem Biophys.*, **409**, 200 (2006).
6. Zhigacheva, I. V., Evseenko, L. S., Burlakova, E. B., Fattakhov, S. G., and Konovalov, A. I. . "*Influence of organophosphorus plant growth regulator on electron transport in respiratory chain of mitochondria*". Doklady Akademii Nauk, **427**(5), 693–695 (2009).
7. Zhigacheva, I. V., Fatkullina, L. D., Rusina, I. F., Shugaev, A. G., Generozova, I. P., Fattakhov, S. G., and Konovalov, A. I. *Antistress properties of preparation Melaphen.* Doklady Akademii Nauk, **414**(2), 263–265 (2007).
8. Zhigacheva, I. V. et. al. *Fatty acids membrane composition of mitochondria of pea seedlings in conditions of moistening shortage and moderate cooling.* Doklady Akademii Nauk, **437**(4), 558–560 (2011).
9. Zhigacheva, I. V. et. al. *Insufficient moistening and Melafen change the fatty-acid membrane composition of mitochondria from pea seedlings.* Doklady Akademii Nauk, **432**(1), 124–126 (2010).
10. Kostin, V. I., Kostin, O. V., and Isaichev, V. A. Research results concerning the application of Melafen when "cropping". *Investigation state and utilizing prospect of growth regulator "Melafen" in agriculture and biotechnology.* Kazan, pp. 27–37 (2006).
11. Alekseeva, O. M., Yagolnik, E. A.,. Kim, Yu. A., Albantova, A. A., Mil, E. M., Binyukov, V. I.,Goloshchapov, A. N., and Burlakova, E. B. "*Melafen action on some links of intracell signalling of animal cells*" International Conference "Receptors and the intracell signaling". Article in collection of works, Puschino, pp. 435–438 (May 24–26, 2011).

12. Alekseeva, O. M., Krivandin, A. V., Shatalova, O. V., Kim, Yu. A., Burlakova, E. B., Goloshapov, A. N., and Fattakhov, S. G. No lipid microdomains destruction, but stabilization by melafen treatment of dimyristoilphosphatidylcholine liposomes» // *International Symposium "Biological Motility: from Fundamental Achievements to Nanotechnologies*", Moscow region, Pushchino, Russia, article in collection of works, pp. 8–12 (May 11–15, 2010).
13. Alekseeva, O. M., Shibryaeva, L. S., Krementsova, A. V., Yagolnik, E. A., Kim, Yu. A., Golochapov, A. N., Burlakova, E. B., Fattakhov, S. G., and Konovalov, A. I. *The aqueous melafen solutions influence to the microdomains structure of lipid membranes at the wide concentration diapason*. Doklady Akademii Nauk, **439**(4), 548–550 (2011).
14. Tarakhovsky, Yu. S.,. Kuznetsova, S. M., Vasilyeva, N. A., Egorochkin, M. A., and Kim, Yu. A. Taxifolin interaction (digidroquercitine) with multilamellar liposomes from dimitristoyl phosphatidylcholine. *Biophysicist*, **53**(1), 78–84 (2008).
15. Alekseeva, O. M., Krivandin, A. V., Shatalova, O. V., Rykov, V. A., Fattakhov, S. G., Burlakova, E. B., and Konovalov, A. I. *The Melafen-Lipid- Interrelationship Determination in phospholipid membranes*. Doklady Akademii Nauk, **427**(6), 218–220 (2009).
16. Binyukov, V. I., Alekseeva, O. M., Mil, E. M., Albantova, A. A., Fattachov, S. G., Goloshchapov, A. N., Burlakova, E. B., and Konovalov, A. I. The investigation of melafen influence on the erythrocytes in vivo by AFM method. *Doklady Biochemistry and Biophysics*, **441**, 245–247 (2011).
17. Alekseeva, O. M., Erokhin, V. N., Krementsova, A. V., Mil, E. M., Binyukov, V. I., Fattachov, S. G., Kim, Yu. A., Semenov, V. A., Goloshchapov, A. N., Burlakova, E. B., and Konovalov, A. I. *The investigation of melafen low dozes influence to the animal malignant neoplasms in vivo and in vitro*. Doklady Akademii Nauk, **431**(3), 408–410 (2010).
18. Rizkina, I. S., Murtazina, L. I., Kiselyov, J. V., and Konovalov, A. I. *Property of supramolecular nanoassociates*, formed in aqueous solutions low and ultra-low concentrations of biologically active substance. DAN, **428**(4), 487–491 (2009).
19. Osipenkova, O. V., Ermokhina, O. V., Belkina, G. G., Oleskina, Yu. P., Fattakhov, S. G., and Yurina, N. P. Effect of Melafen on Expression of *Elipl* and *Elip2* Genes Encoding Chloroplast Light-Induced Stress Proteins in Barley. *Practically Biochemistry and microbiology*, **44**(6), 701–708 (2008).
20. Zamai, A. C., et al. Conference Reception and intracellular signalization. *The ATP influence to the tumor ascetic cells at different stages of cellular growth*. Puschino, pp. 48–51 (2005).
21. Zinchenko, V. P., Kasimov, V. A., Li, V. V., and Kaimachnikov, N. P. Calmoduline inhibitor of R2457I induces the short-time entry Ca^{2+} and the pulsed secretion ATP in cells of ascetic Ehrlich carcinoma. *Biophysicist*, **50**(6), 1055–1069 (2005).
22. Pedersen, S. F., Pedersen, S., Lambert, I. H., and Hoffmann, E. K. P2 receptor-mediated signal transduction in Ehrlich ascites tumor cells. *Biochim Biophys Acta.*, **1374**(1-2), 94–106 (September 23, 1998).
23. Cornet, M., Lambert, I. H., and Hoffman, E. K. Relation between cytosceletal hypoosmotic treatment and volume regulation in Erlich ascites tumor cells. *J. Membr. Biol.*, **131**, 55–66 (1993).
24. Artalej, O. A. and Garcia-Sancho, J. Mobilization of intracellular calcium by extracellular ATP and by calcium ionophores in the Ehrlich ascities tumor cells. *Biochem. Biophys. Acta*, **941**(**9**), 48–54 (1988).
25. Di Virgilio, F., et al. Nucleotide receptors: an emerging family of regulatory molecules in blood cells. *Blood*, **97**, 587–600 (2001).
26. Nagy, P. V., Fehér, T., Morga, S., and Matkó, J. Apoptosis of murine thymocytes induced by extracellular ATP is dose- and cytosolic pH-dependent. *Immunol Lett.*, **72**(1), 23–30 (2000).

20 On the Effects of PABA on Germination

S. A. Bekusarova, N. A. Bome, L. I. Weisfeld, F. T. Tzomatova, and G. V Luschenko

CONTENTS

20.1 INTRODUCTION

Para-aminobenzoic acid (PABA) by well-known activator of growth and development was applied for treatment of ears of winter wheat and winter barley. The varieties of these cultures are destined for North Caucasian region. Also panicles of selection samples of millet (the varieties are introduced in North Ossetia) were subject to treatment by PABA. The PABA was dissolved in acetic acid but not in water as usual. Advantages of that mode are described. These methodologies allow to accelerate a breeding process, when often is necessary to sow in year, when was gathered harvest. In the process of selection for the speedup study of material, it is necessary to get the sufficient amount of seed in the earliest possible dates. For the speedup estimation of plant breeding material it is necessary to get the sufficient amount of seed in the earliest possible dates. To that end conducted sowing of inflorescences without threshing from the plants of cereal cultures selected for a selection. The method of sowing by years was tested during two years in Tyumen research station of N. I. Vavilov Research Institute of Plant Industry on more than 100 samples of winter wheat from world collections. The productivity of winter form of cereals depends on a set of biotic (pathogenic microorganisms) and abiotic (temperature, amount of precipitation etc.) factors. We obtained positive results for winter hardiness and resistance to snow mold which main pathogen is *Microdochium nivale* (Fr.) Samuels and I. C. Hallett (= *Fusarium nivale* Ces. ex Berl. and Voglino).

The PABA discovered by J. A. Rapoport as modifier of metabolic processes already in 1940s. He showed on the example of experimental object, *Drosophila*, that PABA evokes positive changes of nonhereditary character (i.e., it is not a mutagen) in the organism development [1]. Rapoport [2] proposed a scheme of relations between genotype, ferments, and phenotype: genes → their hetero catalysis (on the substrate of ribonucleic acid (RNA) molecule) → messenger RNA (mRNA) → mRNA catalysis (substrate of amino acids) → ferments → their catalysis (different substrates) → phenotype.

Works on the application of PABA in the agriculture continue and deepen. The experiments performed in the Republic of North Ossetia [4-9] showed, that the addition of PABA to the nutritive substances (potassium humate, irlit, leskenit, corn extract, juice of ambrosia, melted snow water, and others) gives a positive effect. The treatment with PABA seeds of pea and honey plant sverbiga east (*Bunias orientalis*) before planting [4], extra nutrition of clover [5], seed of leguminoze grasses [6] stimulates the germination plants in a greater degree, then nutritive substances without PABA. A positive result was got at treatment of sprouts of potato by mixture of the melted water with juice of ambrosia, leskenit, and PABA [7]. This method resulted in the increase of harvest of potato and decline of disease of Fusarium. Addition PABA in nutrient medium for treatment seed of triticale [8] results in the increase of maintenance of protein in green mass. Protracted treatment by PABA of handles of dogwood [9] increased engraftment of grafts. The PABA was used for the receipt of potato without viruses [10].

In the joint research of scientists of the Institute of Biochemical Physics RAS and the North Caucasus Research Institute of mountain and foothill agriculture, it was shown that the treatment by PABA of potatoes tubers with the subsequent enveloping in an ash [11] increase yield of potato. Treatment of seed and seedlings of vegetable cultures by solution of PABA in mixture with boric acid and permanganate potassium [12] improves resistance to diseases of young plantlets and increases harvest of carrot, beet, cucumbers, and tomatoes, seeds and seedlings of vegetables.

After treatment by PABA of binary mixture of seed of winter wheat and winter vetch [13] the productivity and quality of green feed increases at mowing of mixture in a period from the beginning of exit in a tube of wheat and beginning of budding of vetch to forming of grain of milky ripeness of wheat.

In the conditions of the northern forest steppe of Tyumen region, where beside of fertile soils occur saline, jointed impact of salt solutions, and PABA (0, 01% solution) on the seeds of three barley varieties with low salinity resistance considerably increased the salinity resistance of germs independently on NaCl concentration [14]. The positive results were obtained under spraying of inflorescences of mother plants of barley by the solutions of PABA before realization of crossing. In a series of hybrid combinations the exceeding over control of length and width of flag leaf, number of leaves by plant and plant height was observed [15]. Spraying of inflorescences of four amaranth samples by PABA solutions increased seed productivity, the concentration of 0.02% showed to be the most effective [15].

The effect of PABA on germination, germination energy, and winter hardiness of cereals, namely ears of winter cereals and panicles of millet species is studied. In contrast with most of studies, where PABA was dissolved in the hot water following the method developed by Rapoport with collaborators [3], PABA was dissolved in an acetic acid. The combination of PABA and an acetic acid creates an acid media and allows preventing of a set of fungi illnesses simultaneously with the preservation of genotype of samples under multiplying.

The results of experiments performed on the experimental base of the North Caucasus Research Institute of mountain and foothill agriculture where perennial crops of legumes herbs were precursor.

20.2 MATERIAL AND METHODS

Cereals, the most widely applied in the agricultural production and selections were applied as material. The following cultures were tested, of yield of 2009: winter wheat varieties Ivina, Vassa, Batko, Don 107, Kollega, Kalym, and winter barley variety Bastion – the varieties designed for North Caucasian region. The following varieties introduced in North Ossetia: selection samples of millet: Japanese millet (*Echinochloa frumentacea*), panic (*Setaria italica Panicumitalicum*) Italian millet (*Setaria italica*).

The PABA represents a fine-grained powder, it easily and completely dissolves in 3% solution of an acetic acid without heating. We dissolved one tea spoon of dry PABA powder (10 g) in a small volume (20–25 ml) of 3% solution of an acetic acid. Then, the obtained mixture in 1 l of tap water dissolved under room temperature. Thus, the PABA concentration of 0.1% is obtained. In order to obtain the concentration of 0.2% dissolve 20 g of dry powder.

Soaked ten samples for the inflorescences without trashing of cereals with mature seeds, namely ears of cereals and panicles of millet species, in the solutions of PABA (0.1 or 0.2%) during 2–2.5, 3–4, and 5–6 hr. Then the inflorescences were planted in the soil in the open field with the distance of 20–25 cm between samples. Untreated inflorescences, soaked in water during 3–4 hr, were applied as control (control I). When preparing control II dissolved PABA in hot water and soaked inflorescences during 2–2.5 hr.

In tables the variants of experiments were marked: in variant "0" is presented Control of I, where inflorescences were soaked in water (without treatment by PABA), in variant "1" is presented Control of II, where inflorescences were soaked in PABA dissolved in hot water, variants 2–5 are a soakage of inflorescences in PABA dissolved in an acetic acid of concentrations of 0.1–0.2% through 2–6 hr.

Simultaneously, placed thrashed seeds for germination in a Petri dish with PABA dissolved in hot water (control I) or in an acetic acid (control II). For the control II the data averaged for PABA concentrations of 0.1 and 0.2% are presented because their difference was insignificant.

20.3 DISCUSSION

The higher indexes of germinating capacity, energy of germination at all cultures, and winter hardness cold of winter crops are marked - winter wheat and barley in variants (2–5 in Figure 1) after treatment of PABA, dissolved in an acetic acid, as compared to control variants without treatment of PABA (variant 0 control of I without PABA) and at dissolution of PABA in hot water (variant 1 control of II) in analogical variants (see Tables 1–3, Figure 1) were got. All indexes at a soakage in PABA in an acetic acid most long time 5–6 hr (variant 5) as compared to analogical treatment during 3–4 hr (variant 4) has a tendency to go down (see a Table 1–3 and Figure 1). Maybe in this case, braking of development of plantules takes place [10].

This work executed in connection with the necessity of speedup creation of new varieties steady to the ecologically diverse conditions and simultaneously responsive on the additional fertilizing. The work is important from the economic point of view. The sowing of inflorescences without threshing considerably simplifies work of breeder both in a scientific plan and in organizational: the methods of selection are simplified, reproduction of the best plants is accelerated for a selection and forming of new varieties, and diminish expenses of labor. Usually at the unfavorable condones of growing, at introduction of plants, at the high doses of mutagens or during distant hybridization of cultural plants with wild sorts fall down number in inflorescences, germination of seeds, and viability of plants. At sowing of inflorescences, ears, or panicles possibility to distinguish the most perspective families on the number of germinating seed in them appears with mature grain. From every separate inflorescence it is possible to collect more seeds for the receiving of posterity, the terms of estimation of material and terms of reproduction of valuable selection samples grow short. A method especially touches the freshly material, when at preparation to sowing of winter crops in the year of harvest time not enough for threshing of ears and additional treatment of grain. Such quite often happens in those districts, where short vegetation period, for example, in the Tyumen area.

20.4 RESULTS

The PABA concentration of 0.1 or 0.2% in an acetic acid was sufficient for the penetration of substance in the embryo when treating cereal inflorescences. High concentrations (9–11%) inhibit a height and development of plants [10].

The germination of threshed seeds soaked in PABA dissolved in an acetic acid, appeared 3–4 days earlier than when soaked in water. On seedlings from seed germinated in an aqueous solution of PABA developed fungal microflora and they died. When dissolved in an acetic acid PABA seedlings persisted during long time, they continued to grow.

The germinating capacity of seeds in inflorescences under soaking in PABA, dissolved in an acetic acid, considerably exceeded germinating capacity in control variants (Table 1). The effectiveness of treatment by PABA solutions depended on its duration and concentration of solutions. The same trend concerning germination energy was observed for all five cereals under study.

TABLE 1 Germinating capacity of seeds of winter grain crops and millet species under different ways of treatments of cereal inflorescences by PABA.

Variant of experiment	Winter wheat		Winter barley		Japanese millet		Panic		Italian millet	
	%	+/%	%	+/%	%	+/%	%	+/%	%	+/%
0. Control I without treatment: soaking in water for 3–4 hr	62	0	75	0	67	0	72	0	68	0
1. Control II: soaking in PABA (0.1–0.2%) dissolved in hot water for 2–2.5 hr	74	12/16	84	9/11	76	9/12	82	10/1.2	72	4/6
2. Soaking in PABA (0.1%) dissolved in an ac. a. for 2–2.5 hr	82	20/24	93	18/19	87	20/23	92	20/22	84	16/19
3. Soaking in PABA (0.2%) dissolved in an ac. a. for 2–2.5 hr	86	24/28	95	20/21	92	25/27	92	20/22	87	19/24
4. Soaking in PABA (0.2%) dissolved in an ac. a. for 3–4 hr	96	34/35	98	23/23	95	28/29	95	23/23	92	24/25
5. Soaking in PABA (0,2%) dissolved in an ac. a. for 5–6 hr	82	20/24	86	11/13	90	23/26	90	18/20	86	18/21
LSD_{05}	4.5		2.4		2.8		2.1		2.2	4.5

Note. LSD - the least substantial difference; ac. a. = an acetic acid

The middle percent of wintering as compared to control of I (water without PABA) increases with the increase of concentration of PABA and durations of treatment. However, at more long continued soakage of plants during 5–6 hr (variant 5) increase in relation to both controls was less, than at a soakage during 3–4 hr (variant 4). At a soakage in water solution of PABA of 0.1% increase observed also, but on more low level, then at dissolution in an acetic acid.

The energy of germination, presented to the Table 2, exceeded control without treatment of PABA in all variants of dissolution of PABA dissolved in an acetic acid and in a variant with treatment of PABA, dissolved in hot water (control of II, variant 1), and was higher, than in control without addition of PABA (zero variant) at all five investigated cultures. Exceeding of indexes above control of I in control of II was below, than in variants 1–5.

TABLE 2 Germination energy of seeds of winter grain crops and millet species under different ways of treatments of cereal inflorescences by PABA.

Variant of experiment	Winter wheat		Winter barley		Japanese millet		Panic		Italian millet	
	%	+/%	%	+/%	%	+/%	%	+/%	%	+/%
0. Control I without treatment: soaking in water for 3–4 hr	54	0	68	0	53	0	60	0	54	0
1. Control II: soaking in PABA (0.1–0.2%), dissolved in hot water 2–2.5 hr	65	11/17	73	5/7	62	9/15	73	12/16	63	8/13
2. Soaking in PABA (0.1%), dissolved in an ac. a. 2–2.5 hr	72	55/76	81	13/16	75	22/29	81	22/31	73	19/26
3. Soaking in PABA (0.2%), dissolved in an ac. a. 2–2.5 hr	80	27/34	84	16/19	79	26/33	87	26/30	76	22/29
4. Soaking in PABA (0.2%), dissolved in an ac. a. 3–4 hr	85	32/38	89	21/24	83	30/36	89	29/33	79	25/32
5. Soaking in PABA (0.2%), dissolved in an ac. a. 5–6 hr	78	25/32	80	11/14	78	26/33	84	24/29	75	21/28
LSD_{05}		3.4		2.1		2.6		2.9		2.5

Winter hardness of winter cereals (Table 3) - wheat and barley in all variants with treatment of PABA dissolved in an acetic acid (variants 2–5) and in control of II (dissolution is in hot water) was higher than in control of 1 (without PABA). As well as other experiments indexes at dissolution of PABA in hot water in control II were below, than in other variants (3–5).

TABLE 3 Winter hardiness (% of plants overwintered) of winter cereals under different ways of treatments of ears by PABA.

Variant of experiment	Winter wheat		Winter barley	
	%	+/%	%	+/%
0. Control I without treatment: soaking in water for 3–4 hr	78	0	84	0
1. Control II: soaking in PABA (0.1–0.2%), dissolved in hot water 2–2.5 hr	82	4/4.8	88	4/4.8
2. Soaking in PABA (0.1%), dissolved in an ac. a. 2–2.5 hr	86	8/9.3	95	11/22.6
3. Soaking in PABA (0.2%), dissolved in an ac. a. 2–2.5 hr	90	12/13.3	97	13/13.4
4. Soaking in PABA (0.2%), dissolved in an ac. a. 3–4 hr	98	20/20.4	98	14/14.5
5. Soaking in PABA (0.2%), dissolved in an ac. a. 5–6 hr	88	10/11.4	92	8/8.7
LSD_{05}		3.1		0.8

The dynamics of activating of phenotype under act of PABA is evidently shown on a histogram (Figure 1), reflecting the middle indexes of levels of increases energy of germination and winter hardness. The tendency of growth of middle indexes is evidently.

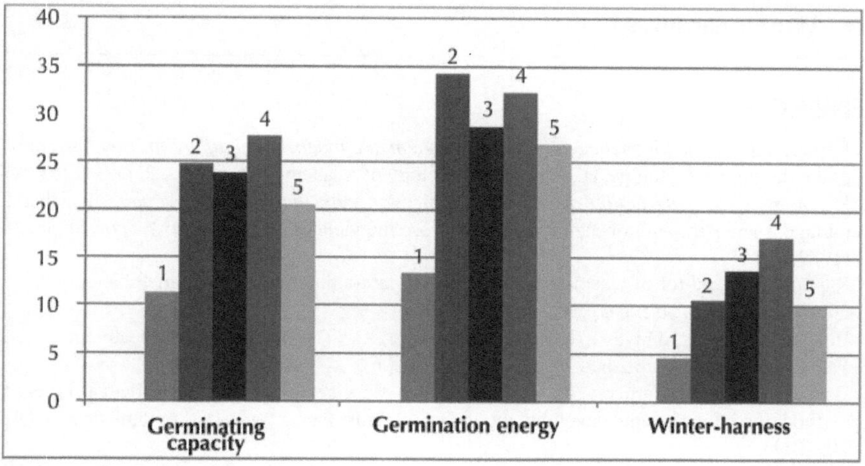

FIGURE 1 Comparison of averages (totally on all cultures, %) of germinated capacity, energies of germination of seed, and winter hardness of plants at the different methods of treatment of ears of winter crops and panicles of kinds millet by PABA in the variants of experiments from 1 to 5 (see Tables 1–3) by comparison to control of I (without treatment, in tables - a zero variant). Ordinate axis %.

20.5 CONCLUSION

(1) The method of acceleration of plant breeding process offers at sprouting of mature seed in inflorescences of cereal cultures, ears of grain growing, and panicles of millet. A method plugs the soakage of inflorescences in PABA (concentrations of 0.1% and 0.2%) dissolved in an acetic acid.

(2) Dissolution of PABA in an acetic acid is more effective, than dissolution in hot water (control II). Activating of phenotype at soakage in PABA, dissolved in an acetic acid, in variants (3–5) was higher than in control II – at soakage in PABA dissolved in hot water.

(3) Soakage of inflorescences during 3–4 hr in PABA a dissolved in an acetic acid, gives the best result. More protracted soakage - 5–6 hr some reduces by all variants.

(4) Concentrations of PABA applied to the treating of inflorescences were higher than in the cited works in which treated trashed seeds.

KEYWORDS

- **Cereals**
- **Germinating capacity**
- **Germination energy**
- **Para-aminobenzoic acid**
- **Winter hardness**

REFERENCES

1. Rapoport, I. A. *Phenogenetic analysis of dependent and independent differentiation*. Proceedings of the Institute of Cytology, Histology, Embryology of Academy of Sciences, **2**(1), 3–128 (1948).
2. Rapoport, I. A. "*The action of PABA in connection with the genetic structure*" in: Chemical mutagens and para-aminobenzoic acid to increase the yield of crops. Moscow, *Science*, pp. 3–37 (1989).
3. Rapoport I.A., editor of Chemical mutagens and para-aminobenzoic acid in enhancing the productivity of crops. *Science*, p. 253 (1989).
4. Bekuzarova, S. A., Abiyeva, T. S., and Tedeeva, A. A. The method of pre-treatment of seeds. Patent № 2270548, Published on (February 27, 2006).
5. Bekuzarova, S. A., Farniyev, A. T., Basiyeva, T. B., Gaziyev, V. I., and Kaliceva, D. N. The method of stimulation and development of clover plants. Patent № 2416186. Published on (April 20, 2011).
6. Bekuzarova, S. A., Shtchedrina, D. I., Farnoyev, A. T., and Pliyev, M. A. Method of additional fertilizing of leguminous grasses. Patent № 2282342. Published on (August 27, 2006).
7. Ikaev, B. V., Marzoev, A. I., Bekuzarova, S. A., Basayev, I. B., Bolieva, Z. A., and Kizinov, F. I. The method of treatment of pre-plant shoots of potato tubers. Patent № 2385558. Published on (April 10, 2010).
8. Bekuzarova, S. A., Antonov, O. V., and Fedorov, A. K. Method of increase of content of protein in green mass of winter triticale. Patent № 2212777. Published on (September 27, 2003).
9. Cabolov, P. H., Bekuzarova, S. A., Tigiyeva, I. F., Tadtayeva, E. A., and Eiges, N. S. The method of reproduction of dogwood drafts. Patent № 2294619. Published on (March 10, 2007).
10. Shcherbinin, A. N. and Soldatova, T. B. The nutrient medium for micropropagation of potato. Patent № 2228354. Published on (May 10, 2004).
11. Eiges, N. S., Weisfeld, L. I., Volchenko, G. A., and Bekuzarova, S. A. Method of pre-treatment of tubers of potatoes. Patent № 2202701. Published on (October 02, 2007).
12. Eiges, N. S., Weissfeld, L. I., Volchenko, G. A., and Bekuzarova, S. A. The method of pre-treatment of seeds and seedlings of vegetable crops. Patent № 2200392. Published on (March 20, 2003).
13. Eiges, N. S., Weissfeld, L. I., Volchenko, G. A., Bekuzarova, S. A., Pliyev, M. A., and Hadarceva, M. V. The method of receiving of feeds in green conveyer. Patent № 2330410. Published on (August 10, 2008) Bull. № 22.
14. Bome, N. A. and Govorukhin, A. A. The effectiveness of the influence of para-aminobenzoic acid on the ontogeny of plants under stress. *Bulletin of Tyumen State University. Tyumen*: Tyumen State University Publishing House, № 2. S. 176–182 (1998).
15. Bome, N. A., Bome, A. Ja., and Belozerova, A. A. Stability of crop plants to adverse environmental factors. *Monograph*, Tyumen State University Publishing House, Tyumen p. 192 (2007).

21 A Study on Damage of Chromosomes Activity in Seedlings of *Crepis Capillaris*

Larissa I. Weisfeld and G. E. Zaikov

CONTENTS

21.1 INTRODUCTION

In this chapter, we attempt to consider mechanism of origin of rearrangement of chromosome in the different phases of mitotic cycle. We are showing the appearance of rearrangements of chromosome under the influence of antineoplastic medicament phosphemid- di- (ethyleneimine)-pyrimidine-2-amino phosphoric acid, named mutagen. Seeds *Crepis capillaris* and fibroblasts of human and mouse were treating by this mutagen. In seedlings on metaphase plates were analyzing rearrangements of chromosomes. The rearrangements were analyzed at different terms after appearance of seedlings named arisings. Also chromosomal rearrangements were examined at anaphases and telophases of fibroblasts.

Scientists attracted attention to the chemical compounds that where cause heritable changes. It remains unsolved mechanism of their effects on the chromosome. The action of many compounds is similar with ionizing radiation, they are cause mutations of genes, disruptions of cell division, and rearrangement of chromosomes.

The ethyleneimine and its derivatives alkylate deoxyribonucleic acid (DNA) and proteins [1]. They induce mutations as it was shown in various model objects (*Drosophila*, higher plants, fungi, bacteria, viruses, and others) and breakage of the chromosome apparatus [2]. Chemical compounds that cause mutations and breakage of chromosomes usually are called "chemical mutagens".

The study of chemical mutagens is a large role belongs to works of J. A. Rapoport [3]. He is discoverer phenomenon of the chemical mutagenesis. He revealed out supermutagens and discovered of possibilities their application in the breeding of crops and in other areas of agricultural production. J. A. Rapoport organized synthesis of supermutagens. Every year, he organized since 1959 the All-Union Conferences for Scientists and Breeders on chemical mutagenesis and its application in agriculture. These meetings served in those years a good genetic school, especially for young breeders, who were trained and the unscientific method of Lysenko. Based on the methodology developed by Rapoport and with the help of mutagens, which he distributed free of charge, breeders created the source material of crops and introduced new varieties [3]. A series of investigations on the application of the chemical mutagen ethyleneimine and mechanism of its action in winter wheat was conducted and is ongoing, N. S. Eiges [6, 7], which is the follower of J. A. Rapoport.

The most effective and affordable way to analyze the mutagenesis is a cytogenetic method: the study of rearrangements (aberrations) of chromosomes observed in dividing cells (mitosis and karyokinesis), and disorders in passing of the mitotic cycle.

The mechanism damage of the chromosomes and their relationship to the mitotic cycle and synthesis DNA at various biological objects in the cells of the meristem of plants onions, *Vicia faba*, *Crepis*, and so on, embryonic cells of animals and humans *in vitro*, at microbes, viruses, and other objects.

The ionizing radiation damages the chromosomes immediately after irradiation at all stages of the mitotic cycle "undelayed" effect. In the cells that came into mitosis from G_2 phase and S, arise aberrations of chromatid type (nature), in cells that come in mitosis from phase G_1 (pre-synthesis) occur aberrations of chromosome type (double bridges, paired fragments at anaphase).

Under treatment by chemical mutagens of asynchronous cell cultures, no rearrangements detect during the first 2–4 hr (depending on the duration of G_2 phase at different objects). This phenomenon is named "delayed" effect. The chromosomal rearrangements appear later after the entry to mitosis of cells treated at the beginning or during DNA synthesis (phase S) or before phase S (named pre-synthetic phase G_1). Chromosome rearrangements are usually analyzed in the ana-telophases or meta-phases. A large number of chromosomes in many objects complicate the identification of chromosome aberrations in metaphase plates. In this case fragments of chromosomes or broken bridges cannot be identified. In this case, scientists analyze the anaphase and early telophases (ana-telophase method).

For the estimate of environmental contamination the method of analyzes of ana-telophases is sufficient. An example is the estimation of pollution in the industrial area of the city of Staryj Oskol on the meristem of birch [8].

In Moscow 1960–1970s, extensive cytogenetic studies of chemical and radiation mutagenesis were carried N. P. Dubinin, his colleagues and followers. The work was begun in the laboratory of radiation genetics of Institute of biophysics of Academy of Science the USSR and continued in Institute of general genetics. A large number of articles and monographs about cytogenetic effects of ionizing radiation or chemical compounds were published [9]). N. P. Dubinin formulated the idea of the mechanism of action of chemical mutagens, mutagens cause potential changes in the chromosomes at all stages of the mitotic cycle, which are realized in a number of cell generations. He called it "chain processes" in mutagenesis [10-12].

At the same time period staff of the laboratory under the direction of B. N. Sidorov and N. N. Sokolov (Andreev V. S., Generalova M. V., Grinih L. I., Durymanova S. E., Kagramanyan R. G., Protopopova E. M., Shevchenko V. V., and others) were conducted extensive work on the induction of chromosomal damage, their localization in the chromosomes, their association with the phases of the mitotic cycle and DNA synthesis after treatment of seedlings by ethyleneimine, tio-TEF, radiation, and other mutagens on the model object *Crepis capillaris*. They studied chromosome aberrations in metaphase plates, which make it possible to take into account the polyploidy of metaphases.

In the early 1960s, working at N. P. Dubinin, I investigated the cytogenetic effect of alkylating agent phosphemidum (lat. synonym phosphemid, phosphasin) di-(ethyleneimid)-pyrimidyl-2-amidophosphoric acid (Figure. 1).

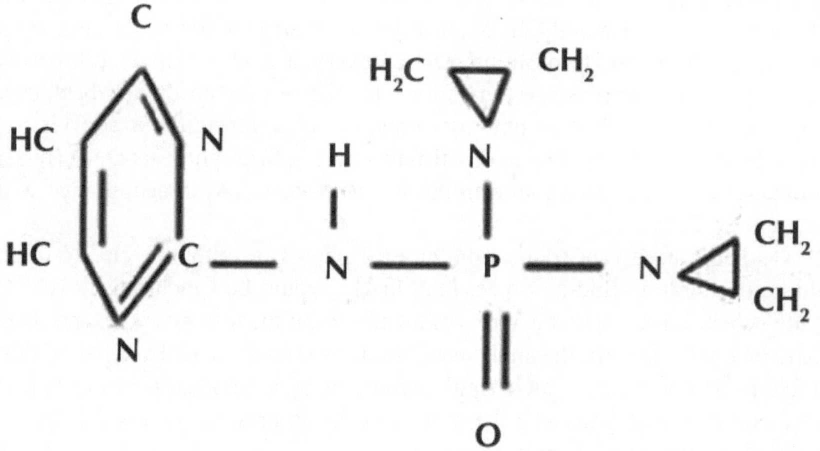

FIGURE 1 Phosphemidum (syn. Phosphasin and phosphemid).

This compound is interesting because it consists of pyrimidine base and three molecules of ethyleneimine. It was assumed that phosphemid will cause a lot of damages in DNA synthesis and thereby would inhibit tumor growth. The drug was synthesized in the laboratory of V. A. Chernov et al. [13], the All-Union Scientific Research Chemical-Pharmaceutical Institute (now the Center for Chemistry of Drugs), and was referred to N. P. Dubinin for cytogenetic studies. The drug is a white crystalline powder, dissoluble in water and alcohol. Chernov et al. [13, 14] have shown that phosphemid (named phosphasin) inhibits the growth of tumors in rats, mice, rabbits, but at the same time it causes leukopenia, leukocytosis, and suppressed erythropoiesis. But currently the drug used in medicine for the treatment of certain tumors such as leukemia, and lymphoma. In the early work with phosphemid (1963–1964), it was important to approach the mechanism of its antineoplastic action. It was assumed that due to pyrimidine bases and ethyleneimine groups of drug will be directly to affect DNA during a synthesis and thereby destroys actively dividable tumor cells.

The work was carried out on primary culture of embryonic tissues the mouse and human fibroblasts [15, 16]. We analyzed chromosomal aberrations in cells at the stages of anaphase or telophases (ana-telophases), after treatment culture by phosphemid in a concentration of 1×10^{-4}M.

It was shown that phosphemid delays entry of cells into mitosis and inhibits the mitotic activity of fibroblasts during the total period of culture growth. The waves of fall and of rise of mitotic activity generally are repeating waves of mitotic activity in control (without treatment), but at a lower level. The average frequency of mitoses made 54% from control. Lesions spindle was not observed, as a rule, but in individual cells were visible clumping of the chromosomes in the form of "stars". At later stages of fixing the number of nuclei was decreased by 3–4 times, which on indicates cell death or on the loss of their contact with surface of glass. At consideration the types of chromosomal rearrangements, it became clear that in the ana-telophases appear mainly chromatid type of rearrangements. At later stages of fixation after 26 hr growth of culture were observed double bridges (chromosomal type). They could have arisen as a result of doubling of the chromatid bridges in the second cycle (as a result their passing to one of poles during the first division) and as a result of the breaking-fusion of chromosomes before synthesis DNA (phase G_1) with the subsequent doubling during the synthesis of DNA, in accordance with the thesis of N. P. Dubinin.

It was necessary to set connection with the phases of mitotic cycle. To do this, it would be desirable to find an object that, firstly, would be synchronized (or at least partially synchronized) with the terms of the phases of the mitotic cycle, and secondly, it would be convenient for the analysis of chromosomes in metaphase. To do this, it is desirable to find an object which, firstly, would be synchronized (or at least partially be synchronized) with phases of the mitotic cycle, and secondly, was suitable for the analysis of chromosomes in metaphase.

In this regard *Crepis capillaris* (L.) Wallr. serves an ideal object. Mikhail S. Navashin studying chromosomes in the metaphase plates of seedlings published a number of works about taxonomy of genus *Crepis* [17]. It was showed emergence of a large number of chromosome rearrangements in aging seeds.

At metaphase plate *Cr. capillaris* clearly identified three pairs of homologous chromosomes (Figure. 2). Analysis of rearrangements here is not in doubt.

It was assumed that the results of experiments with ionizing radiation, the cells of seeds (not seedlings!) *Cr. capillaris* are in phase G_1. The phase of G1 in germinal cells of seed of *Crepis* is heterogeneous, as was shown [18] that first mark cellular nuclei appear in seedlings through 10 hr after treatment of grain of thymidine H3.

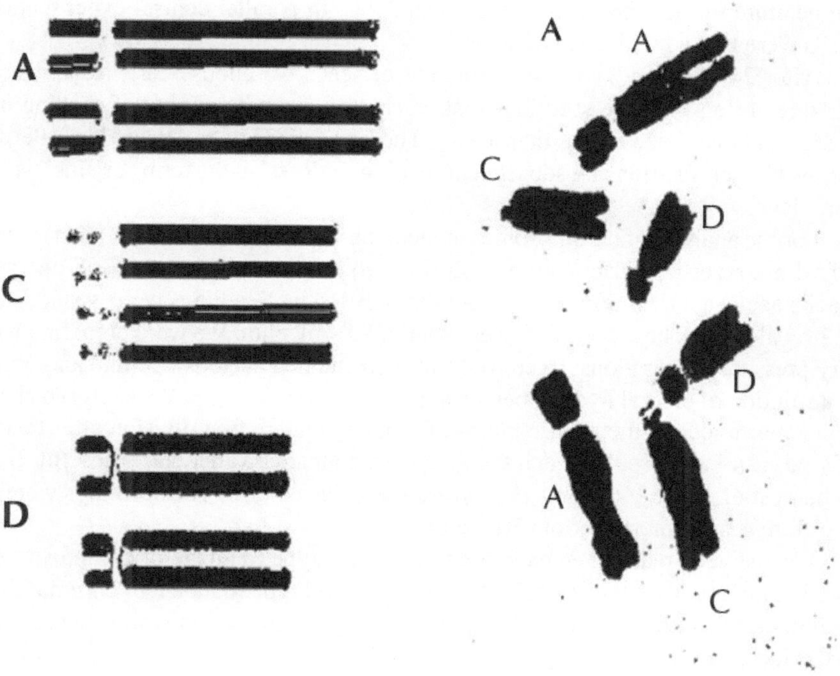

FIGURE 2 Karyotype *Crepis capillaris*.

As far as the seeds germinate at first in mitosis comes the cells from the phase of S then from the phase of G1. If the chemical mutagen interacts with the chromosome before the start of germination that is before the start of DNA synthesis in phase G1, then in metaphases of germinating seedling must appear rearrangements of chromosome type. The data of our experiments, show that after treatment by phosphemid of seeds in the 2*n*-methaphases appear rearrangements of chromatid type.

21.2 MATERIALS AND METHODOLOGY

Air-dried seeds of *Crepis capillaris* of crop of 1967 were analyzed in 1968: April (8 months of storage after harvest), June (10 months of storage), and July (11 months of storage). In the control and in the experiments used distilled water. The chromosome aberrations were analyzed in seedlings (meristem of root tip).

A certain amount of seeds (100 pieces) were treated in an aqueous solution phosphemid of concentrations: 1×10^{-2}M (22.4 mg was dissolved in 10 ml water), $2 \times 10^{-2}M$ (22.4 mg was dissolved in 50 ml) or 2×10^{-3}M (2.24 mg dissolved in 50 ml of water). The treatment was carried out at room temperature (19–21°C) for 3 hr. Then the seeds were washed with running water for 45 min. The washed seeds were placed in Petri dishes on filter paper moistened with a solution of colchicine (0.01%). Seeds were germinating in a thermostat at 25°C, but in July 1968 because of hot weather the temperature in the thermostat could reach 27°C. In parallel control experiments, the seeds were treated with aqua distillate.

After 24, 27, and 31 hr, after soaking of seeds we choose "arisings". The term "arising" refers to those seedlings that emerge after the beginning of soaking of dry seeds and have a size of less than 1 mm. These seedlings were selected in a Petri dish for further germination and subsequent fixation. In Russian the term "arising" is called "proklev".

For the analysis of chromosomes in metaphase plates, root tips (0.5–1cm) were cut off with a razor and placed in the solution: 96% ethanol three parts and one part of glacial acetic acid. Solution poured out through 3–4 hr. Seedlings were washed 45 min in 70% alcohol. These seedlings were kept in 70% alcohol. We were preparing temporary pressure preparations: fixed root tips were stained acetous carmine and crushing in a solution of chloral hydrate between the slide and cover slip. We analyzed chromosome aberrations in metaphase plates of seedlings in the first division after treatment of seeds ($2n$-karyotype). In each seedling were counted up all metaphases. Intact seeds of harvest of 1966, 1967, and 1969 years served as control. The seedlings were fixed at different time intervals from 3 to 24hr.

We analyzed rearrangements of chromosomal type (damage of chromosomes before beginning of synthesis of DNA) and chromatid type (damage of chromatids with beginning of synthesis of DNA). Mitotic activity in the seedlings was determined on two criteria:

(1) On the criterion of the number of metaphase plates, depending on the number of nuclei in seedlings in the control and experiment (after seed treatment phosphemid). We used the seedlings after 2, 4, 6, 8, 10, and 20 hr after the "arising". In each seedling, we counted between 500 and 1,000 nuclei, in an each seedling counted from 500 to 1,000 nuclei.
(2) On the criterion of the number of seedlings with metaphase to all watched seedlings at all stages of fixation.

In all the experiments was estimated standard deviation from the mean.

21.3 DISCUSSION AND RESULTS

The natural level of mitotic activity, control was estimated by the criterion of the number of metaphases in seedlings of the 18,000 counted nuclei metaphase plates were 2.83 ± 0.12%.

Table 1 shows data within 3 years of the evaluation of natural mitotic activity on the criterion of the number of metaphases in all investigated roots, in different years (1966, 1967, and 1969) ranged 90–99%.

TABLE 1 The natural level of chromosome rearrangements in the 2*n*-meristem cells of *Crepis capillaris* after soaking the seeds in water and germinating in 0.01% solution of colchicine. Harvest of 1966, 1967, and 1969 years.

Year of harvest	Month, year studies	Number of investigating seedlings		Metaphases		Rearrangements of chromosomal type	
		Σ	with meta-phases	Σ	with rear-rangements, % ±	Σ	% ±
1966	XII,1966	25	25	1125	0.44 ± 0,198	0	-
	I, 1967	27	27	2120	0.24 ± 0,105	2	-
	III, 1967	43	42	1692	0.24 ± 0,118	0	-
Average:		*95*	94/99.0%	4937	0.31 ±0,008	2	0.04 ± 0.029
1967	IV, 1968	49	48	1169	0.77 ± 0.256	0	-
	VI,VII,1968	20	14	350	1.14 ± 0.569	3	0.86 ± 0.492
	III, 1969	20	18	1316	2.96 ± 0.468	24	1.73 ± 0,369
Average:		89	80/89.9%	2835	1.62 ± 0.214	27	0.81 ± 0,155
1969	IV,1970	33	32	1384	1.59 ± 0.006	21	1.52 ± 0.329
Total (1996-1969гг):		217	206(94.9 ± 1.49%)	9156	1.05 ± 0.134	50	0.55 ± 0,080

Note: Single metaphase with numerous damages of the spindle and chromosomes we do not take into account.

On the average, over the years and months, were studied 217 seedlings. Of these, 94.9% contained metaphases. The frequency of metaphases with rearrangements in 1966 was in the range 0.24–0.44%. In 1967 there was a clear tendency to increase the frequency of rearrangements, as they are stored. Mitotic activity in 1967 was about 90%. Number of rearrangements of chromosomal type was various and small. For all the years discovered 50 such changes (see Table 1). It should be noted a tendency to increase their number during the aging of seeds (harvest 1967, 19 months of storage). The average level of metaphases with rearrangements in different years was less than 2%, rearrangements of chromosomal type in average of 0.55%, the largest number of them was in 1967.

As the count the nuclei in the seedlings after treatment by phosphemid at concentrations of 1×10^{-2}–5×10^{-4} M of 68 500 counted nuclei were 743 metaphases, an average of 1.08 ± 0.06%, thus phosphemid suppressed mitotic activity more than doubled (see above 2.82% in control).

After exposure to seeds by phosphemid the average for all concentrations and of fixations analyzed 17,513 seedlings with metaphases, among them were discover

2,306 metaphases with rearrangements (13.17%). Alterations of chromosomal type made less than 1–0.113%. This magnitude is similar to the natural frequency. At a concentration of phosphemid of 2×10^{-3}M mitotic activity was on the average 51.5% (Table 2), that is two times lower than in the controls (94.9%) (Table 1).

TABLE 2 Aberrations of chromosomes in 2n-meristem cells of *Crepis capillaris* after 24 and 27 hr from the start of treatment of seeds in a solution of phosphemid 2×10^{-3}M (2. 24 mg, 50 ml of water). Fixation through 3–8 hr after arising. Seeds were germinating in 0.01% solution of colchicine. Harvest of 1967 year. The analysis in April 1968.

Time, in hours		Number of investigating seedlings			Metaphases		Rearrangements of chromosomal type
from soaking up to «arising»	from soaking up to fixation	Σ	with metaphases Σ	with metaphases %	Σ	with rearrangements, %±	
24	3	57	24	42.1	741	9.4 ± 1.08	0
	6	56	32	57.1	1270	17.2 ± 1.06	1
	8	54	36	66.7	1815	14.0 ± 0.82	0
27	3	47	16	34.0	373	20.6 ±2.10	1
	5	19	12	63.2	313	15.3 ± 2.04	0
Total		233	120	51.5 ± 3.28	4512	14.8 ± 0.53	0.043 ± 0.031%

Data of Table 2 show that at the weak concentration of preparation 2×10^{3} with the increase of term of fixation in every "arising" mitotic activity increases, and frequency of alterations diminish.

At max concentration 1×10^{-2}M (Table 3), we registered the large number of rearrangements nearly 60% through 27 hr after arising in term 24 hr, of which 16.4% were mitosis with multiple rearrangements (it is marked an asterisk). On average, significant increase the number of strongly damaged metaphases was 13.3 ± 2.9%. Rearrangements of chromosomal type at this dilution also did not find.

Average frequency of mitotic activity of phosphemid in a concentration of 1×10^{-2}M at the same quantity of seeds, as in Table 2, was lower almost twice 25% (Table 3). Average number of metaphases with rearrangements increased nearly 22 *versus* 14.8%.

Thus, the average level of metaphases with chromatid aberrations was increasing with increasing of dose of preparation.

TABLE 3 Rearmaments of the chromosomes in the 2*n*-meristem cells of *Crepis capillaris* seedlings at 24 and 27 hr from the start of treatment of seeds in a solution of phosphemid 1 × 10^{-2}M (22.4 mg, 10 ml water). Fixation through 3–24 hr after arising. Seeds were germinating in 0.01% solution of colchicine. Harvest 1967. The analysis: April 1968.

Time, in hours		Number of investigated seedlings		Metaphases		
from soaking up to «arising»	from soaking up to fixation	Σ	with metaphases	Σ	with rearrangements	
					Σ	%±
24	3	21	6	148	25	16.9 ± 3,09
	6	19	3	365	24(2*)	6.6 ± 1,23
27	3	20	2	73	7(2*)	9.59 ± 3,47
	24	12	7	185	110(18*)	59.46 ± 3,62
Total		72	16/25.0 ± 5.14%	771	166(22*)	21.53 ± 2,85

Note: (*) Number of greatly damaged metaphases with multiple rearrangements.

At max concentration (1 × 10^{-2}M), we registered the large number of rearrangements 59.46% through 27 hr after arising, of which 16.4% were mitosis with multiple rearrangements (it is marked an asterisk). On average, significant increase the number of strongly damaged metaphases was 13.3 ± 2.9%. Rearrangements of chromosomal type do not arising in spite of increasing concentration of preparation.

During storage of untreated seeds and after subsequent treatment of seeds, namely in June and July of 1968 patterns of mitotic activity and frequencies of rearrangements were different, despite the use of the same concentration of phosphemid: 2 × 10^{-2}M (Figures 3(a), (b), and 4(a), (b)). Mitotic activity was above than in April. With increasing time from arising to fixation after 12 hr frequency of seedlings with mitoses close to 90% and was slightly below of control.

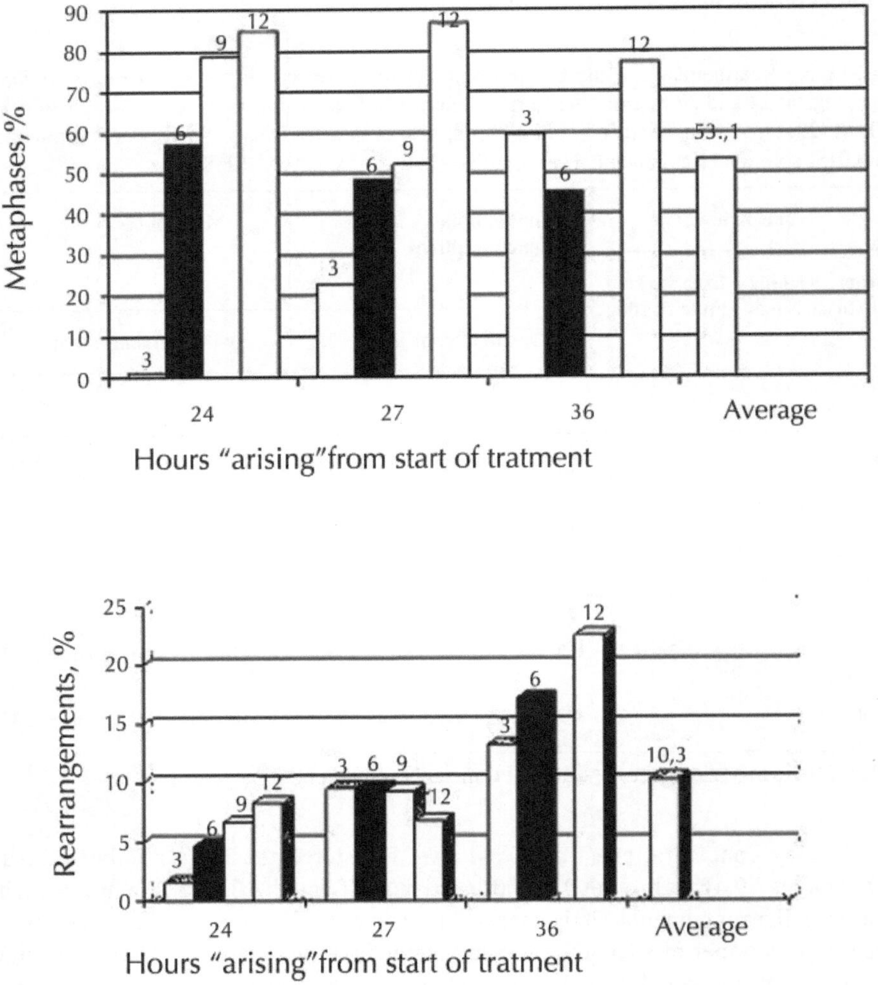

FIGURE 3 *Crepis capillaris*: the mitotic indexes in seedlings (a) and rearrangements of chromosomes in metaphase (*b*) after 24, 27, and 36 hr from the start of seed soaking in a solution of phosphemid 2×10^{-2}M (22.4 mg, 50 ml of water). Fixation: through 3–12 hr after "arising". Seeds germinate in 0.01% colchicine. Seeds are of harvest 1967. Analysis: June 1968.

Through 24 and 27 hr from the start of soaking of seeds the mitotic activity steadily was increasing to depending on term of fixation in each "arising". At a later arising 36 hr (see Figure 3a) or 31 hr (see Figure 4a) mitotic activity was fluctuating at a highest level as compared as at the early stages of arising and terms of fixation from arising. These data on Figures 3 and 4 are statistically significant. The frequency of metaphases with rearrangements increased steadily in the fraction of the 24 hr from, arising from 3 to 12 hr (see Figure 3b) and from 3 to 9 hr (see Figure 4(b)).

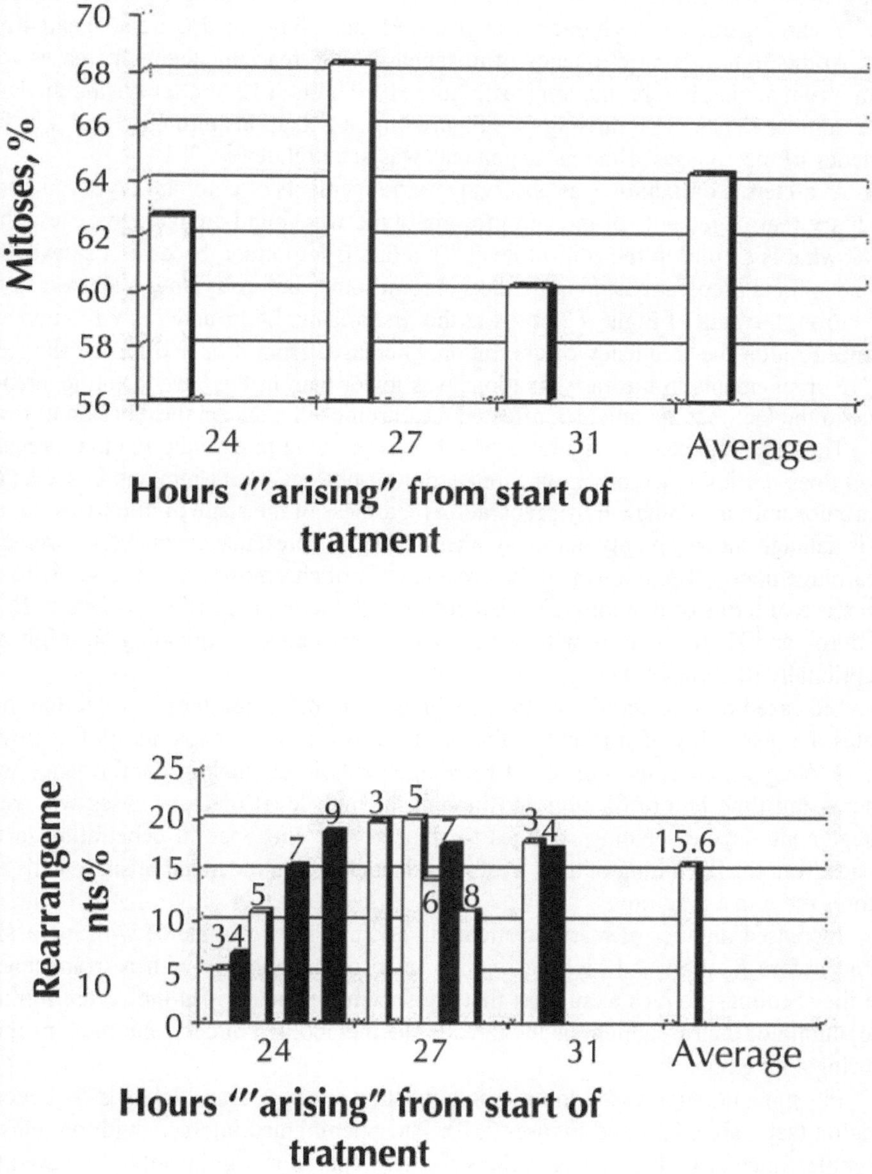

FIGURE 4 *Crepis capillaris*: the mitotic indexes in seedlings (a) and rearrangement of chromosomes in metaphase 2n-meristem cells (b) after 24, 27 and 36 hr from the start of seed soaking in a solution of phosphemid 2 × 10^{-2}M (22,4 mg, 50 ml of water). Fixation: through 3–9 hr after seedling. Seeds were germinating in 0.01% colchicine. Seeds are of harvest 1967. Analysis: July 1968.

Through 24 hr after the soakage of seeds the frequency of metaphases with rearrangements was growing steadily with time of fixing after "arising", average frequency of rearrangements was higher through 27, 31, and 36 hr (see Figure 3(b) and 4(b)).

At the 36 hr arising frequency of metaphases with rearrangements increases with time from arising before fixation (see Figure 3b). Through 12 hr after arising the level was almost 23% in 31 hr arising (see Figure 4b), at a fixations through 3 and 4 hr frequency of metaphases with rearrangements was at a high level 17–18%.

At all terms of fixation was observing rearrangements of chromatid type, and only solitary rearrangements of the chromosomal type was found an average of less than 1%, what is similar to the control level. This fact is important, because it shows, that the chemical preparations do not breaks chromosomes before synthesis DNA.

From the data of Figures 3 and 4 at the first arising, 24 hr under any fixation and concentration the frequency of aberrations increased from 3 hr and more. The level of rearrangements in the early fixations was lower than in the later them, apparently due to the fact that the mutagen affected the chromosome the smallest period of time.

The facts reflected in Figures 3 and 4, it is possible to explain: (1) that preparation does not influence on chromosomes during phase G1, but remains in seeds; (2) chromosomes are damaged by preparation regardless of the stage of mitotic cycle, but this damage shows up only during synthesis of DNA (presence exclusively chromatid rearrangements). The number of the broken areas of chromosomes increases with the increase of terms of fixation; (3) existence of both factors is possible. Works of B. N. Sidorov and N. N. Sokolov will be described, convincingly explaining an origin exceptionally of chromatid alterations.

Increased mitotic activity of the seedlings with increasing terms of fixation indicates the possibility of washout of the preparation from the cells during the growth of seedlings. Increasing number of rearrangements in arisings in conformance with increasing time before fixation and kipping of high level of metaphases with rearrangements in the seedlings obliges to suggest that phosphemid penetrating in the seeds from the beginning of their treatment, is included in the metabolism of cells and stores there in a long time.

Increased number of rearrangements in arisings in conformance with increasing time before fixation and kipping of high level of metaphases with rearrangements in the seedlings obliges to suggest that phosphemid penetrates in the seeds from the beginning of their treatment, is included in the metabolism of cells and remains there during a long time.

Phosphemid suppressed mitotic activity then stronger than higher is its concentration (see Table. 2, 3 and Figures 3, 4). Phosphemid also interacts with proteins of spindle, since at higher concentrations of the preparation significantly increased the number of heavily damaged mitoses (see Table 3). Often such metaphases form a sort of star at the center of the cell. A similar pattern sometimes was observed in the culture of fibroblasts. In addition to these disruptions, in some cells we have seen all the chromosomes were fragmented.

Mitotic activity relatively of the number of metaphases and of number of non dividing nuclei through 2–10 hr after arising in the control amounted to 2.11% at

16,000 nuclei. After treatment of the seeds the number of metaphases in the seedlings over the same period (2–10 hr) was lower and depended on the concentration of phosphemid. At a high concentration of 1×10^{-2}M at an average around 16,000 nuclei was 0.73% of metaphases, with preparation concentration 5×10^{-3}M 0.82% of metaphases around 26,500 nuclei, at a concentration of 5×10^{-4}M observed 1.42% of metaphases around 21,000 nuclei. However, by 20 hr after the start of treatment increased the frequency of mitoses both in control 3.40%, and in the experience at concentration of phosphemid 5×10^{-3}M 2.38%. These data also reflect the decline of mitotic activity with the increase of concentration of phosphemid and her increasing with reduction of concentration of preparation.

Mutagen may be remains in the seeds in conjunction with other cellular proteins, by that braking advancement of phases of mitotic cycle, thus influencing on the chromosomes longer and therefore stronger is damaged the synthesis of larger number of loci of chromosomes.

In the 1960–1970s, outstanding scientists N. N. Sokolov, B. N. Sidorov [19-21] realized a series of studies on effects of ethyleneimine on seedlings *Cr. capillaris*. They cultivated seedlings in a solution of colchicine for five cell generations. They found in tetraploid and higher polyploidy cells rearrangements of chromatid type "not multiplied" under influence of colchicine.

The authors explain this phenomenon is the fact that the mutagen is saved in the cells and there are new rearrangements. [20] Seedlings were treating of ethyleneimine: these seedlings were washed in running water within 2 hr. From these through 48 hr were preparing "thin gruel". Intact seedlings of *Cr. capillaris* were treating by that thin gruel. In these seedlings treated by thin gruel were appearing rearrangements of chromatid type.

The authors suggested that ethyleneimine formed active secondary mutagens, connecting with the components of the cell, including with nucleic acids. [21], the author's *in vitro* added ethyleneimine to amino acids: glycine and histidine, to the hexamine (hexamethylenetetramine), to vitamins: thiamin (vitamin B_1), nicotinic acid.

Consequent treatment of seedlings of these preparations did not cause aberrations. The frequency of them was at the level of control. Ethyleneimine (concentration 0.05%) caused about 15% of rearrangements, while in mixture with thiamine caused significantly more rearrangements (+22.27%). Treatment by ethyleneimine in mixture with nicotinic acid, glycine, and histidine showed even some protective effect. Treatment by ethyleneimine together with adenine, guanine (derivative of purine), cytosine (derivative of pyrimidine) showed absence of effect or a small excess of rearrangements above the level of rearrangements of pure ethyleneimine.

The received significant excess frequency of rearrangements under the influence of mixtures: uracil + ethyleneimine gave an increase nearly 19%, thymine + ethyleneimine caused nearly 61% of rearrangements (+37.25%). Guanine and cytosine in a mixture with ethyleneimine gave an insignificant action. In the mixture with thio-TEPA (three ethyleneimine groups), only thiamine gave significant excess frequency of rearrangements (+25.97%) over control (ethyleneimine). After treatment of seedlings by a mixture of ethyleneimine with thymine, the excess frequency rearrangements not found. All the rearrangements were chromatid type. The authors explain this phenomenon of the formation of secondary mutagens in cells. Thusly, these experiments [19-21] were showing that mutagens do not cause aberrations of chromosomes before or after phases of the mitotic cycle G2, G1, and causes only chromatid aberrations effecting on the chromosomes in the course of DNA synthesis. Modern microscopy suggests that mutagen penetrating into the cell is remaining in "space around a chromosome" [22]) in bonding with proteins or DNA, but its effect is revealing when mutagen passes through phase of DNA synthesis and becomes discovered in the form of chromosome rearrangements during mitosis. The same can be evidence of the phenomenon of fragmentation of chromosomes, strong destruction of mitosis and the spindle during treatment by high doses of the alkylating agent (in our case by phosphemid) in culture of fibroblasts of human and mouse or in germinal cells of seeds *Cr. capillaris*. Perhaps the same principle is working in the course of "chain process" of Dubinin. Someday, 4D microscopy will reveal the mechanism of interaction chromosomes with the chemical mutagens.

Understanding the mechanism of chemical mutagenesis in the present time is fundamentally important in view of the global contamination of the surrounding nature by different chemicals that damage the genetic structure of organisms.

21.4 CONCLUSION

(1) We have shown that in cultivated fibroblasts phosphemid was suppressing mitotic index, induces rearrangements of chromosomes.
(2) In seedlings *Crepis capillaris* phosphemid also causes inhibition of the mitotic cycle. The average number of metaphases on the number of nuclei in seedlings after treatment phosphemid decreased twice.
(3) Phosphemid after treatment of seeds *Cr. capillaris* causes rearrangements in the cells of seedlings, regardless of age treated seeds, but depending on the concentrations of drug. The greatest number of rearrangements occurs when using the highest concentration of the preparation.
(4) After treatment of the seeds *Cr. capillaris* by phosphemid were found rearrangements only chromatid type in seedlings. The number of rearrangements of chromosomal type was at the level of controls or smaller. This means that the preparation works as well as other chemical mutagens, that is chromosomes break during DNA synthesis.
(5) Phosphemid in treatment seeds *Cr. capillaris* showed heterogeneity of germinal cells in the seeds during the G1 phase. The frequency of chromosomal rearrangements varies depends on the time between fixation and "arising".
(6) At high concentrations ($1 \times 1^{-2}M$) phosphemid caused the destruction of mitotic spindle and multiple fractures of the chromosomes.

KEYWORDS

- **2*n*-Meristem cells**
- **Chemical mutagenesis**
- **Chromosome rearrangements**
- **Colchicines**
- **Ethyleneimine**
- **Fibroblasts**
- **Karyotype**
- **Mitotic activity**

REFERENCES

1. Ross, W. Biological alkylating agents, Fundamental chemistry, and design of compounds for selective toxicity. 1962. Translation A. Ja. Berlin's. (Ed.). Medicine Moscow, London p. 259 (1964).
2. Loveless, A. *Genetic allied effects of alkylating agents*. N. P. Dubinin's (Ed.). Nauka, Moscow and London, p. 255 (1966) (1970).
3. Stroeva, O. G. *Josef Abramowitz Rapoport*. Nauka, Moscow, p. 215 (1912–1990).
4. Josef Abramowitz Rapoport – scientist, warrior, citizen. Essays, memoirs, materials. Compiled by O. G. Stroeva. Nauka, Moscow p. 335 (2003).
5. Kihlman, B. A. Aberrations induced by radiomimetic compounds and their relations to radiation induced aberrations In *Radiation-induced chromosome aberrations*. London and New York p. 260 (1963).
6. Eiges, N. S. Characteristic features of chemical mutagenesis method I. A. Rapoport and its use in breeding of winter wheat. Penza, pp. 14–17 (2008).
7. Eiges, N. S., Vaysfel'd, L. I., Volchenko, G. A., and Volchenko, S. G. Some aspects of the securities chemomutant characters a collection of winter wheat and characterization of these characters. *International teleconference number 1: Basic Science and Practice*, Retrieved from http://tele-conf.ru/nasledstvennyie-morfologicheskie-kletochnyie-fakto/aspektyi-ispolzovaniya-tsennyih-hemomutantnyih-priznakov-kollektsii-ozimoy-pshenitsyi-i-ih-har-ka-chast-1.html (January, 2010).
8. Kalaev, V. N., Butorina, A. K., and Sheluhina, O. Yu. Assessment of anthropogenic pollution areas Staryj Oskol on cytogenetic parameters of birch seed family. *Ecological genetics*, **4**(2), 9–21 (2006).
9. Dubinina, L. G. *Structural mutations in the experiments with Crepis capillaries*. Nauka, Moscow, p. 187 (1978).
10. Dubinin, N. P. Some key questions of the modern theory of mutations. *Genetika*, (7), 3–20 (1966).
11. Dubinin, N. P. *Unresolved issues of modern molecular theory of mutations*. Proc. USSR, (2), 165–178 (1971).
12. Dubinin, N. P. *Unresolved issues of modern molecular theory of mutations*. Proc. USSR, (3), 333–344 (1971).
13. Chernov, V. A. Cytotoxic substances in chemotherapy of malignant tumors. *Medicine,* Moscow, p. 320 (1964).
14. Chernov, V. A., Grushina, A. A., and Lytkina, L. G. Antineoplastic activity of phosphasin. *Pharmacology and Toxicology*, **26**(1), 102–108 (1963).

15. Weisfeld, L. I. Cytogenetic phosphasin effect on human cells and mice in tissue culture. *Genetika*, (4), 85–92 (1965).
16. Weisfeld, L. I. Influence phosphasin on the duration of phases of the mitotic cycle of human cells and mice in culture. *Genetika*, 4(7), 119–125 (1968).
17. Navashin, M. S. *Problems karyotype and cytogenetic studies in the genus Crepis*. Nauka, Moscow, p. 349 (1985).
18. Protopopova, E. M., Shevchenko, V. V., and Generalova, M. V. Beginning of DNA synthesis in seeds *Crepis capillaries*. *Genetika*, (6), 19–23 (1967).
19. Sidorov, B. N., Sokolov, N. N., and Andreev, V. A. Mutagenic effect of ethyleneimine in a number of cell generations. *Genetika*, (1), 121–122 (1965).
20. Andreev, V. S., Sidorov, B. N., and Sokolov, N. N. The reasons for long-term mutagenic action of ethyleneimine. *Genetika*, (4), 28–36 (1966).
21. Sidorov, B. N., Sokolov, N. N., and Andreev, V. A. Highly active secondary alkylating mutagens. *Genetika*, (7), 124–133 (1966).
22. Rubtsov, N. B. Human chromosome in four dimensions. *Priroda*, (8), 1–8 (2007).

Index

F

N